Geography of Technology Transfer in China

A Glocal Network Approach

East China Normal University Scientific Reports
ISSN: 2382-5715

This book series reports valuable research results and progress in scientific and related areas. Mainly contributed by the distinguished professors of the East China Normal University, it will cover a number of research areas in pure mathematics, financial mathematics, applied physics, computer science, environmental science, geography, estuarine and coastal science, education information technology, etc.

Published

Vol. 17 *Geography of Technology Transfer in China: A Glocal Network Approach*
by Chengliang Liu (East China Normal University, China)

Subseries of Symposia and Topic Studies

Published

Vol. 16 *Forty Years of Algebraic Groups, Algebraic Geometry, and Representation Theory in China: In Memory of the Centenary Year of Xihua Cao's Birth*
edited by Jie Du (University of New South Wales, Australia), Jianpan Wang (East China Normal University, China) and Lei Lin (East China Normal University, China)

More information on this series can also be found at https://www.worldscientific.com/series/ecnusr

(Continued at end of book)

East China Normal University Scientific Reports | Vol. 17

Geography of Technology Transfer in China

A Glocal Network Approach

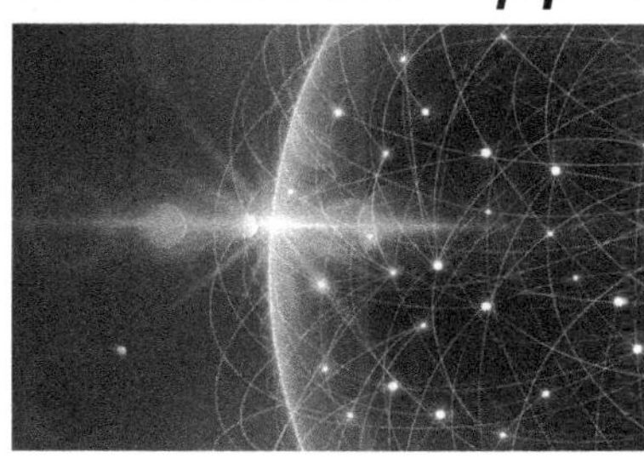

Chengliang Liu
East China Normal University, China

World Scientific

EW JERSEY • LONDON • SINGAPORE • BEIJING • SHANGHAI • HONG KONG • TAIPEI • CHENNAI • TOKYO

Published by

World Scientific Publishing Co. Pte. Ltd.
5 Toh Tuck Link, Singapore 596224
USA office: 27 Warren Street, Suite 401-402, Hackensack, NJ 07601
UK office: 57 Shelton Street, Covent Garden, London WC2H 9HE

Library of Congress Control Number: 2023031045

British Library Cataloguing-in-Publication Data
A catalogue record for this book is available from the British Library.

East China Normal University Scientific Reports — Vol. 17
GEOGRAPHY OF TECHNOLOGY TRANSFER IN CHINA
A Glocal Network Approach

ISBN 978-981-127-495-4 (hardcover)
ISBN 978-981-127-496-1 (ebook for institutions)
ISBN 978-981-127-497-8 (ebook for individuals)

For any available supplementary material, please visit
https://www.worldscientific.com/worldscibooks/10.1142/13371#t=suppl

Desk Editors: Nambirajan Karuppiah/Lai Ann

Typeset by Stallion Press
Email: enquiries@stallionpress.com

PREFACE

When we used to glance over the research on innovation diffusion and knowledge flow, there has already been a great volume of outcomes toward different innovation actors in various regions. Academics in China and abroad have focused on the technology transfer studies on core innovation actors such as multinational corporations, universities, and research institutes, along with industry–university–research communities. However, the majority of results are derived from economics and management science, and there are relatively few studies that examine the spatial concerns of technology transfer and national innovation issues highlighted by it from a geographical perspective. Therefore, our edition of *Geography of Technology Transfer in China* primarily aims at discovering the geographical character of technology transfer and its determinants from the interaction of global and local scales, which involves the study of innovation geography, economic geography, urban, and regional economics.

In essence, our choice of China is due to its role of innovation growth pole under the current background which is a suitable objective to depict a classic increasing model of emerging economies. In the era of the knowledge-based economy, China's economic development has entered a new economic mode from factor-driven growth to innovation-driven development, and innovation has emerged as the dominant driver to promote national and regional economic growth. To accelerate the transfer and transformation of innovative technologies, the State Council has made the clarity that

China must urgently strengthen the distribution and construction of a national technology transfer system in accordance with the laws of technological innovation, technology transfer, and industrial development. With the help of national innovation policies and local innovation milieu, China has shown a great increase in both technology generation and transfer since 2000s. Therefore, the empirical research on China's technology transfer has precious value in indicating a successful innovation system and innovation spillover pattern to the developing countries that have a serious locked-in effect or locate in lower reaches of global innovation chain.

The chapters selected in this book create a clear structure of a series of studies. The contents are organized from easy to complicated, as well as from theoretical to empirical. The research content begins by constructing an analytic framework for urban technology transfer network. It abstracts the study target from innovation elements to cities of technology transfer, and develops an initial conceptual framework for urban technology transfer network. In addition, it further puts forward the dual-pipeline framework to analyze the mechanism of internal technologies and external technologies on local innovation performance (Wang *et al.*, 2023b), which enriches the perspective of current innovation geography research and opens a new field of study. In the course of examining the urban system of technology transfer in China, the research scale has been refined with the interaction of both the global and local (named as glocal network); in this way, the following empirical studies depict the spatial evolution by integrating international, national, and regional scales at first. Then this book deeply analyzes the spillover effect of the technology transfer network and also its determinants, which can be regarded as a bidirectional study. On the other hand, the research methodology emphasizes not only the combination of geographical visualization and econometric analysis but also the combination of big data and small data mining analysis, as well as bibliometric analysis and traditional literature review.

Finally, we hope the outcomes in this book will raise more debates on the geography of science and technology (S&T) innovation between scholars. Due to the context of S&T sovereignty, the competition among countries has targeted S&T which is endorsed to be the necessity for nations and regions to cope with external constraints to achieve long-term sustainable development. Therefore, the voice on innovation development may forge ahead with new ideas and understanding for especially developing countries to achieve a breakthrough when faced with a locked-in predicament. We would also be very grateful to anyone who provides further ideas and research on the geography of technology transfer in different regions, especially in those late-developing countries led by technology catch up.

In the publishing process, this book has been jointly supported by the National Social Science Fund of China (NSSFC) (No. 21ZDA011) and the Shanghai Shuguang Talent Program (No. 19SG22).

I would like to thank my research group who have helped me to prepare this book efficiently. Junxian PIAO, a Ph.D. candidate and former Master's student from the Bartlett School of Planning in the University College of London, took the effort of translating the book into English. Ms. Shanshan YAN, Ms. Caicheng Niu, and Mr. Mingming Guan, who have received their Master's degrees, participated in most of the data collection, analysis, and content writing. My four Ph.D. candidates and three Master's undergraduate students enhanced the theoretical and empirical analysis. Specifically, Mr. Weisheng MAO oversaw the theoretical background and framework. Mr. Yuan LI designed the figures on the transnational technology transfer. Ms. Bangjuan WANG renewed the models of hubs and their hinterworlds and helped in designing figures in various chapters of the book. Mr. Jiange LIN revised the models on the determinants of technology transfer network. Mr. Xiangjie Liu was responsible for the research methods design, the section on domestic technology transfer network as well as designing its figures. Ms. Xue Luo collected the data material on the transportation network section.

Additionally, I would also like to express my appreciation to other undergraduates such as Shuqi Sun, Tong Liu, Xue Lan, and Qixiang Li, who helped me in redrawing figures and data collection.

ABOUT THE AUTHOR

Chengliang Liu is a professor and supervisor of economic geography, distinguished professor (class B) of geography, the vice director of the Institute for Global Innovation and Development at East China Normal University (ECNU), Shanghai, China. He earned his Ph.D. in human geography from Central China Normal University and was a visiting scholar at Louisiana State University in the US. He has won five subnational-level research awards and four government-funded talent projects. His research interests focus on geographical complexity of glocal innovation network and transport network by integrating Global Innovation System (GIS), spatial econometrics, along with methods of complex network and other complexity science. His work has been supported by seven national grants including a major project of the National Social Science Fund of China (NSSFC) and two general programs of the National Natural Science Foundation of China (NSFC). He is also on the editorial board of several academic journals including *Chinese Geographical Science*, *World Regional Studies*, and *Regional Sustainability*. He has published four monographs and over 100 refereed articles signed first or corresponding author. These works are distributed in famous geography journals (such as *Geoforum*, *Journal of Transport Geography*,

Computers Environment and Urban Systems, *Environment and Planning A*, etc.) indexed by SSCI, SCI, and CSSCI. Moreover, he is devoted to teaching world regional geography and serves as a teacher in charge of several national-level courses. He has been honored as the *Teaching Master in Curriculum-based Values Education* by the Ministry of Education of China.

CONTENTS

LIST OF FIGURES

LIST OF TABLES

PART I
Introduction

CHAPTER 1

NATURE OF INNOVATION NETWORK

1.1 Rising Discussions and Challenges

Recent decades are often characterized as the era of globalization — never before has such a web of linkages and interconnections existed (Acs & Preston, 1997), which is experienced by a worldwide system of invention and innovation (Nam & Barnet, 2011). Competitive advantages have shifted from access to financial and physical capital to knowledge-based resources and human capital (Audretsch *et al.*, 2014). With the increased globalization of science and technology (S&T), there is increasing political and academic discourse concerning geographical laws of innovation activities, providing new insights into the spread of S&T (Nam & Barnet, 2011; Gui *et al.*, 2019b). It has — without any doubt — evolved into one of the mainstream research topics in Economic Geography and Regional Science (Duan, 2020). Any comprehensive analysis is likely to include the spatial dimension of technology transfer, the local clusters of innovation activities under regional innovation system, the interaction between innovation actors as well as the innovation activities in various industries. Existing evidence overwhelmingly demonstrates that innovation activities have broken through the borders of regions, organizations, and industries. The trans-local flow of knowledge and technology is regarded as the main feature of regional opening innovation, which indicates the globalization of technology has become increasingly important in the shifting global economy (Audretsch *et al.*, 2014).

Research on innovation has evolved from regional innovation system (Cooke, 1992) to global innovation networks (Ernst, 2009), which focus either on local or global scales, emphasizing the internal or external flow of innovation actors. However, this academic approach has also overlooked the multiscaled interaction known as glocalization, which occurs at global, national, regional, and local levels (Mackinnon *et al.*, 2019). It is necessary to analyze the flow of innovation actors and innovation path from the glocalization or glocal perspective, which can provide new ideas about their interconnection and interaction among different scales — a topic of heated discussion today (Tavassoli & Carbonara, 2014; Gertler & Levitte, 2015; He & Mao, 2015). In addition, the concepts of "global" and "local" in the term "glocal" are not two fixed scales, but rather two terminals proceeding through a series of complex interactions (Dicken, 1994; Si *et al.*, 2016). This book aims to illustrate not only a clear understanding of innovation networks in specific "global" and "local" scales but also discover the landscape dynamics of their interaction, which differs from previous studies.

Innovation studies have also sparked discussion toward the openness of innovation system. Chesbrough (2003) demonstrated that enterprises leveraging abundant external resources tend to have higher innovation efficiency, leading to the concept of "innovation openness" (Wang & Sun, 2009). Additionally, innovation does not solely come from the accumulation and reorganization of preexisting knowledge and ideas (Aghion & Howitt, 1992; Weitzman, 1998). It is also essential to introduce and absorb external knowledge and technology to avoid "path dependence" or "locked-in effect" (Nooteboom, 2000a; Boschma, 2005; Martin, 2010; Trippl *et al.*, 2018), which can drive regional innovation performance and knowledge production (Bottazzi & Peri, 2003; Moreno *et al.*, 2005; Miguélez & Moreno, 2013). However, external knowledge may not be always effective, and only related but not identical knowledge can significantly promote innovation. Recently, this idea has been demonstrated by scholars regarding the influence of related external knowledge on local innovation through technological trade and patent cooperation (Boschma & Iammarino, 2009; Miguelez & Moreno, 2018). However, there is still a lack of empirical analysis from a technology transfer perspective.

In reality, actors involved in S&T innovation display both extensive global flows and strong local concentration, which contribute to the formation of a glocal innovation network. This perspective suggests that technology globalization has led to an unprecedented transfer of innovation actors at a transnational scale, with greater scope, speed, and spillover effects in the global innovation network. Moreover, innovation actors also follow the rules of regional agglomeration and localization, which leads to their high concentration in a small number of cities and urban areas that become S&T hubs of a local innovation network. Consequently, a "hub-and-network" structure is embedded into the glocal innovation network through "global pipelines" of knowledge flows.

Under this context, the topic of "technology transfer" has generated significant interest among academic researchers and policymakers (Bozeman, 2000), although it is strongly considered at a regional level (Audretsch *et al.*, 2014), revealing its spatiotemporal complexity challenges for scholars and planners. Firstly, there is still much work to be done in constructing a theoretical and interdisciplinary framework to advance the understanding of technology transfer. Secondly, outlining its dynamic patterns and networking organizations geographically can be challenging due to the concurrent nature of many processes. Thirdly, quantitative measurement of the determinants of technology-transferred network requires a deep understanding of its self-organizational processes.

Technology transfer is regarded as an important impetus for technological progress and economic growth, acting as a "bridge" between them (Farris, 2007; Yang *et al.*, 2021). It is widely recognized as a fundamental pipeline for local integration into the glocal innovation network (Yang *et al.*, 2021), and a driving force for emerging latecomers to catch up with industry leaders (Romer, 1986; Lim, 2007). Regions or countries that occupy a prominent position in the field of technology possess a large potential for economic growth. For those aspiring to achieve technological independence and leadership, it is important to adapt to their technology landscape and integrate into the glocal innovation network through technology channels or pipelines. Although glocal technology networks are a relatively new phenomenon, the literature on geographical structure and dynamics of technology transfer network is still in its early stages in both theoretical and empirical terms (Chaminade *et al.*, 2015). Furthermore, there is also a fierce debate on the current situation, in which the globalization of technology transfer is being challenged, thus setting barriers for some regions to participate in the network. For instance, the rise of trade protectionism, technological sovereignty, and multidimensional competition has intensified the S&T "war".

Despite the increasing number of documents on this topic, there still remains a paucity of evidence on the structures and dynamics of technology transfer in emerging innovative countries such as China. As a post-developing nation in terms of technology (Atkinson, 2012; Yang *et al.*, 2021), China is currently not only falling short in translating its own research into applications (Phillips, 2016) but also experiencing profound changes driven by the globalization of technology, as well as facing great threats from more aggressive competition. To achieve continuous independent innovation and sustainable development in a complex international context, China's policymakers need to understand the channels to accelerate the progress. Research on output and input channels can promote Chinese cities to actively integrate into the global innovation network and play a leading role in the national (or regional) innovation system by discovering China's specific technology progress path. So far, limited attention has been paid to the landscape complexity of technology transfer networks within emerging countries like China (Liu *et al.*, 2018a; Yang *et al.*, 2021).

This is an important critical thinking that this book aims to express and depict.

1.2 Objective and Motivation

Despite the abundance of research in this field, there is still a need for a better depiction of the geographical complexity related to technology transfer network. The vision of this book is to bridge the research gap and promote effectively China's integration into the glocal network of technology transfer, as previously mentioned. To achieve this goal, the book uses patent transaction and application data from the Patent Information Service Platform of *China National Intellectual Property Administration* and the Knowledge Service Platform of *Wanfang Data*, and leverages qualitative and quantitative analysis methods to pursue the four objectives:

First, it aims to establish a theoretical framework of the "global–local" network among innovators as a function of technology transfer, based on a range of classic theories including innovation system theories, technology transfer theories, and emerging theories in economic geography.

Second, the book portrays the spatiotemporal characteristics of intercity technology transfer networks from transnational, domestic, and regional perspectives through social network analysis. It also discovers different types of diversities and similarities of transferred technology as well as the spatial evolution pattern of urban patent application through relatedness diversity and unrelatedness diversity models in the cities.

Third, this book demonstrates the spillover effect of the technology transfer network based on the "dual-pipeline" framework. It aims at discovering whether the technological relatedness and technological similarity of a city's existing knowledge base to its domestic and transnational technologies contribute to the development of local innovation capacity.

Fourth, the book also aims to identify the determinants of the technology transfer volume. It searches for the socioeconomic indicators and investigates their degree of influence on the technology transfer network. Meanwhile, it also explores the influencing mechanism from another type of network to the technology transfer network. The topic around the influence of transportation network focuses on the transportation accessibility and its specific range, which can deeply explore the rational configuration of the transportation network for an adequate flow of technologies.

Given the focus of the book with the value it can bring to China and even the world, it can be divided into practical and academic contributions. In terms of practical contributions, the findings can provide theoretical advice for the implementation of China's Innovation-driven Development Strategy and improvement of local capacity. Currently, the innovation capacity and regional innovation system in Chinese cities are constantly developing (Figure 1.1), which has been regarded as a strong driving force for China's socioeconomic development to achieve strength in science and industry. This book aims to establish the theoretical framework of the "global–local" interaction to reflect China's intercity technology flow and regional innovation system. For instance, it describes the spatial pattern of the transnational, domestic, and regional technology transfer networks, as well as the hubs in the network, which makes full use of innovation hubs' leading roles in technology

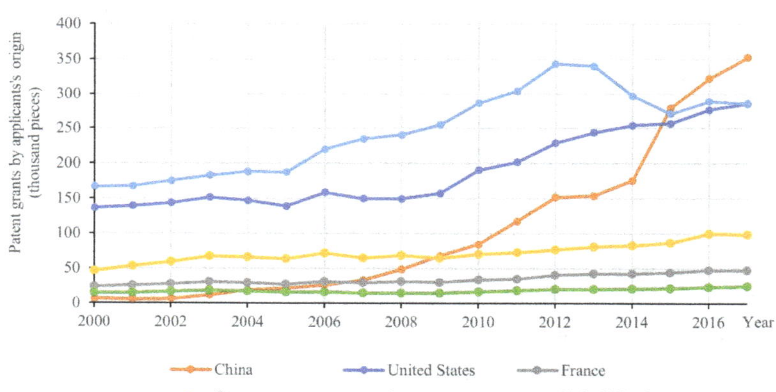

Figure 1.1 Comparison of PCT applicants between China and other major developed countries from 2000 to 2017.

Source: Redrawing from Du & Duan (2021).

transfer and diffusion. On this basis, we discover the difference in spillover effect in two different scales (transnational and domestic) from the perspective of external technology's correlation to provide a rational way for regions to improve local innovation capacity expressed in policy advice.

This book also aims to promote a rational flow of technology actors between regions, and further improve local innovation milieux and strengthen external technology linkages. China is increasingly playing a prominent role in global innovation and engaging in more frequent knowledge and technology exchanges both domestically and internationally. However, current technology transfer and diffusion in China are strongly concentrated on the three urban agglomerations such as the Beijing–Tianjin–Hebei Region (BTH), the Yangtze River Delta Region (YRD), and the Guangdong–Hong Kong–Macao Greater Bay Area (GBA), which accounted for about one-half of the total amount of patent applications in 2015 (Ye *et al.*, 2020). This concentration implies that the technology gap between the regions is still enormous (Figure 1.2). The analysis of influential factors on intercity technology transfer in this book can provide policy advice on promoting a more well-distributed and efficient-and-rational technology flow to improve local innovation milieux and technological strength. By studying technology diversity and relatedness, we can identify the technology demand of various cities, enabling them to identify their external technology fields with high relatedness and strengthen their external knowledge communication by importing technology. The benefits of stronger external technology linkages will be reflected in improved local innovation capacity and economic development, ultimately enhancing competitiveness.

When considering the theoretical contributions, the book establishes a theoretical framework of glocal technology transfer network and reveals the evolutionary rules and spillover effect of China's technology transfer network. These discoveries address a research gap in the theory of innovation networks. Over the past few decades, researchers have focused on spatial borders and observed different types of innovation networks at global and local scales. However, this

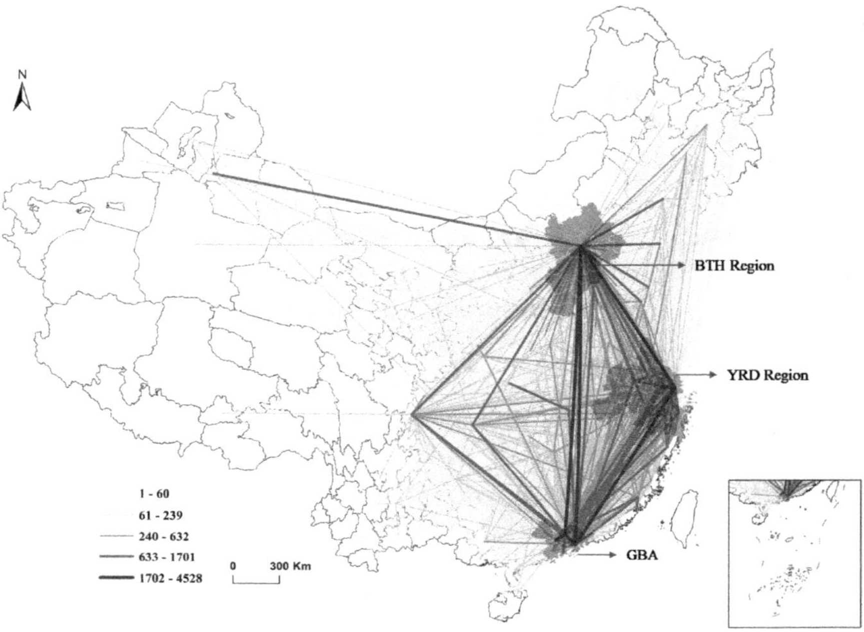

Figure 1.2 Spatial disparity of China's intercity technology transfer network from 2009 to 2018.

Source: Redrawing from Liu *et al.* (2019a).

book takes a novel approach by examining the "global–local" interaction, also called the "glocal" perspective. Furthermore, previous studies have lacked a direct analysis of technology actors, especially in comparative studies of technology transfer. The framework constructed in this book, which uses big data mining technology, spatial analysis, and complex network analysis, directly demonstrates the interrelation and influential factors of intercity innovation network at the transnational, national, and regional scales. It also uncovers the spatial evolution of diversity and relatedness of transferred technology, as well as the spillover effect from the perspective of relatedness. This approach reveals the path of technology innovation in Chinese cities in two dimensions: technology volume and categories, which enriches the theoretical research of global innovation network and regional innovation system in the geography of innovation.

1.3 Overview of the Chapters

Notwithstanding the increasing awareness of technology transfer networks, the spatial complexity of their dynamic structures remains underreported (Ma & Xu, 2022). Therefore, the chapters in this book are logically designed to provide a comprehensive and comparative analysis of the multiscalar geography of technology transfer networks in China. It consists of 5 parts and 10 chapters, covering the research background, theoretical framework, research methodology, spatial evolution, network effect, determinant factors, and policy implications of technology transfer networks (refer to Figure 1.3).

Part II of the book introduces the theoretical framework and methods related to technology transfer network. Chapter 2 provides a general literature review that supports the theories used in the subsequent chapters. It covers the theories of innovation system, technology transfer activity, and technology transfer network based on the studies of relational economic geography, evolutionary economic geography, and innovation geography. The chapter aims to provide an overview of the technology transfer network from a system perspective of innovation elements, structure, and mechanism. Chapter 3 includes a comprehensive introduction to the methods and data used in the following sections. At the beginning of this chapter, it explains the complex network establishment and the indicators to measure the importance of a node and linkage in the network. The chapter also introduces the technology diversity measurement, which evaluates the technology complexity of a city, thus enabling the discovery of the relationship between the imported technologies and local knowledge base. Finally, some regression models and Exponential Random Graph Model (ERGM) of different types of indicators are used to evaluate the spillover effect and determinants of the network. Its detailed description and the improvement of the model are stated in the following related chapters.

Based on the theories of complex network and technology transfer network, Part III conducts an empirical and dynamic analysis of China's technology transfer network at transnational, domestic, and

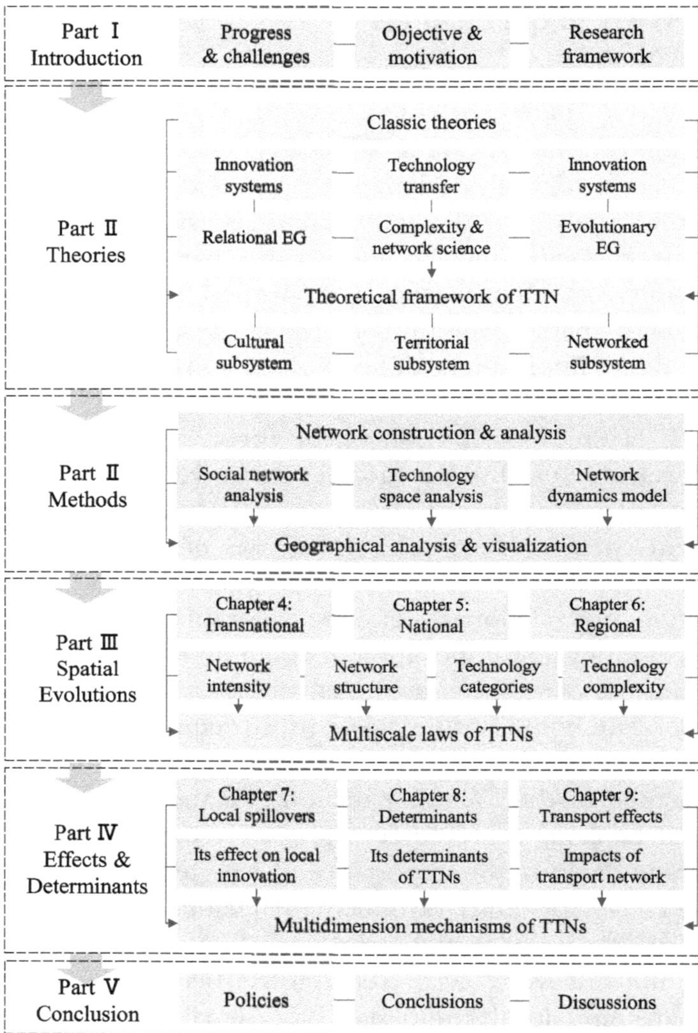

Figure 1.3 Research framework of this book.

regional levels. Chapter 4 focuses on the transnational scale, while Chapter 5 focuses on the domestic scale, both of which illustrate the evolution of technology transfer landscape. This provides insight into the relationship between China and the rest of the world, as well as intercity cooperation within China. Furthermore, by using technology complexity calculation, the technology sectors and fields analyzed in these chapters reveal how external technologies are related to the local knowledge base. Consequently, it clarifies the advantageous technology fields of a certain region or city and provides a clear regional division of labor. In contrast to the previous two chapters, Chapter 6 describes the evolution of domestic technology transfer at the regional scale. It not only depicts the evolution of node cities within the region but also the relationship with other domestic cities, revealing the interaction between two networks at different scales.

Part IV emphasizes the spatial spillover effect and determinants of technology transfer network. The spillover effect can be regarded as the impact of network on local innovation performance and capacity, which in turn affects the determinants of network. Therefore, this part aims to explore the two-way relationship between technology transfer network and local innovation capacity. Chapter 7 focuses on the effect of technology transfer network on local innovation capacity using the Ordinary Least Squares (OLS) regression. Chapter 8 uses the ERGM and linear regression model to identify the potential local innovation elements that influence technology transfer network. What's more, at the end of Chapter 8 and through Chapter 9, the determinants are extended to other networks, which is the first attempt to examine the relationship between two networks. Chapter 8 analyzes the domestic technology transfer network in relation to the transnational technology transfer network, while Chapter 9 analyzes the high-speed transportation networks and their impact on the technology transfer network.

Part V is the final part of this book, and Chapter 10 presents the policy recommendations for the establishment and improvement of China's technology innovation network based on the findings from the previous chapters. The political implications are discussed from three dimensions, namely the national perspective, regional perspective, and actors' perspective. The aim is to foster a sustainable innovation system from a relatively comprehensive aspect. At the end of Part V, the concluding chapter summarizes the main findings of the book and highlights the limitation of the research, which can be addressed in future studies.

PART II

Theoretical Advances and Methodology

CHAPTER 2

THEORIES OF TECHNOLOGY TRANSFER NETWORK

2.1 Classic Innovation Theories

Since the 1990s, there has been rapid advancement in technological and industrial changes, leading to increased attention on innovation theories across interdisciplinary disciplines such as economic geography, technology management, innovation economics, and technological sociology.

Simultaneously, there has been great progress in research on the spatial system of innovation activities and organizations (Teirlinck & Spithoven, 2008), resulting obviously in the formation of many schools of thought such as the technological change (Nelson & Phelps, 1966; Perez & Soete, 1988; Gomulka, 1990), innovation milieu (Camagni, 1991; Storper, 1997; Maillat, 1998a, 1998b), geography of innovation (Feldman, 1994; Bathelt & Henn, 2014; Polenske, 2007), and innovation systems (Edquist, 1997; Carlsson *et al.*, 2002), etc.

Innovation system (or network) theories are essentially the result of the integration and development of different innovation modes, such as non-linear innovation, collaborative innovation, network innovation, institutional innovation, and open innovation.

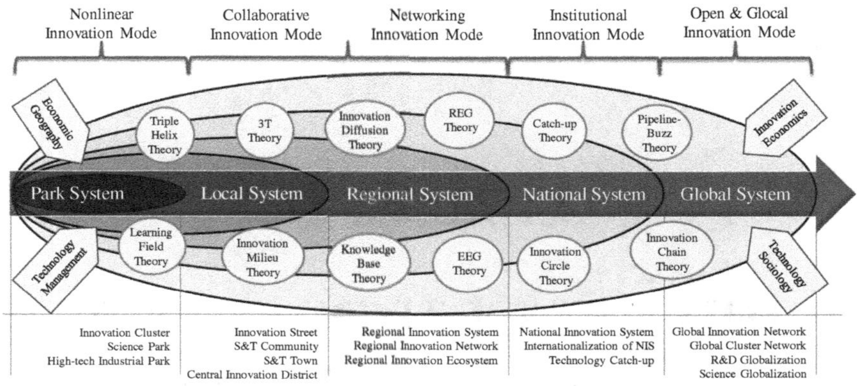

Figure 2.1 Theoretical dynamics related to innovation geography.

Consequently, the theoretical framework for the spatial system of innovation network has been continuously constructed and developed, ranging from local industrial clusters to global cluster networks, from local innovation networks to global and glocal innovation networks, and from local innovation systems to global innovation systems (Figure 2.1).

A large number of studies have shown that the spatial linkages of innovation networks are manifested in agglomeration-and-diffusion and global-and-local interactions, leading to the formation of two major schools of research. The first is the "institutional or cultural turns" in economic geography (i.e., the "new regionalism"), which focuses on cultural embeddedness, institutional thickness, innovation milieu, and cooperation or competition among firms, universities, and R&D institutions (Cooke & Morgan, 1998a). This school is represented by the national innovation system, regional innovation system, and regional innovation ecosystem (Freeman, 1987; Lundvall, 1992; Cooke *et al.*, 1997). The second is the "relational or evolutionary turns" in economic geography (Boschma & Lambooy, 1999; Yeung, 2005), which focuses on networked structures and dynamics of innovation networks, and is represented by global innovation network (Liu *et al.*, 2013) and glocal innovation network (Si *et al.*, 2016; Binz & Truffer, 2017).

2.1.1 *Innovation system theories*

Innovation is recognized as a networked system (Liu *et al.*, 2013), with the boundaries of "Systems of Innovation" (SIs) identified spatially/geographically, sectorally, and functionally (Edquist, 1997). As a result, SIs have been developed and studied from four perspectives: Technological systems of innovation (Carlsson *et al.*, 2002), Spatial systems of innovation (including global, transnational, national, regional, etc.) (Freeman, 1987; Lundvall, 1992; Cooke *et al.*, 1997; Binz & Truffer, 2017), Sectoral systems of innovation (Breschi & Malerba, 1997), and Industrial clusters (Porter, 1998) and Ecosystems of innovation (Moore, 2006) (Figure 2.2). These theories have been further expanded to include the innovation network theories at different scales, including the global innovation network (Liu *et al.*, 2013), global cluster network (Bathelt & Li, 2014; Turkina *et al.*, 2016), regional innovation network (Cooke, 1996), glocal innovation network (Si *et al.*, 2016), etc.

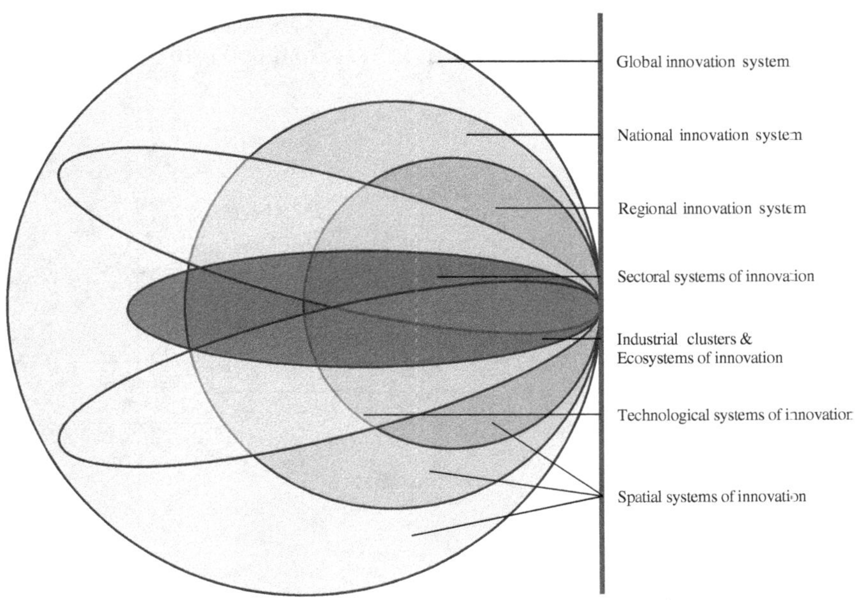

Figure 2.2 Structure of SIs.

Source: Own illustration based on Asheim *et al.* (2011)

2.1.1.1 *National innovation system*

The intellectual foundations of the National Innovation System (NIS) or the National System of Innovation (NSI) can be traced back to List's (1841) national system (List, 2011) and Schumpeter's (1912) innovation theory (Schumpeter, 2020). Since then, theoretical research on NIS has generally embraced List's institutional and national perspective and Schumpeter's knowledge and technology perspective (Zheng & Si, 2006).

The term "NIS" was first introduced by Freeman (1987), who focused on various institutional networks involved in the initiation, import, modification, and diffusion of new technologies in both private and public sectors. Lundvall (1992) later defined it as a set of interrelated innovation elements that interact in the process of new knowledge production, diffusion, and application related to economic activities. Metcalfe (1995) further illustrated NIS as a system of interconnected institutions, including educational institutions, research institutions, intermediating service institutions, industrial firms, and government institutions, which is involved in new knowledge creation, storage, application, and transfer (Figure 2.3).

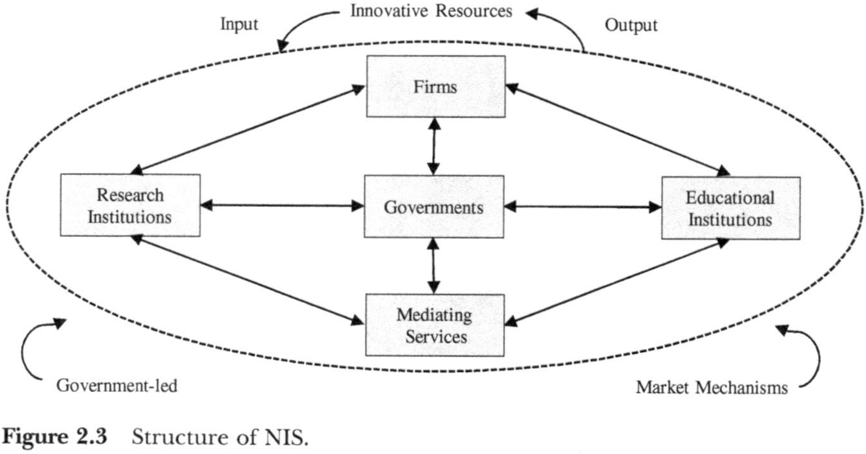

Figure 2.3 Structure of NIS.

Source: Redrawing from Freeman (1987).

There is a consensus that NIS is a unique cluster of public and private institutions that, individually or collectively, promote the creation and diffusion of new technologies. This includes the production, storage, and transfer of knowledge, skills, and technologies (Nelson, 1993; OECD, 1997; Liu *et al.*, 2013; Hui, 2017), as defined by several scholars and developed countries (Meyer-Krahmer & Schmoch, 1998; Furman *et al.*, 2002).

Openness is a fundamental characteristic of national innovation systems during the process of internationalization (Carlsson, 2006). The internationalization of NIS mainly involves the flow of knowledge, talent, and information with host country organizations through the internationalization of R&D by transnational corporations. The internationalization of innovation systems and organizational networks, the internationalization of the flow and allocation of innovation resources, and the internationalization of the interaction of innovation actors and innovation activities are the three main interrelated ways (Liu *et al.*, 2015). The increase in the number of innovation actors, the expansion of the boundaries of innovation element flows, and the resulting complexity of innovation networks are important features of the internationalized evolution of NIS (Niosi & Bellon, 1994) (Figure 2.4). When it comes to the evolution stage, there are three main stages of NIS internationalization, that is, regional innovation network to national innovation system to open NIS, with significant differences in terms of innovation elements, institutions, functions, and NIS internationalization. These can be clearly divided into three stages: dependent, catching-up, and autonomous (Table 2.1).

2.1.1.2 *Regional innovation system*

The notion of a Regional Innovation System (RIS) was formally introduced by Philip Cooke in 1992 (Cooke, 1992). RIS is a dynamic concept, and scholars' perspectives on its definition, components, functions, and characteristics vary widely (Asheim & Isaksen, 2002; Buesa *et al.*, 2006). However, they commonly emphasize the regionalism, local embeddedness, and institutional cohesion of RIS

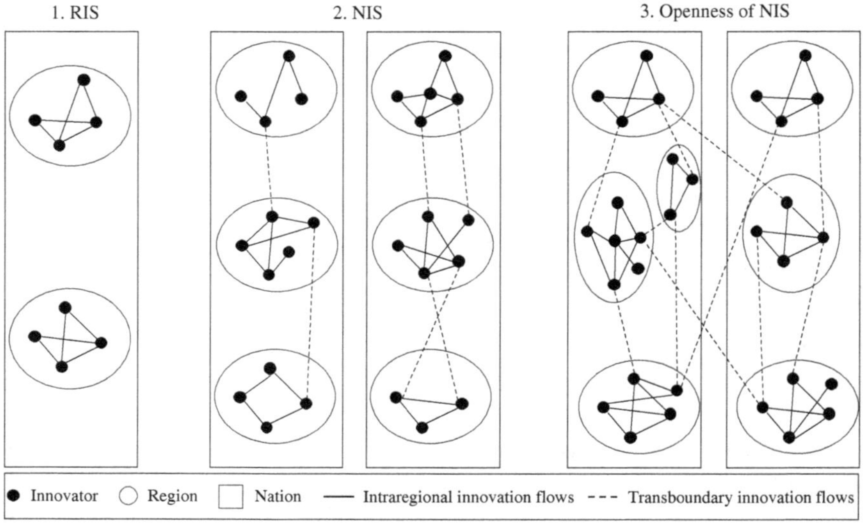

Figure 2.4 Evolution of NIS internationalization.
Source: Redrawing from Niosi and Bellon (1994).

(Fu, 2015). It is widely believed that financial capital, collective learning and trust among innovators, cultural embeddedness, and institutional thickness are crucial in promoting the performance of RIS, which may act as anchor and dynamo within the development of RIS (Cooke & Morgan, 1998a; Cooke *et al.*, 2004) (Figure 2.5). This can better explain the stickiness of innovation spaces such as the "Third Italy" (Miao, 2006), as well as path-dependence and technological lock-in (Martin, 2010).

In general, RIS is defined as a regional organizational system comprising two subsystems. One is a knowledge generation and diffusion subsystem (or *exploration* subsystem), including public research institutions, educational institutions, technology and workforce intermediating institutions, among others. Another is a knowledge application and exploitation subsystem (also known as the *exploitation* subsystem) consisting of competitors, contractors, collaborators, and consumers. Additionally, their socioeconomic and cultural setting is established through vertical and horizontal networking ties (Autio, 1998; Cooke, 2014) (Figure 2.6).

Table 2.1 Internationalization characteristics of NIS.

Period	Dependent	Catching-up	Autonomous
Element	• Indigenous allocation of domestic human, material, and market resources and foreign capital and technology • Lack of initiative to allocate global innovation resources	• Accumulating and utilizing local innovation resources while tapping into global capital, technology, and talent • Active and proactive participation in global innovation networks • Fostering multinational corporations in the nation to deploy innovation elements globally	• Effective allocation of innovation resources on a global scale • Effectively attracting and harnessing global innovation talent • Playing a leading role in global innovation networks • Multinational corporations in the nation are highly competitive globally
Institution	• Weak production capacity for new knowledge and technology • Lack of effective linkages between knowledge production, diffusion, and application linkages • Lack of national strategies and targets for autonomous innovation • Innovation policies focusing on the domestic • Weak intellectual property protection • Introduction of foreign investment mainly in low technologies	• A national strategy and objectives for autonomous innovation, with a focus on shaping a national innovation system adapted to an autonomous innovation strategy • Innovation networks to improve the interaction of knowledge production, diffusion, and application • Internationalization as a component of innovation policy • Increased protection of intellectual property rights • Introduction of foreign investment mainly in medium to high technologies	• Forming a national innovation system of independent innovation • Forming a strong social network for the production, spread, and diffusion of new knowledge and technologies • Strong influence and radiance in global innovation network • Internationalization as a strategic goal and means of innovation • Strong protection of intellectual property rights • Introduction of foreign investment mainly in high technology

(*Continued*)

Table 2.1 (*Continued*)

Period	Dependent	Catching-up	Autonomous
Function	• Local industries lack international competitiveness and are mostly based in the domestic market • Participation in global production networks in a subordinate manner, with a focus on processing production • At the lower end of the value chain in the global division of labor	• Fostering new knowledge in several fields and production capacity of new technologies through technological learning • Formation of internationally competitive brands in several industries • Proactive participation in global innovation networks and active cultivation of advantages in strategic emerging industries • Strive to move from the low end to the middle and high end of the value chain in the global division of labor	• Local industry with strong international competitiveness • Strong capacity to produce and apply new knowledge and technology • Forming internationally competitive brands with intellectual property advantages and leading the development of several industries • At the high end of the value chain in the global division of labor

Source: Liu *et al.* (2015).

Based on its structural landscape and organizational mechanisms, RIS can be categorized into different types with a diversity of typologies. In terms of governance structure, RIS includes Grassroots (locally driven, bottom-up, informal coordination), Network (mixed national and local leadership), and Dirigiste (state-led, top-down, formal coordination) (Cooke & Morgan, 1998a). The governance actors are grouped into entrepreneurship, public–private sector partnerships, and developmental government, respectively. From the perspective of business innovation, RIS is identified as Localist (led by small- and medium-sized enterprises, SMEs), Interactive (strong ties among different innovators), and Globalist (dominant

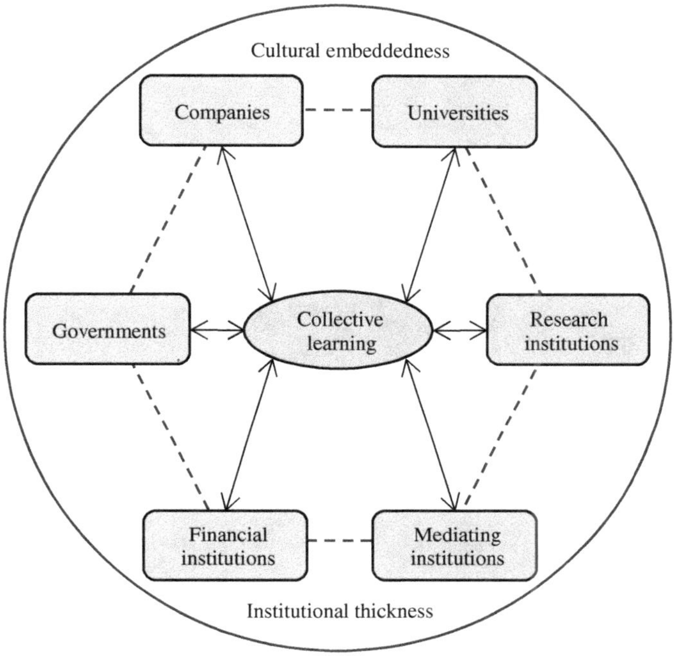

Figure 2.5 Internal linkage of RIS.

transnational corporations) (Cooke & Morgan, 1998a) (Table 2.2). In terms of the in-between roles of its intermediaries, RIS can be expressed as institutional regional innovation system (or IRIS) and entrepreneurial regional innovation system (or ERIS) (Cooke, 2014). According to its knowledge organization, flow, and stimulus, RIS is also consisted of Territorially Embedded Regional Innovation Network (TERIN) (Grassroots RIS, localized learning processes stimulated by geographical, social, and cultural proximity), Regionally Networked Innovation System (RNeIS) (Network RIS, localized and interactive learning stimulated by planned, systemic and networking embeddedness), Regionalized National Innovation System (RNaIS) (Dirigiste RIS, more linear learning and innovation cooperation with actors outside the region based on same education or common experience) (Asheim & Isasken, 2002).

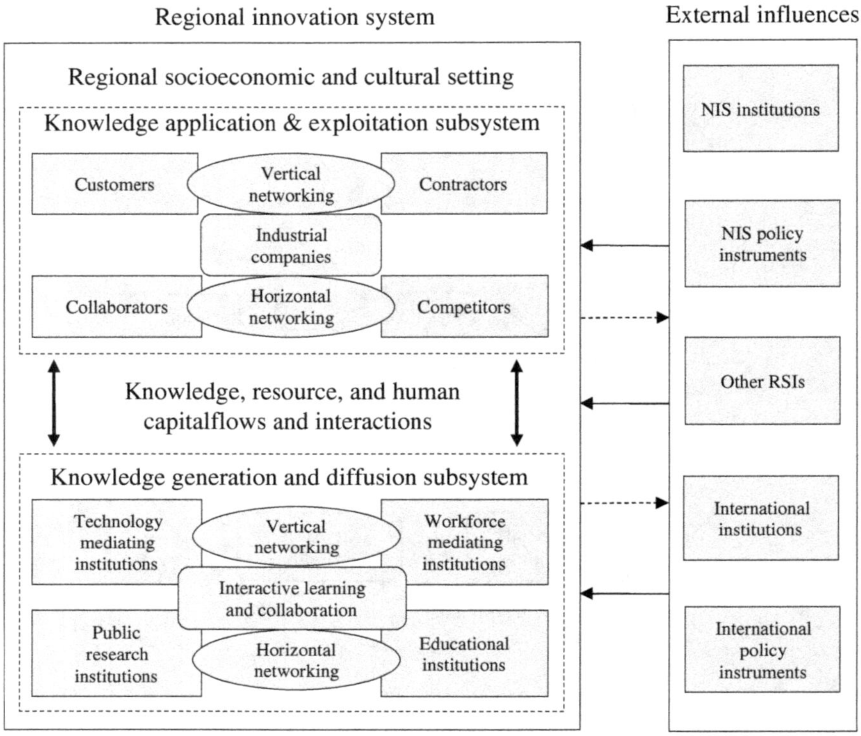

Figure 2.6 Structure of RIS.

Source: Redrawing from Autio (1998).

Table 2.2 Classic types of RIS.

Types		Governance structure		
		Grassroots	Network	Dirigiste
Business Innovation	Localist	Tuscany in Italy	Tampere in Finland	Northeast Japan
	Interactive	Catalonia in Spain	Baden-Württemberg in German	Gyeonggi in Korean
	Globalist	Ontario in Canada	Welsh	Singapore

Source: Cooke & Morgan (1998a).

The evolution of RIS is stage-specific (Liu *et al.*, 2016) and has become a research frontier in economic geography. Many studies have shown that the evolution of RIS is a response to continuous changes in both internal and external innovation settings (Glückler & Doreian, 2016), which manifests as the adjustment of network structure at the macro level and the evolution within network organization at the micro level (Richard & Doh, 2001). Like an evolutionary organism, RIS has a similar life-cycle and self-organizational process, including an initial stage, growth stage, maturity stage, and renewal or decline stage, as well as an innovation ecosystem, owing to "evolution turn" of innovation system in the 1990s.

2.1.1.3 *Innovation ecosystem*

The concept of Innovation Ecosystem (IE) originated from the Business Ecosystem in bionomics and was first proposed by Moore (1993). It was developed with a relational analysis focused on skill sharing and value cocreation through the interactions of competition and cooperation (Bassis & Armellini, 2018).

Mainly categorized as ecological, the IE considers national or regional innovation as a competitive and collaborative process formed by a series of innovative species or communities that adapt to changes and perturbations in the innovation context (Bloom & Dees, 2008). The theory emphasizes the interaction and synergistic innovation between different innovators, referred to as innovation ecosystem species and communities, and their innovation context (called innovation ecosystem soil). It is characterized as holistic, dissipative, complex, and regulatory (Figure 2.7).

The innovation ecosystem develops at different scales: the enterprise innovation ecosystem at a micro scale (Moore, 1993), the industry innovation ecosystem at a meso scale (Benitez *et al.*, 2020), and the regional or national innovation ecosystem at a macro scale (Fukudaa & Watanabe, 2008). Its theoretical framework encompasses four essential areas: its key constructive elements, their typical patterns of element configurations, the life-cycle process, and the nurturing strategies/processes from a company perspective (Rong & Shi, 2014).

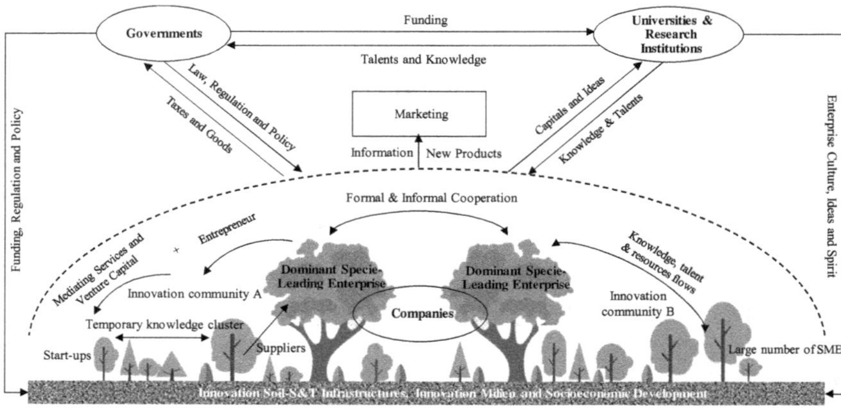

Figure 2.7 Structure of IE.

Source: Redrawing from Hu *et al.* (2016).

The key species of the innovation ecosystem are the innovation "predators", including government (research funding and political options), academia (talent training and R&D), and industry (human and risk capitals), which are conducted as an evolutionary model — the Triple Helix model of academia–industry–government relations (Etzkowitz & Leydesdorff, 1995; Etzkowitz, 2008) (Figure 2.8).

From an innovation chain perspective, the innovation ecosystem serves five major functions: knowledge generation, knowledge application, technology diffusion, technology-driven development, and social transformation, which are similar to those in NIS and RIS. The functional evolution of the innovation ecosystem follows four patterns: genetic, variation, selection, and embedding, and mainly presents four ecological forms, including the "innovation grassland", "innovation nursery", "S&T plantation", and "innovation rainforest" (Figure 2.9). Additionally, the IE exhibits seven major stages, such as initial trickle, surging spring, hot spring, rough sea, torrent river, calm lake, and shrinking pond (Zhang, 2018) (Figure 2.10). Regarding business organizational form, the IE can be categorized into eight paradigmatic stages: modularized technology (e.g., IBM) and flexible organization (e.g., HP), evangelistic-vision community

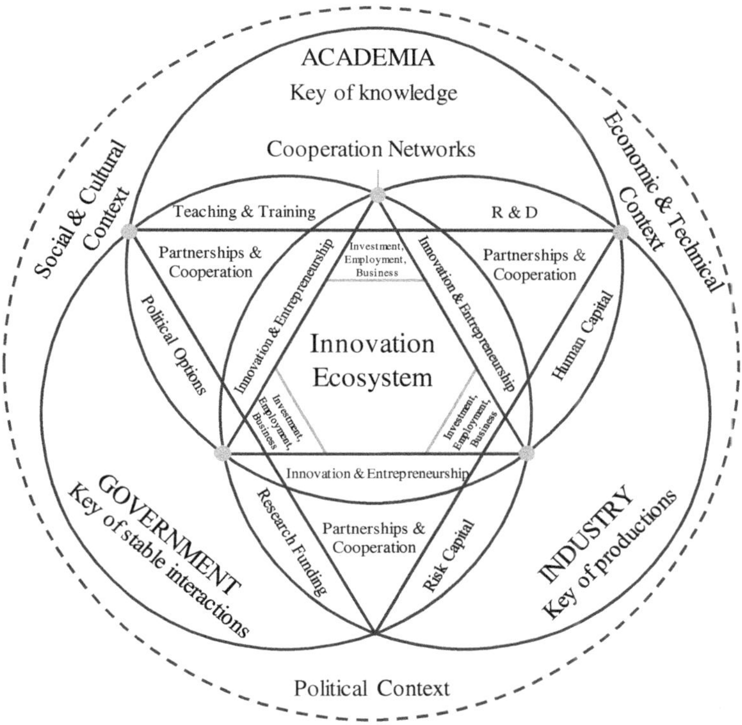

Figure 2.8 Structure of IE.
Source: Redrawing from MIT.

(e.g., Apple), open alternative (e.g., Sun and AT&T), block innovation (e.g., Compaq and IBM), community of volunteers (e.g., IBM and HP), Web 2.0 (e.g., Yahoo! and Google), justice system (e.g., Microsoft and Netscape), organizational ecosystem (e.g., ICT firms), etc. (Moore, 2006).

2.1.1.4 *Global innovation network*

Networks are the basis for interactive learning and innovation (Glückler, 2013; Liu *et al.*, 2013). Due to global production network (GPN), they appear to bring more globalization of innovation into being, which results in the emergence of global innovation networks

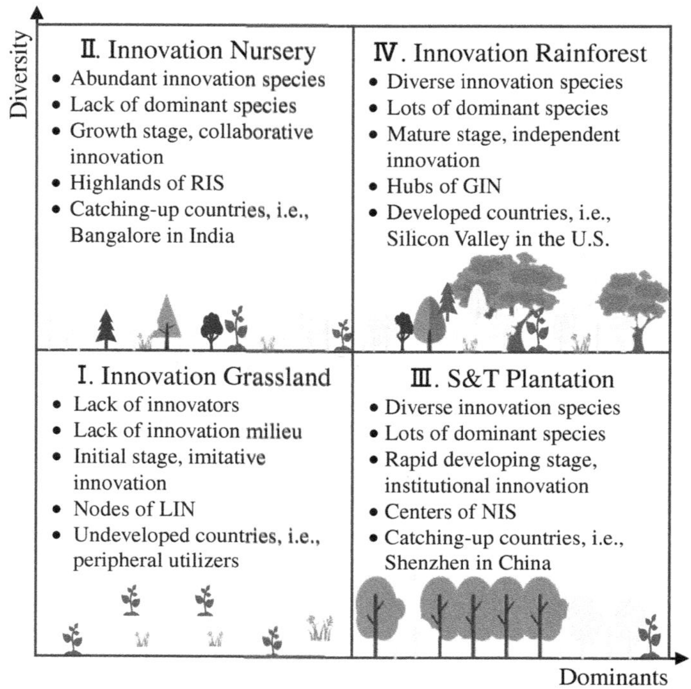

Figure 2.9 Ecological types of IE.

Source: Redrawing from Zhang (2018).

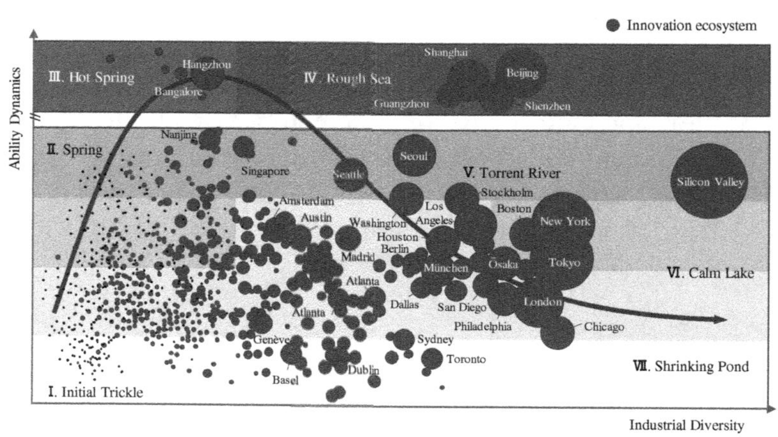

Figure 2.10 Evolution of IE.

Source: Redrawing from Du (2015); Du *et al.* (2022); Zhang (2018).

(Rycrof & Kash, 2004; Ernst, 2002) or globalizations of innovation networks (Laberge, 2011). The concept of the Global Innovation Network (GIN) was early proposed by Ernst (2009). He pointed out that GIN emerges as a natural extension of GPN and is constructed by global corporations ("global network flagships").

GIN shares three significant characteristics with the GPN that preceded them (Ernst, 2002): *asymmetry* (with multinational corporations dominating as network flagships and defining network organization and strategy), a variety of *governance* structures (ranging from loose linkages to highly formal networks), and *knowledge sharing* (aimed at gaining quick access to skills and capabilities at lower-cost overseas locations) (Ernst, 2009). Chaminade *et al.* (2015) further pointed out three specific characteristics of GIN in terms of its geographical spread (global), the nature of the interactions (networks) and the outcomes (innovation). Although its emergence is viewed as a long-standing deepening or add-on function of GPN and global value chain (GVC), GIN has a particular focus on knowledge exchange rather than production (Chaminade *et al.*, 2015).

To understand GIN, scholars have emphasized the networking interactions of different organizations. GIN is defined as a globally organized web of collaborative interactions between various organizations engaged in knowledge production including firms and non-firm organizations (Barnard & Chaminade, 2011). According to innovator's differences, GINs consist of the global technological network driven by transnational corporations, the global scientific network led by universities and research institutions, and the government-led service innovation system (Du, 2015; Si *et al.*, 2016) (as illustrated in Figure 2.11). Among these actors, flagship firms (or "leadership firms") are regarded as the most crucial innovators and constructors of GIN (Archibugi & Michie, 1995; Binz & Truffer, 2017).

Based on Asian cases, GIN can be distinguished into five types: intrafirm networks (offshore stage), intrafirm networks (outsource stage), regional own networks (mostly intrafirm), international R&D networks (public-corporate R&D consortia), informal social networks (specialized knowledge communities) (Ernst, 2009).

Figure 2.11 Main types of GINs.

GIN is not only models of network innovation and networking organizations of globalized firm alignments and their R&D activities in transnational contexts (Rothwell, 1992; Barnard & Chaminade, 2011; Binz & Truffer, 2017) but also acts as a series of important "global pipelines" for local access to external knowledge and technology (Bathelt *et al.*, 2004) and a major driver of regional competitiveness (Bathelt & Henn, 2014).

Additionally, GIN is also a significant result of the internationalization of NIS (Niosi & Bellon, 1994) and the integration of cross-border RIS (Makkonen *et al.*, 2016), which encourages the expansion and renewal of innovation pathways in various innovation systems (Aslesen *et al.*, 2017) (Figure 2.12). As such, GIN serves as a complementing or compensatory mechanism for local innovation system (or local innovation clusters), regional innovation system, and national innovation system (Grillitsch & Nilsson, 2015), which is named as the glocalization of GINs (Figure 2.12).

Recently, scholars have conducted extensive research on the evolution of GIN, revealing a trend from R&D globalization to global R&D network (Saint-Pierre, 1997; Wang *et al.*, 2014), from industrial clusters to global industrial cluster networks

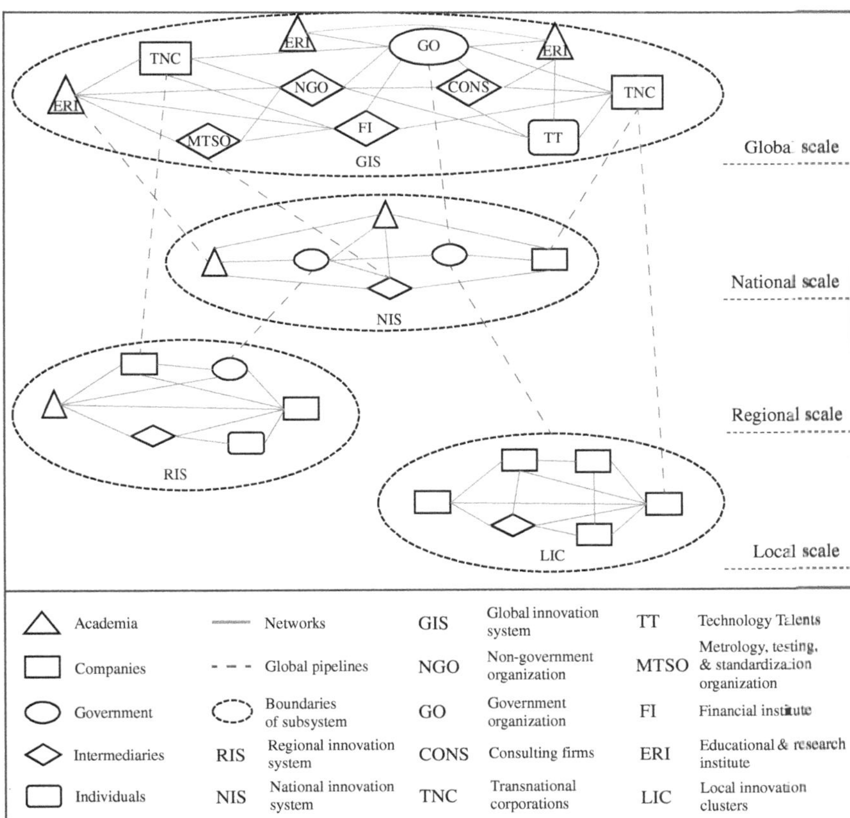

Figure 2.12 Integrating framework of LIC, NIS and RIS.

Source: Redrawing from Binz and Truffer (2017).

(Bathelt & Li, 2016; Turkina *et al.*, 2016), and from globalization of innovation networks to glocalization of innovation networks (Laberge, 2011; Si *et al.*, 2016; Wang *et al.*, 2016). The dynamics of GIN is primarily impacted by endogenous and exogenous mechanisms (Zhao *et al.*, 2019), such as the endogenous effects of network structure and nodal embeddedness (Buchmann & Pyka, 2015), the influence of organizational elements (Boschma *et al.*, 2007; Glückler & Doreian, 2016), the exogenous effects of innovation context (Mckelvey *et al.*, 2003), and the role of multidimensional proximities (Balland, 2012; Caragliu & Nijkamp, 2016; Crescenzi *et al.*, 2016; Liu *et al.*, 2017; Gui *et al.*, 2018a, 2018b). Among them, the network structure and multidimensional proximity are the internal drivers of the network, whereas the network's innovation context are the external determinants of the network.

2.1.1.5 *Glocal innovation network*

Globalization has gone with a dual phenomenon, with the strengthening of regionalism worldwide, referred to as glocalization) (Lysak, 2003; Swyngedouw, 1997, 2004; Swyngedouw & Kaka, 2003). In the context of innovation glocalization, various theoretical frameworks have been further developed from a global and local interaction perspective, such as *Local Buzz and Global Pipelines* (Bathelt *et al.*, 2004), *Local Sticky and Global Ubiquitous* (Asheim & Isaksen, 2002), *Global Cluster Network* (Bathelt & Li, 2014), and *Strategic Coupling* (Yeung, 2016), among others.

Based on these frameworks, Si *et al.* (2016) were the first to propose the concept of a Glocal Innovation Network (GLIN), which has important implications for technology catching-up countries. GLIN is defined as the sum of knowledge network channels of various innovators, which increasingly connect globally scattered innovation resources together (Si *et al.*, 2016). It is viewed as a global network system insisted of a series of local innovation network subsystems connected by global pipelines (Figure 2.13).

The innovation capacity of a specific region is highly related to its network position and network capability in GLIN. According to nodal

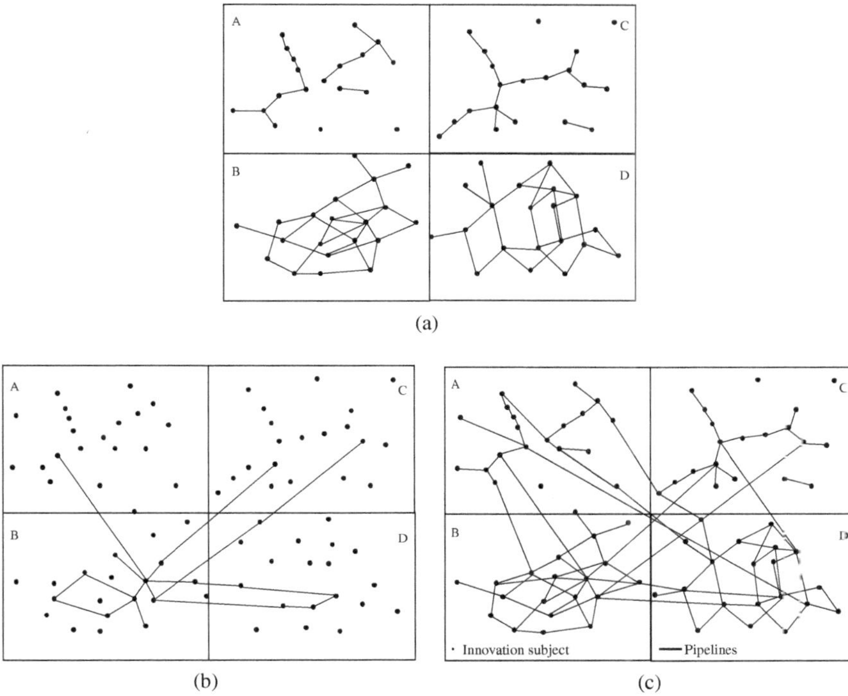

Figure 2.13 Structural comparison of RIS, GIN, and GLIN. (a) Regional innovation system; (b) Global innovation network and (c) Glocal innovation network.
Source: Redrawing from Si *et al.* (2016).

position and concentration, a region in GLIN can be classified into four types: *knowledge-access agglomeration* (direct access to different knowledge pools due to high centrality, such as Beijing in China), *knowledge gateway* (potential to disseminate internal knowledge, such as Hong Kong and Tianjin in China), *knowledge bypassed agglomeration* (a cluster of scientific activities that lacks a leading position, such as Wuhan in China), and *knowledge peripheral utilizer* (less relevant in general) (Liefner & Hennemann, 2011) (Figure 2.14).

GLINs are organized through negotiations between industrial associations and technology alliances and their members (Si *et al.*, 2016). The linkages within GLINs are distinguished into three channels: vertical intra-firm cooperation (upstream and

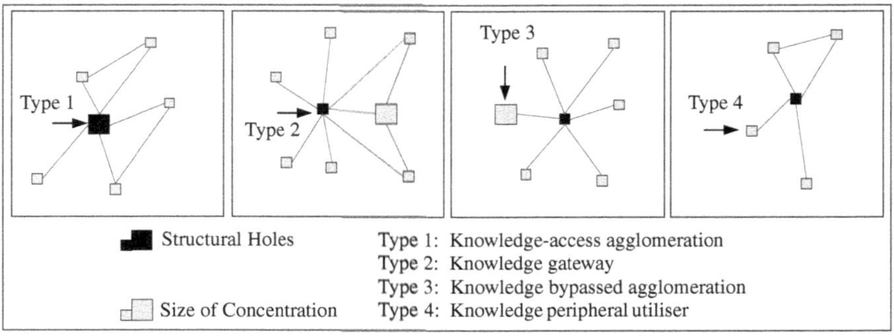

Figure 2.14 Position and concentration of a region in GLIN.
Source: Redrawing from Liefner and Hennemann (2011).

downstream links within global industrial chain and innovation chain), horizontal university–industry–research cooperation, and intra-firm sectoral collaboration (Si *et al.*, 2016). Through these channels, local sticky and global ubiquitous knowledge are transferred, diffused, integrated, and transformed in GLINs (Nonaka & Takeuchi, 1995; Asheim & Isaksen, 2002; Haasis *et al.*, 2018). GLINs not only integrate coded knowledge (trans-border, formal and long-distance technology transactions) and uncoded knowledge (face-to-face, local and informal communication, and learning) (Asheim & Isaksen, 2002) but also transform them mutually through the SCEI model (socialization, externalization, combination, and internalization) (Nonaka & Takeuchi, 1995) (Table 2.3; Figure 2.15).

2.1.2 *Technology transfer theories*

The definitions and theories of technology transfer have been discussed in various ways across different disciplines and for research purposes (Bozeman, 2000). Technology transfer is often used to describe the process of transferred activities from laboratory to marketplace (Phillips, 2002), from technologically developed to less-developed countries (Radosevic, 1999), and from origins to secondary adopters (William & Gibson, 1990). These definitions are respectively referred to as process-embodied, product-embodied,

Table 2.3 Structural characteristics of innovation networks.

Types	Main innovators	Key linkages	Dominant knowledge
Global innovation networks	Intra-firms	Vertical cooperation between headquarters and branches	Explicit and analytic knowledge
Local innovation networks	Inter-firms, firms-institutions	Horizontal cooperation among users, institutions, firms, academia	Tacit and comprehensive knowledge
Glocal innovation networks	Umbrella organizations	Vertical cooperation (industrial associations and technology alliances) and horizontal cooperation (universities, industries, and research institutions)	SCEI of tacit and explicit knowledge

Source: Redrawing from Si *et al.* (2016); Nonaka and Takeuchi (1995).

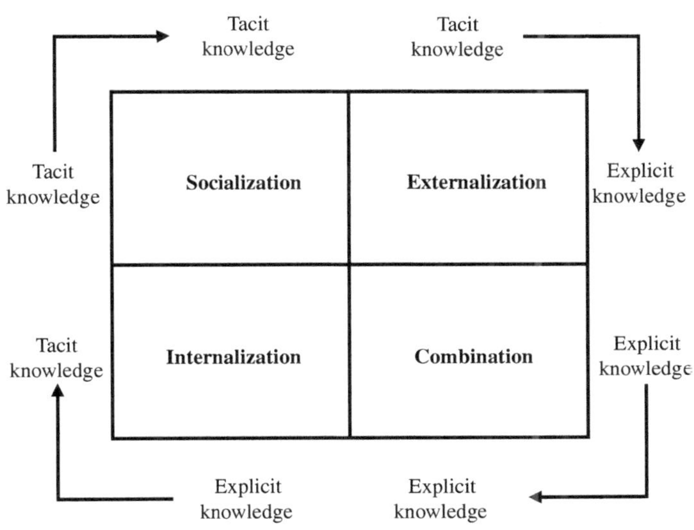

Figure 2.15 SCEI model of knowledge transformation.

Source: Redrawing from Nonaka and Takeuchi (1995).

and person-embodied technology systems (Hall & Johnson, 1970). Based on these three perspectives, several theories related to technology transfer have been continuously developed, such as *International Technology Trade* (product-related approach, noticeably in *Technology Gap* and *Technology Transfer Choice*, etc.), *Product and Technology Life-cycle* (process-related approach), and *Diffusion of Innovations* (person-related approach), among others.

2.1.2.1 *Technology gap*

In this view, the Technological Gap Theory and Model, initially proposed by Posner (1961) to explain the impact of technological change on the pattern of international trade, is discussed. It was further developed by scholars like Vernon (1966) and Basu and Weil (1998). In this context, the technology gap can be seen as how advanced or efficient a country's advanced technology is in comparison with the cutting-edge technologies globally available, resulting in the transfer of technology from its core group (innovation countries) to its periphery group (imitation countries) for the "imitation lag".

Technologically, the lag can be classified into three stages: *demand lag, reaction lag,* and *mastery lag.* The first is the time lag between a new technology being produced by innovators and imitators importing the technology. The second is the time lag for the imitators from importing new technology to adjusting and producing it. While the third is the time interval for the imitators to catch up with the innovators after they begin the production of its new technology (Posner, 1961) (Figure 2.16).

The concept of technological gap has been divided into two categories: transferring gap and cumulative gap (Dosi, 1984). The transferring gap refers to the difference in technological life-cycle stages between developed innovation countries and developing imitation countries, which is also known as the external gap. On the other hand, the cumulative gap refers to the lack of absorptive capacity and proficiency processes within peripheral countries due to a shortage of high-skilled technicians who can effectively accept

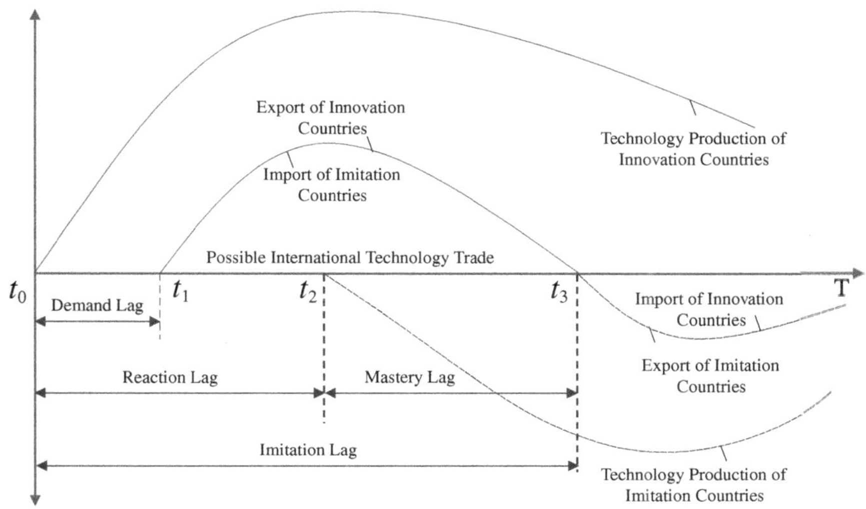

Figure 2.16 Theoretical framework of technology gap.

Source: Redrawing from Posner (1961).

new technologies. This gap is also known as the internal gap. These gaps are illustrated in Figure 2.17 with the vertical axis representing the transferring gap and the lateral axis representing the cumulative gap.

The technological gap has its origins in two drivers: the innovation of leaders and the limitation of latecomers. The application of new technology by technological leaders enlarges the gap between developed and developing countries, while the introduction of external technology fosters technical convergence through technology spillover or diffusion (Fagerberg, 1994). However, the transfer of imported technology does not always work and depends on the absorptive capacity of the latecomers (Fracasso & Marzetti, 2014). Many studies have shown that the technology gap has a two-fold effect on technology transfer. Firstly, there is a substitutive relationship under an oversized technology gap which results in reduced absorptive capacity (Kumar, 1987). This effect limits the spillover benefit of transferred-in technology in host countries, leading to what is known as the "crowding-out effect". Secondly, there is a

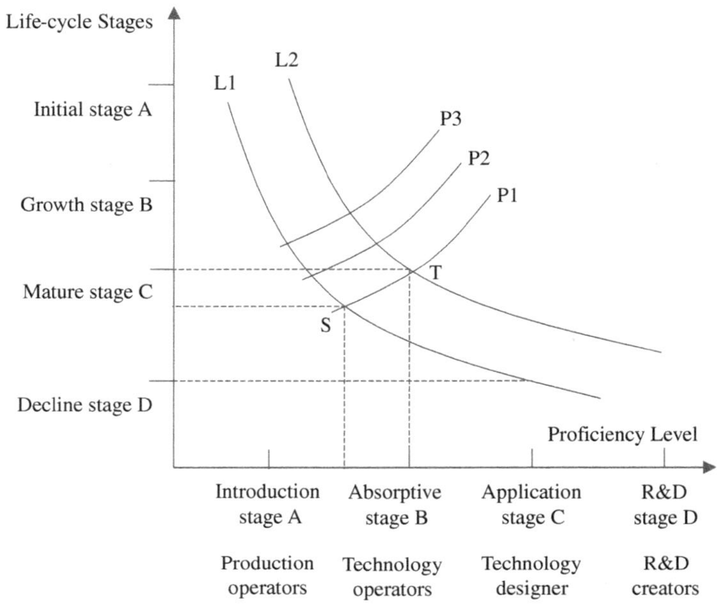

Figure 2.17 Dual types of technology gap.
Source: Redrawing from Kim (1997).

complementary relationship under an appropriate technology gap (Kim & Stewart, 1993). Imported technology is well absorbed and applied to improve the innovation capacity of the host countries, as shown in Figure 2.18.

The relationship between technology gap and technology transfer is dynamic, and can be seen in the periodic inverted U-shaped curve as observed by Driffield *et al.* (2009). When the technology gap is relatively small (Gap$<G_1$), the imported technology transfer may have an obvious crowding-out effect on independent R&D investment of the host country in the short term due to a lack of absorptive capacity. However, when the gap surpasses the minimum threshold G_1, the technology spillover effect on the host country will increase until it reaches the inflection point *B*. As the gap continues to widen toward G_3, the technology spillover effect will gradually diminish until it reaches

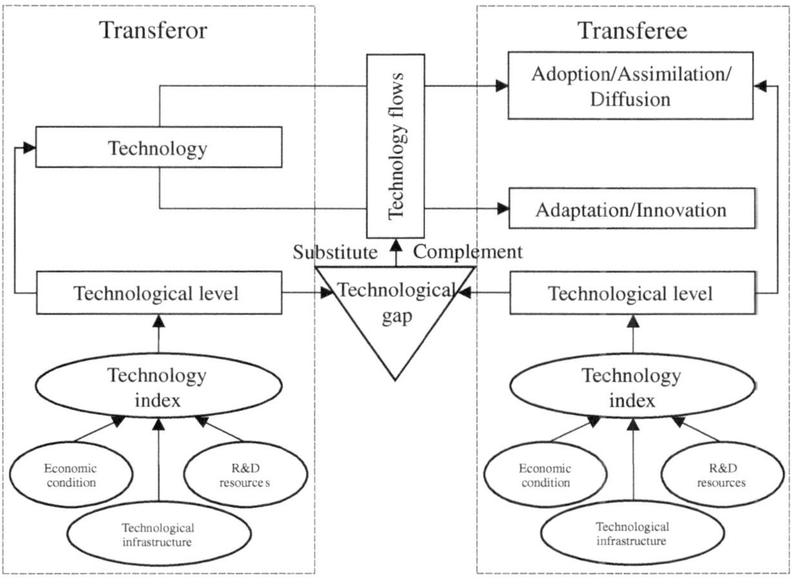

Figure 2.18 Two-fold effects of technological gap on technology transfer.
Source: Redrawing from Sharif and Haq (1980).

its maximum, and the range between G_1 and G_3 is considered to be the appropriate technology gap range. During this period, enterprises in the host country can fully absorb the imported technologies, leading to technological convergence between the two countries due to the higher absorption capacity of the host. However, when the gap is further reduced to exceed G_3 (Gap>G_3), it begins to play a negative role in the technology spillover despite its increasing effect. This is because the reduction of technology selection leads to a decrease in the marginal effect of the technology spillover. Consequently, the technology spillover effect moves from decrement to increment, reaching the turning point *D*. Furthermore, the period between points *E* and *F* is a crucial time window that sees fierce international competition for technology advancement, with point *F* marking the beginning of a new round of technology spillover from innovation forerunners, as shown in Figure 2.19.

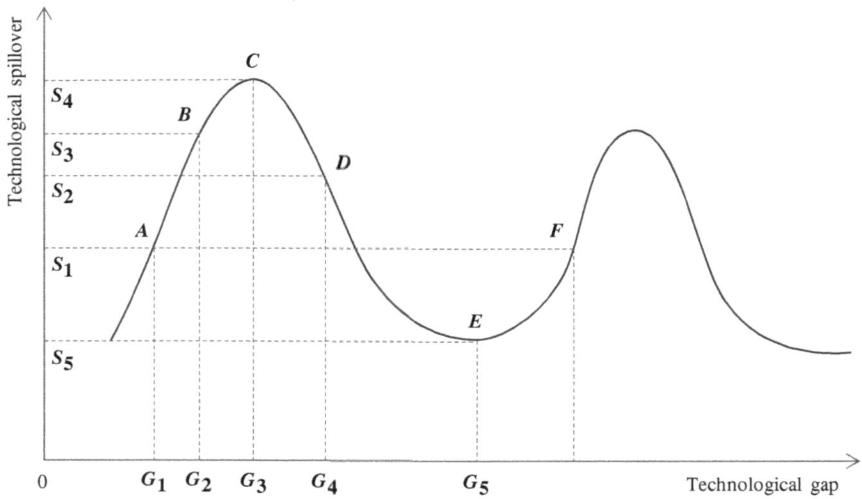

Figure 2.19 Dynamic effects of technological gap on technological transfer.
Source: Redrawing from Driffield *et al.* (2009); Yu (2009).

Moreover, the relationship between technology gap and technology transfer can be seen as a circular coupling process, as depicted in Figure 2.20. In the initial stage, there is a larger technology gap between host and home country enterprises, while the host's absorption capacity is smaller. As a result, there is only a low-level and small-scale technology spillover between them through the "demonstration effect". However, in the maturity stage, greater competitive pressure prompts forerunning enterprises to increase their R&D and human capital investments to enhance their technology absorption capacity. This leads to an enlarging technology competition. On one hand, the follower achieves a breakthrough of its absorptive capacity and technology catch-up, narrows its technology gap with the home country, and enlarges its technology export. On the other hand, the home country has to improve the speed and level of technology transfer, increase R&D investment, and control intellectual property to consolidate its technological leadership. This results in the widening of the gap between them. International competition, industry linkage, innovation milieu, and absorptive

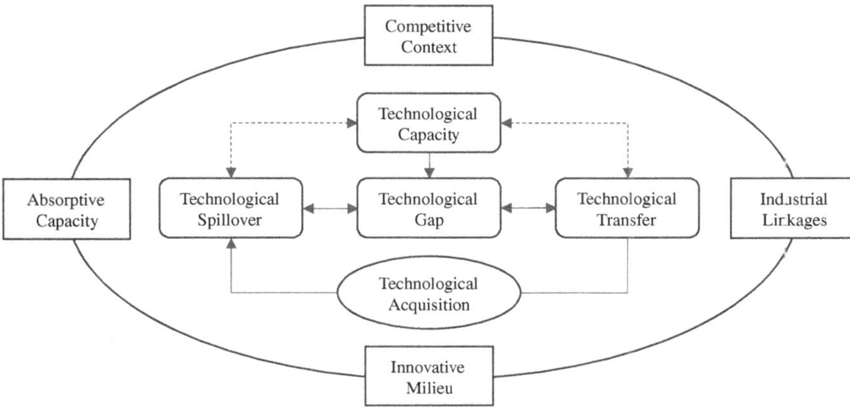

Figure 2.20 Coupling effects of technological gap on technological transfer.
Source: Redrawing from Qi (2013).

capacity are among the crucial external factors influencing these relationships.

2.1.2.2 *Technology transfer choice*

The Technology Transfer Choice Theory was mainly developed by Mansfield (1975), Saito (1979), and Dunning (1977, 1981), to demonstrate the substitution relationship among production trade, foreign direct investment (FDI), and technology transfer.

In order to maximize their profits, transnational corporations often resort to technology transfer when they are unable to obtain monopoly profits from exports and FDI (Mansfield, 1975; Mansfield *et al.*, 1979). International technology transfer is seen as an eclectic choice that depends on ownership advantages (firm-specific competitive advantages), locational advantages (country-specific factors related to market potential and risk), and internalization advantages (costs of a hierarchical mode of operation over an external mode) (Dunning, 1981). Enterprises only choose the mode of technology transfer when it is difficult to form economies of scale through internalizing markets, and when the locational advantage is not significant (Table 2.4).

Table 2.4 OIL paradigm of technology transfer.

Modes	Ownership advantages	Internalization advantages	Locational advantages
Export trade	Yes	Yes	No
FDI	Yes	Yes	Yes
Technology transfer	Yes	No	No

Source: Redrawing from Dunning (1981).

Based on the three comparative advantages, the priorities for enterprises are product exports, FDI, and technology transfer. To continually maximize their benefits, enterprises cycle through these three choices (Saito, 1979). Initially, enterprises with new technology use it to produce new products for export, gradually increasing the rate of profit as the market expands. However, when the product can be produced using local factors of production, the rate of profit falls, and enterprises switch from exporting goods to direct investment to bring the rate of profit back up. As local production and sales increase, the local application of technology leads to imitation in the market, causing the rate of profit for direct investment to shift from rising to falling. At this point, enterprises shift from direct investment to technology transfer to increase the rate of return on the technology (Figure 2.21).

2.1.2.3 *Technology life cycle*

The Technology Life Cycle (TLC) Theory originates from the Product Life Cycle (PLC) Theory proposed by Vernon (1966). This perspective suggests that the life cycle of a new product can be compared to that of a living organism, with distinct phases of innovation and emergence, growth and maturity, and eventual standardization or decline. The duration of each phase is determined by the technological gaps between different countries, which are classified into innovative countries (the most developed), more developed countries, and developing countries (Figure 2.22).

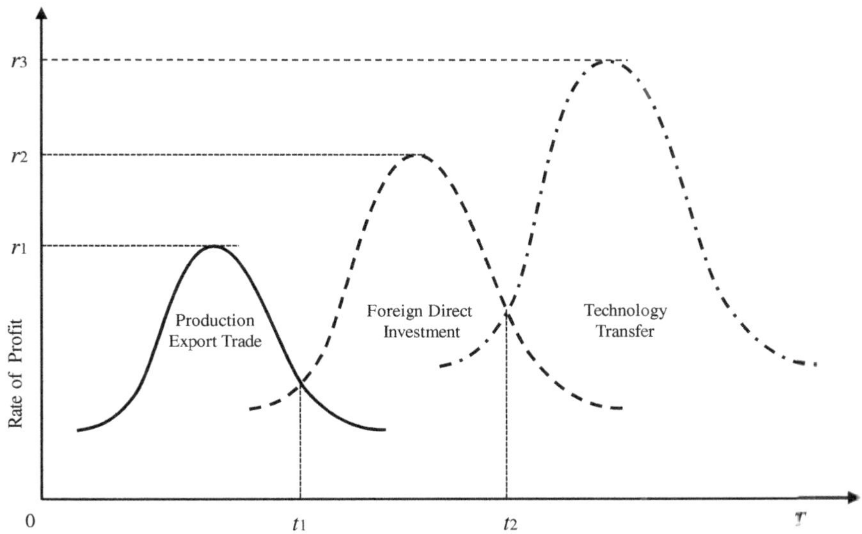

Figure 2.21 Periodic cycle of technology transfer choice.
Source: Redrawing from Saito (1979).

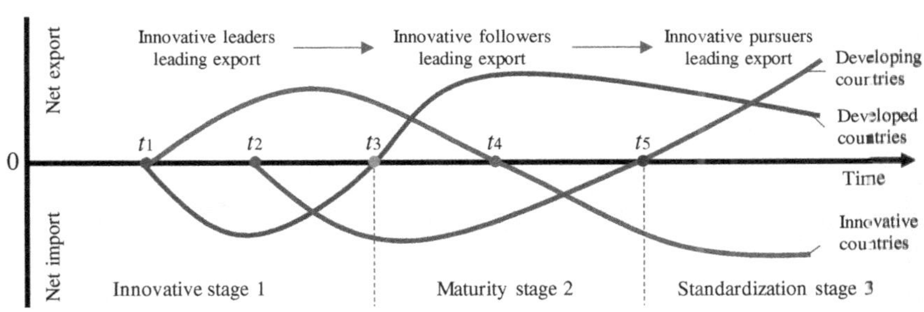

Figure 2.22 PLC of different countries.
Source: Redrawing from Vernon (1966).

The cyclical model is the leading framework for understanding technological dynamics and innovation trajectory over time (Abernathy & Utterback, 1978; Anderson & Tushman, 1991; Murmann & Frenken, 2006). In the evolution of technology changes, there generates long-run fluctuations following S-shaped (or logistic) curve consisted of four changes: (1) *Era of technological discontinuity*: revolutionary, breakthrough, radical, emergent, step-function technologies; (2) *Era of technological ferment*: variation, technical uncertainty, potential users have unclear preferences; (3) *Era of technological standardization*: industry standard and dominant design emerges; (4) *Era of technological incremental change*: retention, evolutionary, continuous, incremental, nuts-and-bolts technologies (Taylor & Taylor, 2012).

In terms of technology sourcing process, TLC comprises three phases which includes the *fluid phase* (external cooperation is an appropriate strategy to face the uncertainty of technologies and markets), the *transitional phase* (companies focus on internal technology sourcing to improve their innovation performance), and the *specific phase* (firms focus more on internal technology sourcing rather than external sourcing due to cognitive distance and knowledge bases) (Albernathy & Utterback, 1978; Stolwijk *et al.*, 2012).

In addition, an alternative perspective on the life cycle of technology is measured by the rate of innovation as a function of time. According to Epicoco (2021) and as illustrated in Figure 2.23, the rate of innovation for a specific technology follows four sequential

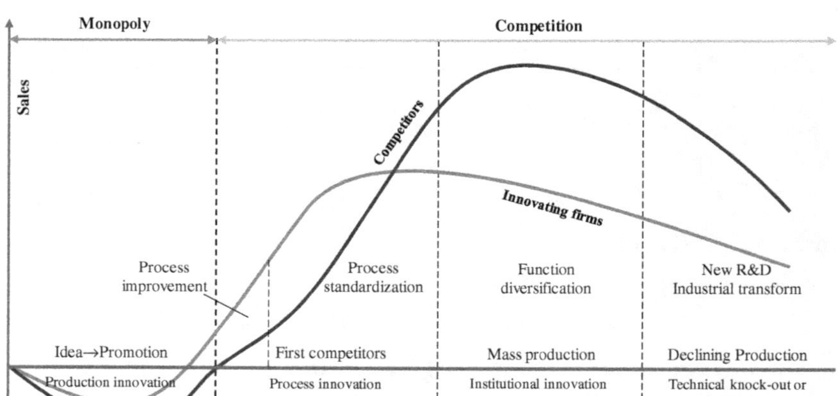

Figure 2.23 Life cycle of technology and production.

Source: Redrawing from Vernon (1966); Saito (1988).

steps: emergence, growth, maturity, and decline. Initially, the rate of innovation slowly increases at the beginning of technology breakthrough. As the technology continues to grow and mature, the rate of innovation steadily increases. During the mature phase, the rate of innovation is maintained at a constant level. Finally, the rate of innovation decreases as the technology becomes less popular and is replaced by other technologies (Lee & Su, 2015).

TLC framework has also been applied in technological and industrial forecasting, most notably in the "Hype Cycle Model" (technology maturity curve) proposed by Gartner in 1995, a leading global forecasting and consulting firm. The model depicts a common pattern that arises with each new technology or innovation, providing a graphic depiction to track its maturity and future potential. It is divided into five distinct phases: technology trigger, peak of inflated expectations, trough of disillusionment, slope of enlightenment, and plateau of productivity (Gartner, 2022) (Figure 2.24).

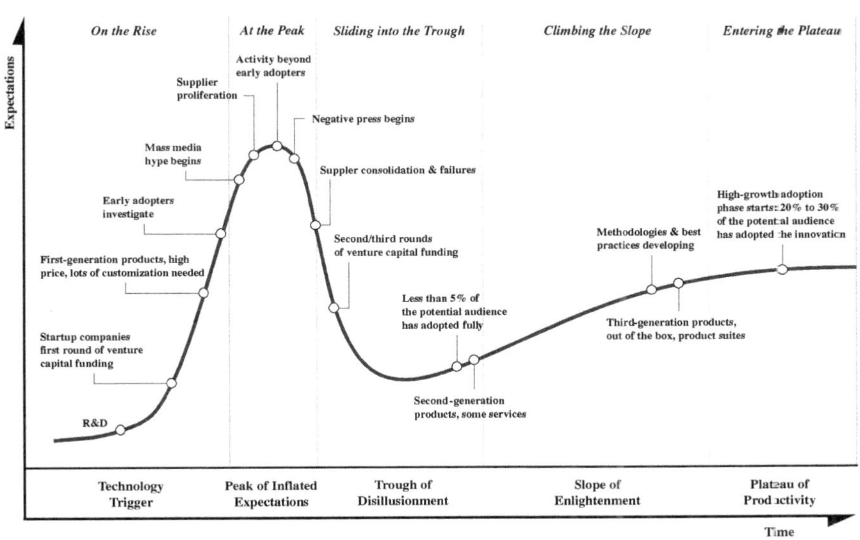

Figure 2.24 Hype cycle curve.

Source: Redrawing from Gartner (2022).

Table 2.5 Indicators of TLC.

Factors		Indicators
R&D profiles	Fundamental research	No. of items in SSCI/SCI databases
	Applied research	No. of items in EI databases
	Development	No. of items in official patents databases
	Application	No. of items in some newspapers databases
	Societal impacts	Issues raised in some business & popular press abstracts
Growth rate		Trends over time in number of items
Technological issues		Technological needs noted
Maturation		Types of topics receiving attention
Offshoots		Spin-off technologies linked

Source: Watts and Porter (1997).

The technology life cycle was evaluated and forecasted by multiple indicators describing the typical sources that report the findings in each stage of R&D, including technology citations (i.e., literature and patent citations or references), technology-related issues (i.e., abstracts & news), technology spin-offs (patent applications, grants, and activities), technology claims, and technology actors (i.e., inventors, assignees, technology developers) (Watts & Porter, 1997; Haupt *et al.*, 2007; Lee *et al.*, 2016; Huang *et al.*, 2022) (Table 2.5). Among them, patent indicators beyond simple counts can deepen one's understanding of TLC (Huang *et al.*, 2022).

2.1.2.4 *Need–resource relationship*

The Need–Resource Relationship Theory, developed by Saito (1979, 1988), posits that technology development and transfer within a societal system of nations, enterprises, or individuals is determined by the "N–R" relationship between their needs (N) and resources (R). The need system comprises a series of elements ranging from individual to social to marketing needs, while the resource system consists of two subsystems: R&D resources and technology stock resources.

The mismatch between needs and resources serves as an impetus for technical development and drives international technology transfer. However, despite the presence of a technology gap, technology transfer does not occur automatically across nations or regions. Rather, there must also be an imbalance between socioeconomic needs and available resources for technology transfer to take place.

The information communication system plays a significant role in balancing the mismatch, allocating the resources, and bridging the relationship. Infrastructure is essential in facilitating this process and can be classified into two types: physical infrastructure (such as R&D institutions, educational institutions, information systems, transfer platforms and centers, transportation and telecommunication facilities, etc.) and institutional infrastructure (including patent systems, incentive systems, support systems, educational systems, R&D market systems, etc.). These infrastructures enable the aggregation of sufficient resources, and reduce failure risks, which further motivates the technology development following ideas→R&D→practicality→commercialization and the technology transfer progress from introduction, to absorption, then to spillover (Figure 2.25).

2.1.2.5 *Innovation diffusion*

Innovation occurs in a special place at a special time. The diffusion of innovation or technology can be categorized into two paradigms: over-time process and spatial pattern.

Temporally, the theory of innovation diffusion was initially proposed by Rogers (1962) and has undergone various revisions since its inception. According to Rogers (1962), the decision-making process for adopting or rejecting an innovation can be described as a progression. Diffusion is therefore considered a social process that involves the communication of an innovation through certain channels, including mass media and interpersonal channels, over time among adopters (Utterback, 1974). Furthermore, the theory also

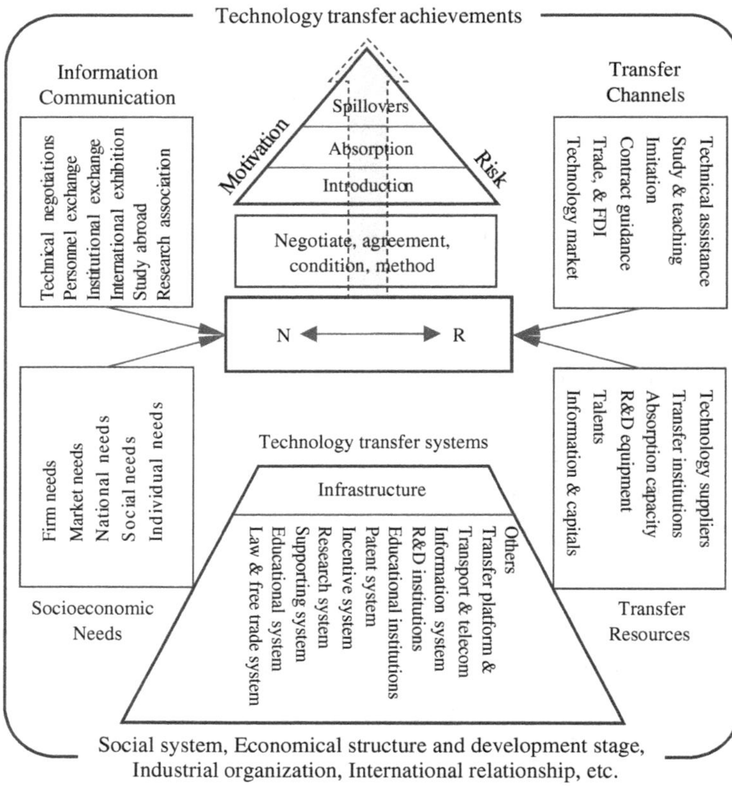

Figure 2.25 NR relationship of technology transfer.
Source: Redrawing from Saito (1979, 1988).

predicts the spread of an innovation through a social system, positing that it follows a sigmoid or S-shaped curve (Figure 2.26).

The S-shaped curve of cumulative adopters results in a bell-shaped distribution of adopters. Through the evaluation of the average time to adopt an innovation, individuals can be classified into five categories: innovators (peers who are venturesome and interested in new ideas), early adopters (gatekeepers who are more oriented toward their local peer networks), early majority (opinion leaders who adopt new ideas just before the average member does), late majority (followers who adopt new ideas after the average member has done so), and laggards (laggers who tend to orient

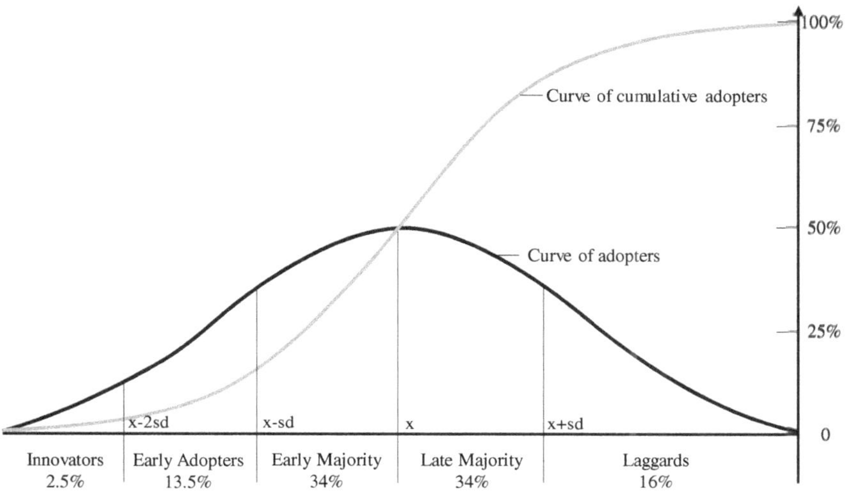

Figure 2.26 Curves of adopters and cumulative adopters.

Note: sd represents the standard deviation of average time.

Source: Redrawing from Rogers (1966).

themselves toward the past as a reference for their decisions) (Figure 2.26).

The diffusion of an innovation is comprises the five steps: knowledge (becomes aware of an innovation), persuasion (develops an attitude toward the innovation), decision (decides whether to adopt the innovation), implementation (starts using the innovation), and confirmation (continues to collect information), which represents a new approach to reduce the uncertainty about an innovation (Figure 2.27).

The rate of innovation uptake is strongly correlated with five attributes of innovations. First, relative advantage refers to the degree to which the innovation is perceived as an improvement over the previous innovations (such as concepts, solutions, etc.) that it replaces. Second, compatibility is the degree to which the innovation is consistent with current values, prior practices, and the requirements of its intended adopters. Third, complexity refers to the difficulty involved in comprehending and implementing the innovation, making it a strong barrier against adoption.

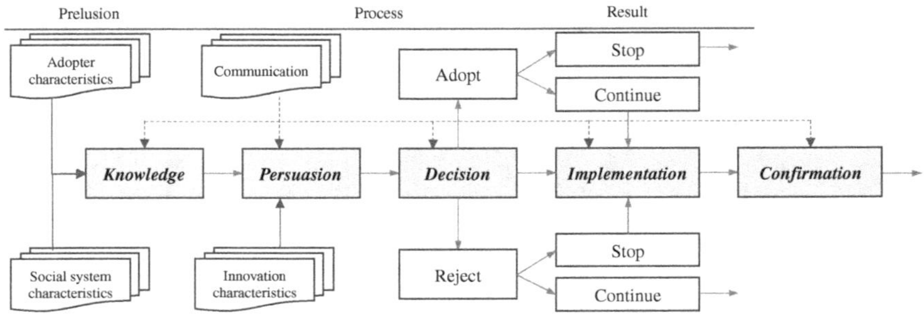

Figure 2.27 Diffusion process of innovation.
Source: Redrawing from Rogers (1962).

Fourth, trialability refers to the extent to which the innovation can be tried on a probationary basis, eliminating any doubts about its effectiveness. Finally, observability refers to the degree to which others can observe the outcomes of an innovation.

The theory of innovation diffusion, which was first introduced by Hägerstrand (1953), also pertains to the spatial diffusion of innovation. Similar to the temporal process, the spatial diffusion of innovation is dynamic and evolves in four stages: emergence, transfer, diffusion or expansion, and prevalence, as illustrated in Figure 2.28. Research has shown that distance and size from earlier adopters are important factors in explaining the spatial diffusion of innovation (Pedersen, 1970; Perkins & Neumayer, 2010). During different stages, the intensity of effective flow of innovation decays with the increasing "distance bias". Based on the two perspectives, spatial organization of innovation diffusion contains four main effects: neighborhood effect (distance decay of diffusion), hierarchy effect (spread down in the hierarchy), axial effect (spread along the infrastructure axis), and aggregation effect (spread to innovation clusters). Accordingly, spatial paths of innovation diffusion were distinguished into four major models: (1) *wave-like diffusion* ("expansion diffusion" spreading from a pole into a neighboring area, following the principle of neighborhood effect) (Casetti & Semple, 1969; Morrill, 1970), (2) *hierarchical diffusion* (innovations spread rapidly down the urban hierarchy, depending on the hierarchy effect) (Pedersen, 1970; Brown & Cox, 1971; Richardson, 1973),

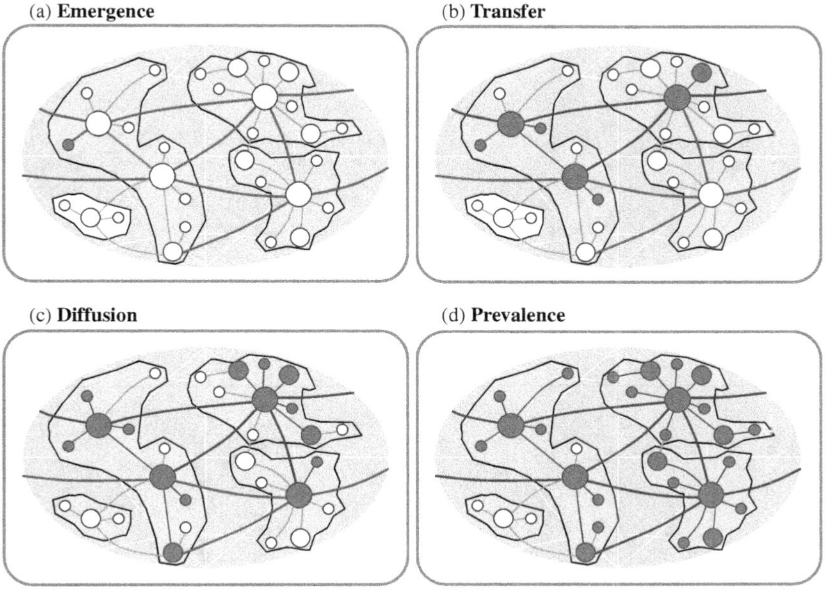

Figure 2.28 Spatial dynamics of innovation diffusion.
Source: Redrawing from Rodrigue *et al.* (2020).

(3) *relocation diffusion* ("leapfrogging diffusion" of innovators that move along the journey as well as the eventual destination), and (4) *axial diffusion* (innovation spread rapidly along some transport corridors based on the axial effect) (Levison *et al.*, 1973) (Figure 2.29). There exists a complementarity among these effects permitting their simultaneous operation and coexistence in innovation diffusion progress (Brown & Cox, 1971).

2.1.3 *Relational economic geography*

Since the mid-1990s, three propositions have been extensively observed: structural *contextuality* (economic agents are embedded in relational contexts) (Granovetter, 1985), *path dependence* of evolution (economic processes are path dependent), and *contingency* of processes (economic processes are contingent in the open-ended strategies of the actors) (Bathelt & Glückler, 2003).

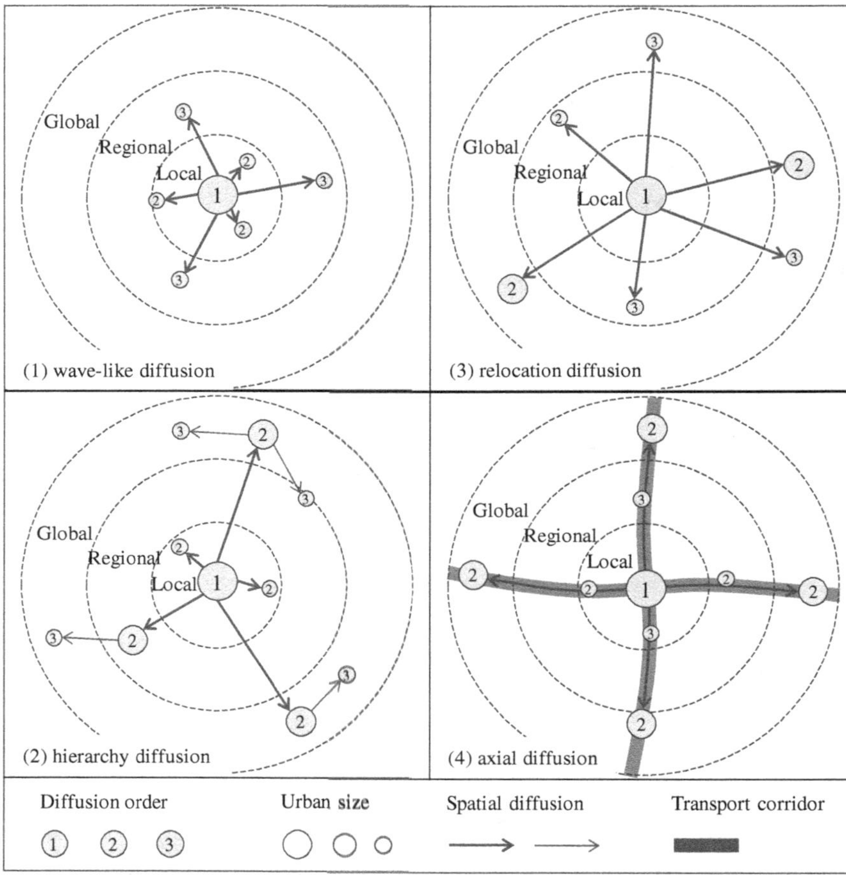

Figure 2.29 Spatial types of innovation diffusion.

The shift toward a relational economic geography (REG) has resulted from a paradigmatic change, commonly referred to as the "relation turn" in economic geography (Bathelt & Glückler, 2003; Yeung, 2005; Sunley, 2008). The basis for the analytical framework employed in REG is centered on organization, evolution, innovation, and interaction (Bathelt & Glückler, 2003) (Figure 2.30). In addition, relational assets, relational embeddedness, and relational scales are considered the "holy trinity" themes in REG (Yeung, 2005) (Table 2.6). Among them, learning fields, global–local

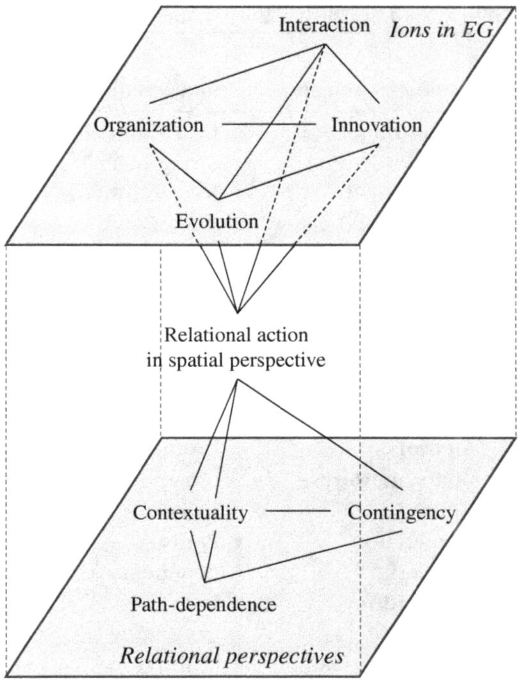

Figure 2.30 Theoretical framework of REG.

Source: Redrawing from Bathelt and Glückler (2003).

interactions, and pipeline buzz provide a conceptual apparatus to explain why and how the relationality of technology transfer matters.

2.1.3.1 *Learning field*

The relational assets approach attempts to explain local development as a spatial outcome of the resurgence of regional economies (Yeung, 2005), which is characterized by Storper (1997) as the "holy trinity" of technology, organizations, and territories (TOT) (Figure 2.31).

Territorialization is often tied to "untraded interdependencies" and theorized to be significantly embedded in networks of relational

Table 2.6 The "holy trinity" framework of REG.

Relational frameworks	Thematic concepts	Spatial manifestations	Antecedents
Relational assets	• Institutional thickness • (Un)traded interdependencies • Agglomeration tendencies • Atmosphere and milieu • Social capital	• Learning fields or regions • New industrial spaces • Industrial districts • Clusters • Marshallian nodes	• Evolutionary & institutional economics • New economic sociology • Organizational analysis • Urban studies • Political studies
Relational embeddedness	• Interorganizational networks • Actor networks • (Global) production networks • Innovation networks • Hybrid & gender relations	• Pipeline buzz • Production of organizational spaces • Path dependence • Hybrid geography • Multiple proximities	• New economic sociology • Organizational analysis • S&T studies • Poststructuralism and feminist studies • Management studies
Relational scales	• Geographical scales as relational constructs • Social relations as scalar constructs • Rescaling and reterritorialization • Glocalization	• Global–local interactions • Scalar geography • Politics of globalization • Urban and regional governance • Social regulations of local labor markets	• Geography • Sociology • Institutional analysis • S&T studies

Source: Redrawing from Yeung (2005).

assets and spatial proximity (Storper, 1997; Yeung, 2005). The relational assets offer a variety of such non-economic factors as organizations (i.e., local rules and institutional thickness) and

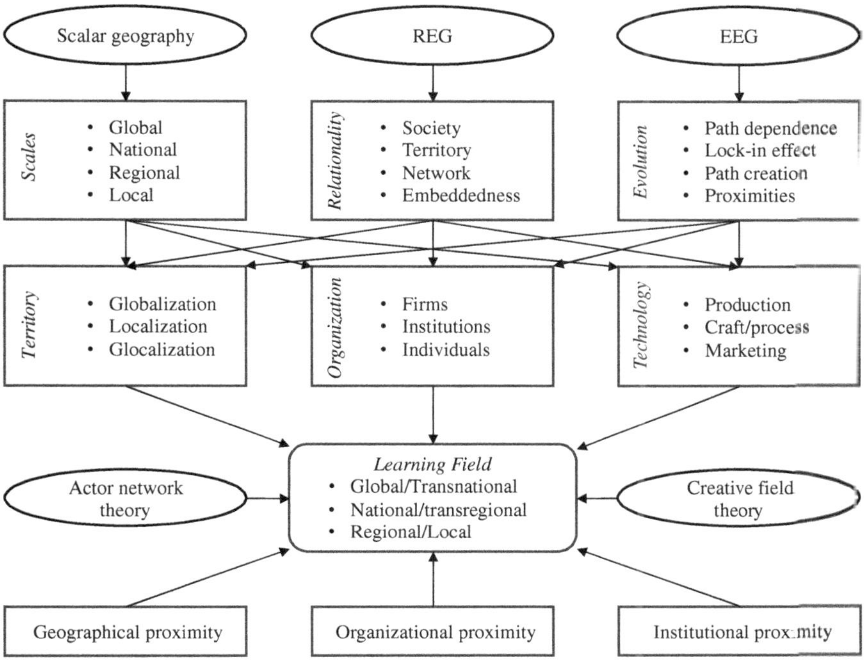

Figure 2.31 Framework of learning field.

Source: Redrawing from Miao *et al.* (2007); Miao and Ai (2009).

technologies (i.e., reflexive knowledge, conventions and technology change) to hamper territorial development (Yeung, 2005).

According to the "holy trinity", Miao *et al.* (2007) further expanded Scott's (1999) concept into a systemic theoretical model termed the "learning field" to explain the collective learning of different firms and strategic coupling of different geographical scales (Figure 2.31) (Miao & Ai, 2009). The learning field origins from concepts such as the "learning region" (Florida, 1995; Asheim, 1996), "technology learning" (Storper, 1997), and "creative field" (Scott, 1999).

It is also named as "ba" of knowledge creation (Nonaka & Konno, 1998; Kostiainen, 2001) and "club" of collective learning (Capello, 1999), which is driven collectively by a set of interactive forces including globalization and localization, external embeddedness and

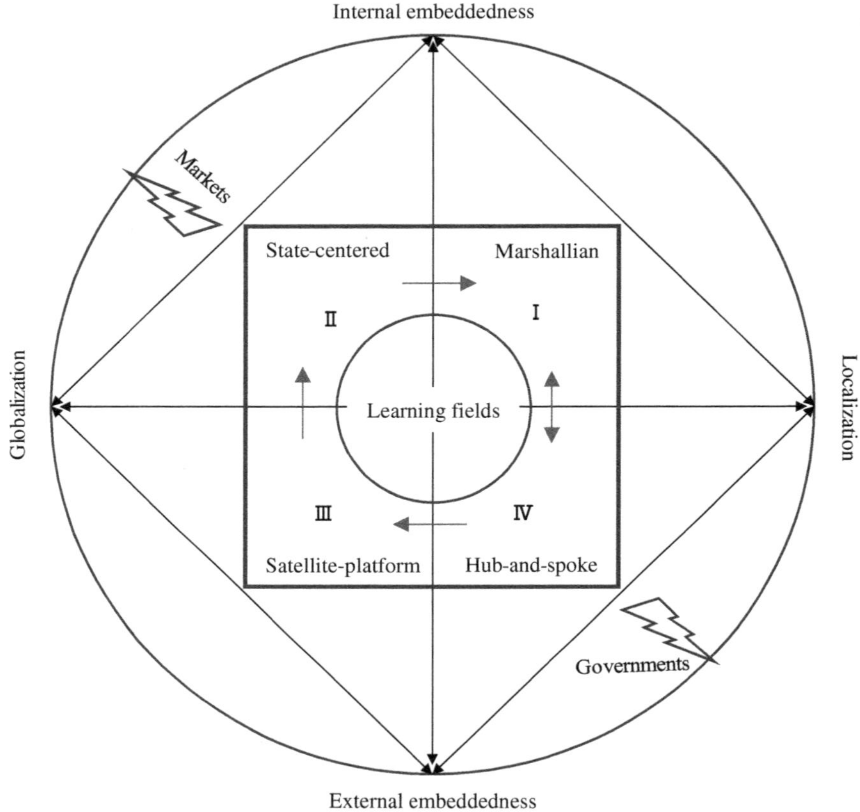

Figure 2.32 Main types of learning field based on its drivers.
Source: Redrawing from Lyu (2013).

internal embeddedness, governments and markets. Consequently, it is classified into the four types: (1) led by localization and internal linkages (Marshallian district), (2) led by globalization and internal linkages (State-centered district), (3) led by globalization and external linkages (Satellite-platform district), and (4) led by localization and external linkages (Hub-and-spoke district) (Lyu, 2013) (Figure 2.32). Moreover, on the basis of its knowledge bases, it is also further defined as different regions as the scientific, the embedded, the creative, and the hybrid (Table 2.7) (Lyu, 2013).

Table 2.7 The main types of learning field based on its knowledge base.

Type	Knowledge base	Proximity	Crucial actors
Scientific	Analytical, know-what, know-why knowledge	Technological proximity	Scientist, universities, research institutions
Embedded	Synthetic, know-how knowledge	Geographical and social proximities	Competitors, customers, suppliers, consult institutions, associations
Creative	Symbolic, know-who knowledge	Geographical, social, and cultural proximities	Competitors, customers, suppliers, consult institutions, associations
Hybrid	Hybrid knowledge	Multiple proximities	/

Source: Redrawing from Lyu (2013).

2.1.3.2 *Global–local interactions*

Instead of focusing on a specific scale, the substantive content in REG has been broadened to social network relations at different geographical scales (Yeung, 2005), viewing the local level in relation to global scales (Perry & Tan, 1998; Yeung & Li, 1999; Bathelt & Glückler, 2003; Zhou & Tong, 2003; Hewitt-Dundas *et al.*, 2005; Cohendet *et al.*, 2018).

The "relational scales" shaping a given territory has become one of the most important themes in economic geography. In REG, geographical scales are reconsidered as one of the relationality natures of socio-spatial system (Yeung, 2005) (Figure 2.33), and assigned as the relational constructions — reterritorialization from the single scale to the relationships within multiple scales (Swyngedouw, 1997, 2004; Sheppard, 2002).

Yeung (2005) suggests that scalar relationality in regulatory governance is not limited to any particular level, but encompasses social networks across various scales (Figure 2.33). This inclusive approach is exemplified by the concept of "global–local tensions" or "global–local interactions" (Dicken, 1994; Massey *et al.*, 1994; Cooke

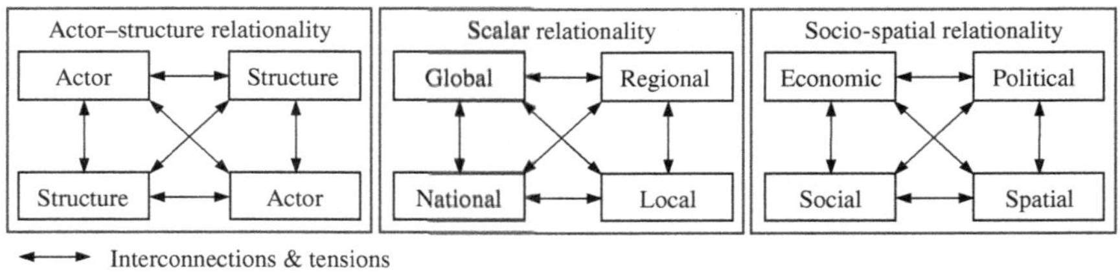

Figure 2.33 Nature of relationality in REG.

Source: Redrawing from Yeung (2005).

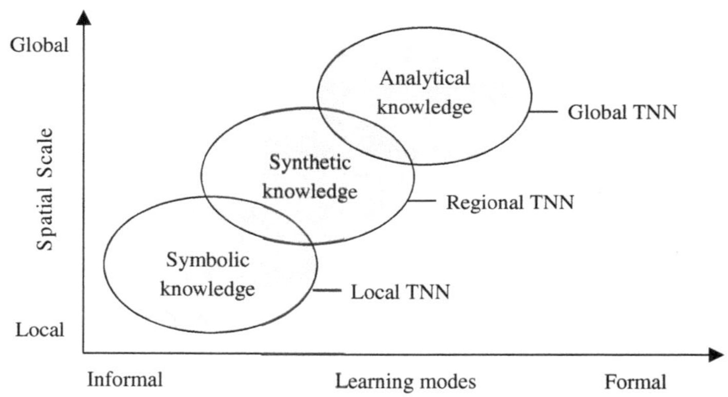

Figure 2.34 Relationship between knowledge base and global–local interaction.

Source: Redrawing from Martin and Moodysson (2011).

& Morgan, 1998b), as well as the notion of "glocalization" (Robertson, 1994). The terms "global" and "local" are not viewed as fixed scales, but rather represent two endpoints of a continuum of scalar interconnections and tensions (Dicken, 1994; Mao & He, 2019a).

The interaction between global and local factors varies depending on the industry, as noted by Asheim and Coenen (2005). This interaction is influenced by the type of knowledge dominant in each industry, such as analytical, synthetic, or symbolic knowledge (Asheim *et al.*, 2007a; Liu *et al.*, 2013) (Figure 2.34). For instance, industries like biomedicine, information, and communication technology rely more on globally transferable analytical knowledge, which can be easily shared across distant locations. On the other hand, synthetic knowledge, which is engineering based and blends codified and tacit knowledge, is mostly exchanged within national or regional technology transfer networks (TTN) led by industries such as food processing, automotive components, and mechanical engineering (Liu *et al.*, 2013). In contrast to both analytical and synthetic knowledge, symbolic knowledge is primarily tacit and local, and relies heavily on face-to-face informal learning processes that are often embedded in trust-based TTN configured locally (Martin & Moodysson, 2011).

Meanwhile, the global–local interaction exhibits a dynamic "glocalization" (Meyer *et al.*, 2012), referring to the twin process whereby: (1) institutional arrangements shifting from the national scale both upward to the global scale and downward to local configurations; (2) economic activities and interfirm networks becoming simultaneously more localized/regionalized and transnational (Si *et al.*, 2016).

The complexity of global–local interactions can be studied through various networks such as innovation networks, production networks, and trade networks. One way to analyze these interactions is through the lens of a cluster network, where the trans-local connections are framed in different ways, including transnational-community, production-network/value-chain, global-pipeline, and global city conceptions. These perspectives focus on individual, organizational, cluster, and city-region scales, respectively (Table 2.8) (Bathelt & Li, 2014). The concept of global pipelines highlights how clusters can be organized as networks within networks, known as global cluster networks (Figure 2.35).

A production network perspective emphasizes that global–local interactions are dominated by leading firms through global value chains, which embed the nexus between firms and territories (Dicken & Malmberg, 2001). These global value chains are organized as a discontinuous trans-local system consisting of local production networks with specialized divisions and cooperation within the context of a global production network (Wen & Zeng, 2005) (Figure 2.36).

Table 2.8 Frameworks of global–local interactions.

Scale	Concept	Key drivers	Main arguments	Limitations
Individuals	• Transnational communities	• Emigrant engineers/ entrepreneurs	• External knowledge transfer/ learning	• Few immigrant groups • Few corporate and FDI
Firms	• Global value chains/ production networks	• Multinational headquarters • Buyers/ producers	• Local clusters as nodes • External learning depends on value-chain	• Vertical interaction • Input–output relation • Secondary spatial relation
Clusters	• Global pipelines	• Trade fair attendees • Business travelers	• Trans-local pipelines as growth resources • Temporary clusters lead to pipelines extension	• Intra-cluster relations • Unexplored FDI-related network
City-regions	• Global cities	• Global producer services • Multinational headquarters	• Concentration in global cities • Control power along urban hierarchy	• Input–output relations • Top city groups

Source: Bathelt and Li (2014).

From a trade network perspective, global–local interactions relate to the structural changes of regions based on both local assets and global exposures (Dauth & Suedekum, 2016). These interactions are embedded in the relatedness between commodity structures and trade networks, which shape regional export specialization (Figure 2.37). The former is defined as the product–product relatedness, while the latter can be captured by either product–region relatedness, indicating market potential, or region–region relatedness, indicating market competition (Mao & He, 2019a).

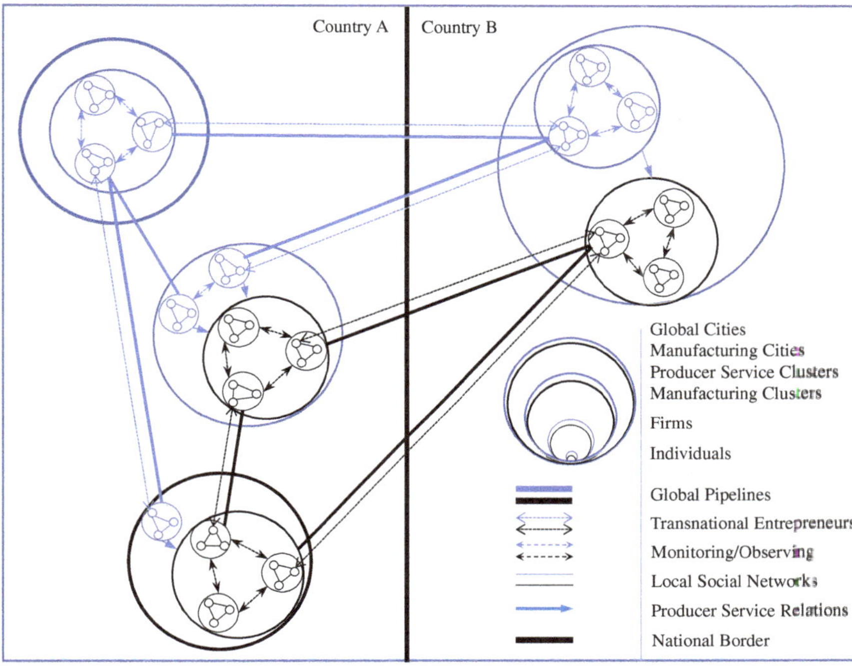

Figure 2.35 Global–local interactions of cluster networks.

Source: Redrawing from Bathelt & Li (2014).

2.1.3.3 *Global pipeline-local buzz*

The interaction between global and local factors is also reflected in the buzz-and-pipeline model of clusters (Bathelt *et al.*, 2004). This model suggests that the growth of a cluster depends on systematic linkages between its internal learning networks, conceptualized as "local buzz", and its external knowledge attainments, referred to as "global pipelines" (Bathelt & Li, 2021). Local buzz on one hand generates in an embedded context of local bonding that induces trust, reduces transaction costs, creates technological spillovers, and shares information (Storper & Venables, 2004). On the other hand, global pipelines play a crucial role in accessing novel and non-redundant information, acting as bridging structures for clusters (Aarstad *et al.*, 2016).

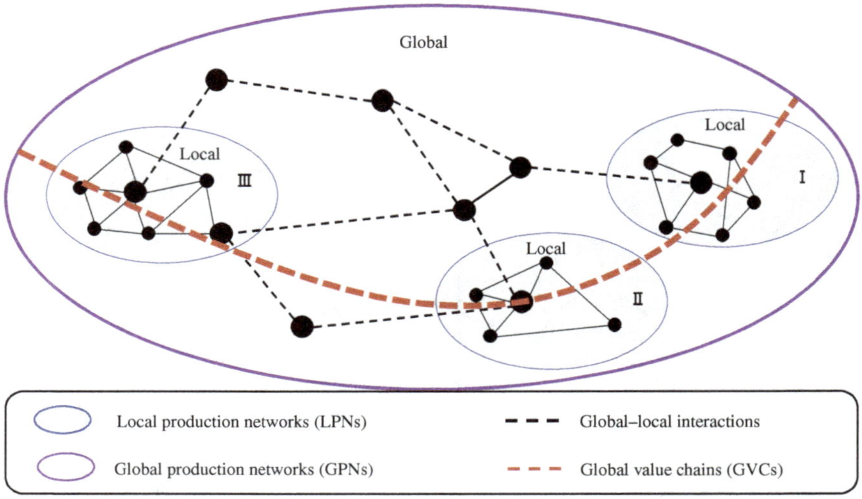

Figure 2.36 Global–local interactions of production networks.

Source: Redrawing from Wen and Zeng (2005).

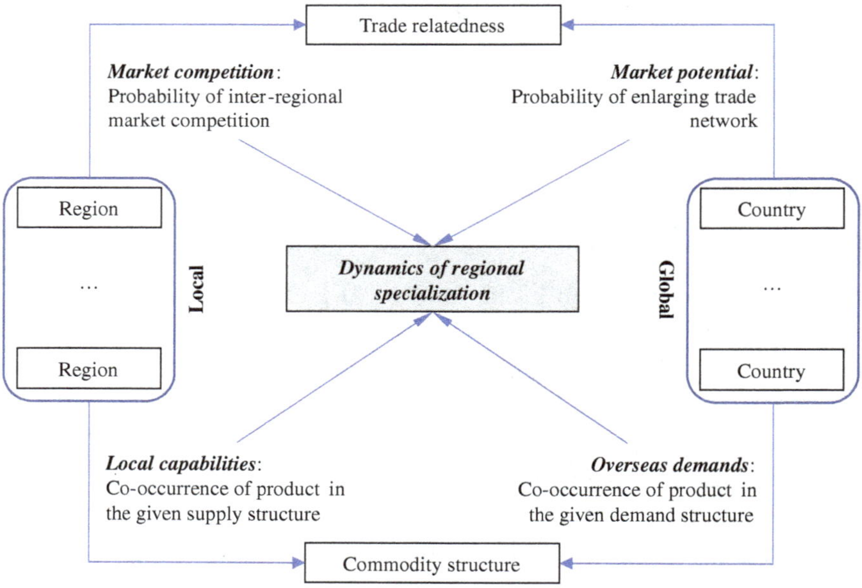

Figure 2.37 Global–local interactions of trade networks.

Source: Redrawing from Mao and He (2019a).

There exists a complementarity between local buzz and global pipelines in facilitating local innovation capacity (Cao *et al.*, 2022). The effect of internal and external technology sourcing on local innovative performance differs throughout the technology life cycle (Abernathy & Utterback, 1978; Stolwijk *et al.*, 2012). Global pipelines have a more positive effect in the early phase. The creation of new knowledge seems to be best viewed as a result of a combination of close (or local) and distant (or global) interactions (Moodysson, 2008; Bathelt & Glückler, 2011; Aarstad *et al.*, 2016; Taira, 2020). The buzz-and-pipeline model suggests that the local information and knowledge ecology is of only limited effect in the absence of trans-local connections (Bathelt *et al.*, 2004, 2018). *Vice versa*, without local buzz, the cluster's external pipelines are also of little use.

Buzz refers to the multilayered information and communications ecology created by face-to-face contacts and co-location of innovators within the same industry or place (Bathelt *et al.*, 2004; Storper & Venables, 2004), as depicted in Figure 2.38. It depends on locally institutional preconditions (or local milieu or atmosphere) and encourages the sharing of values, attitudes, and interpretative schemes, which

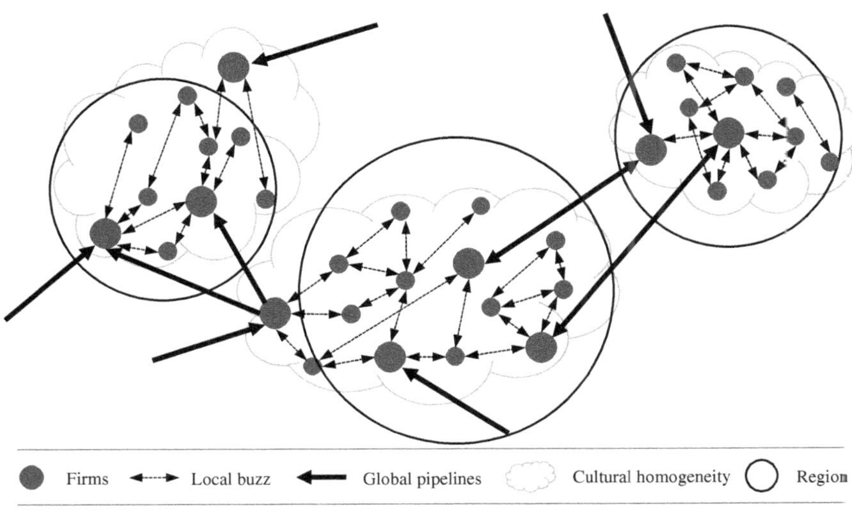

Figure 2.38 Reflexive dynamics of local buzz and global pipelines.

Source: Redrawing from Bathelt *et al.* (2004).

enable local actors to engage in interactive learning and problem solving (Bathelt *et al.*, 2004). While it has an inverted U-shaped relationship with innovation capacity (Cao *et al.*, 2022), external interconnection through "pipelines" is typically more structured, formalized, and standardized, providing local innovators with access to new external knowledge and market environments, thereby positively contributing to the creation of new knowledge. However, the benefit of pipelines relies on the intensity of the buzz. Given that the buzz is sufficiently intense, such new external knowledge may come into productive usage through informal pipelines of communication.

2.1.4 *Evolutionary economic geography*

To address some problem of regional dynamics, key notions of Evolutionary Economics were applied in the field of Economic Geography, such as chance and selection (location decisions of new firms), routines and path dependency (regional adjustment to avoid negative lock-in), and chance and increasing returns (combination of chance and necessity (agglomeration economies)) (Boschma & Lambooy, 1999).

Similar to relational perspectives, evolutionary approaches are based on a critique of conventional research that is mostly static (Bathelt & Li, 2021). In EEG, three approaches deal effectively with the long-term spatial evolution of technology changes (Boschma & Lambooy, 1999), including the cumulative version of Product Life Cycle approach (PLC) (spatial polarization of technology change) (Vernon, 1966; Klepper & Graddy, 1990; see Section 2.1.2), the diffusion version of PLC (spatial diffusion of technology change, see Section 2.1.2), and the Window of Locational Opportunity Model (WLO) (spatial implications of technology change in terms of spatial leapfrogging) (Boschma, 1997) (Table 2.9).

2.1.4.1 *Path dependence*

The concept of path dependence is often considered as "one of the most exciting ideas in contemporary economic geography"

Table 2.9 Three approaches of EEG.

Key notions	Cumulative version of PLC	Diffusion version of PLC	WLO
Nature of technique change	• Cumulative trajectories • Localized, specific, tacit knowledge	• Continuous trajectories • Partly codifiable, semi-appropriable knowledge	• New trajectories • Discontinuous, generic knowledge
Selection	• Strong: specific conditions	• Medium	• Small: generic conditions
Path dependence	• Cumulative causation	• Creative diffusion	• Discontinuous creativity
Chance	• High predictability	• Medium	• Low predictability
Increasing returns	• Localization economies	• Agglomeration economies	• Urbanization economies
Human agency	• Bounded	• Semi-bounded	• Unbounded
Change in spatial matrix	• Small, relatively fixed	• Gradual	• Potentially dramatic

Source: Redrawing from Boschma and Lambooy (1999).

(Walker, 2000; Martin, 2010). It refers to the idea that economic outcomes are dependent on the path of previous outcomes, rather than just current conditions. This notion is closely linked with a place-dependent evolutionary process, where the historical development of a region or industry influences its future trajectory (Martin & Sunley, 2006).

Path dependence was initially developed by economists to explain the processes of technology adoption and industry evolution (David, 1985; Arthur, 1988, 1989). This approach highlights how the first-to-market product or technology can become entrenched as the standard (David, 1985). The theoretical concepts of path dependence have had a significant influence on evolutionary economics (Nelson & Winter, 1982), where they are used to illustrate how historical contingencies shape current and future economic outcomes. With path dependence, both the starting point and "accidental" events (or "noise") can have significant effects on the ultimate outcome (David, 1985; Arthur, 1989; Martin, 2010).

It is subdivided into various types, which can be weak (the efficiency of the chosen path is tied with some alternatives, named as the first-degree path dependence), semi-strong (the second-degree dependence indicates that the chosen path is not the best but not worth fixing), and strong (the third-degree path dependence assumes that the chosen path is highly inefficient, but we are unable to correct it) (Liebowitz & Margolis, 1995).

Path dependence can also be classified into two categories: positive lock-in (also referred to as "self-reinforcement") and negative lock-in. Positive lock-in occurs when local externalities, increasing embeddedness, and interrelatedness generate efficient outcomes, leading to increasing returns and stimulating economic performance. In contrast, negative lock-in occurs when high levels of interrelatedness and embeddedness result in inefficient or inferior feedbacks, leading to a reduction in economic performance due to negative externalities and a hindrance to innovation (Roe, 1996; Martin *et al.*, 2006; Martin & Sunley, 2006, 2007). Based on the works of Liebowitz & Margolis (1995) and Kay (2013), only third-degree path dependence is deemed inefficient, leading to lock-in through market failure.

The path dependence theory has a strong tendency toward historical determinism (David, 1985). It may arise or be hindered by a number of important factors, these may include: *durability of capital equipment, technical interrelatedness, increasing returns, dynamic increasing returns to adoption*, and *imperfect foresight and inefficiency* (David, 1985; Arthur, 1989; Puffert, 2008).

Path creation, which overcomes the hindrance of path dependence, is based on five principles: *technological determination* (which states that the ability to create a new path rests on the prior existence), *mindful deviation* (which states that even when the community has the technical capability to create a new path, entrepreneurs must still actively carry out the task), *real time influence* (which states that all actions have real time consequences on the path in the making), *recognized interdependence* (which states that entrepreneurs cannot work alone, they require other the assistance of others), and *minimal dislocation* (which states that entrepreneurs should be selective about what they translate to outside agents). These principles facilitate the creation of new capabilities, products, and markets or restructuring existing ones resulting in positive outcomes (Pham, 2007). To accomplish this, path creation also involves five interactive building blocks: a "*window of new opportunities*" associated with new techno-organizational paradigms, prospects for new businesses and markets, pressures from external socioeconomic factors, key change events, and the human will to change things (Schienstock, 2011).

Path dependence has significant implications for regional economic development and plays a crucial role in theorizing the evolution of regional economies (Boschma & Frenken, 2006; Scott, 2006; Frenken & Boschma, 2007; Henning *et al.*, 2013). "The 'evolutionary' turn" in economic geography has led to its widespread application to investigate various phenomena, including the persistence of regional differences, the revitalization of old industrial bases, and the self-reinforcing effects of high-tech clusters. Several conceptual frameworks have been proposed to understand the role of path dependence in regional economies, such as the "*path breakthrough*" of old industrial districts (Grabher, 1993; Hudson, 2005), the "*path contingency*" of regional adaptation (Hudson, 2005), the "*local dependence*" or "*place-dependent process*" (Martin *et al.*, 2006), the "*path creation*" of clusters (Garud & Karnøe, 2001; Frenken *et al.*, 2007; Boschma & Iammarino, 2009), and the "*path interdependence*" of clusters (Martin & Sunley, 2008; Schamp, 2008).

The theory of path dependence suggests that we should move beyond the equilibrium-based perspective of neoclassical economics. While David and Setterfield's path dependence evolution models (Figure 2.39) still relied on equilibrium analysis, they

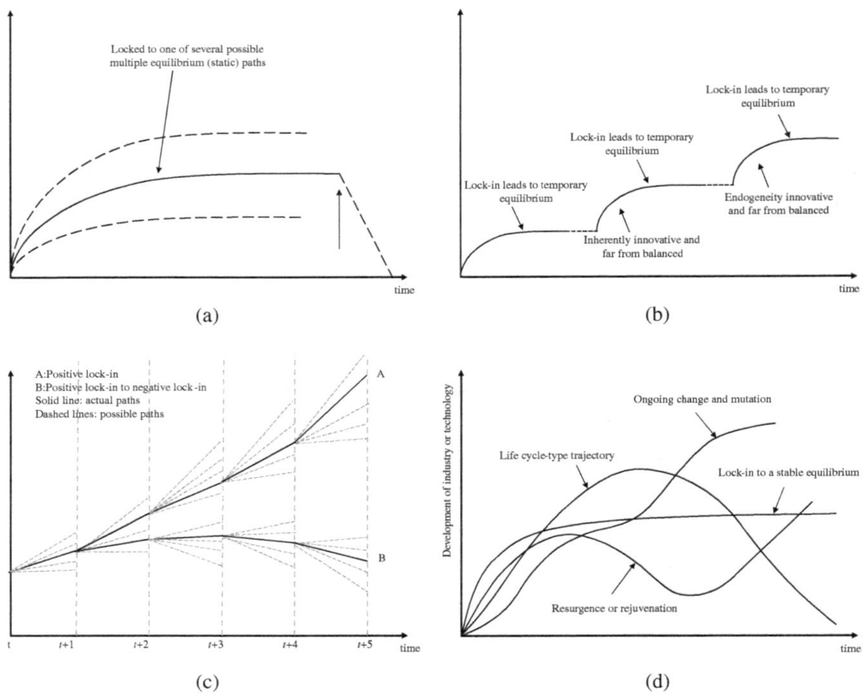

Figure 2.39 Theoretical models of path dependence. (a) David's path dependence model; (b) Setterfield's path dependence model; (c) Martin's path dependence model of regional system (1) and (d) Martin's path dependency model of an industry or technology (2).

Source: Redrawing from David (1985); Setterfield (1997); Martin and Sunley (2006); Martin (2010).

recognized that equilibrium is not a permanent outcome of economic processes. Their models revealed path-dependent processes can lead to subsequent endogenous processes of innovating out of equilibrium. In contrast, Martin's model challenged the concept of equilibrium altogether. He emphasized that technology, industry, and regional systems are open and far from equilibrium, and may never be locked into any stable equilibrium configuration. According to Martin (2010), this understanding highlights the importance of continuously adapting to changing circumstances rather than relying on fixed outcomes. (Figure 2.39d).

2.1.4.2 *Multiple proximities*

As previously mentioned, innovation faces distance-decay effects in the process of spatial diffusion (Caragliu & Nijkamp, 2016). Geographical proximity is central to regional innovation studies (Sun & Liu, 2016). Marshall's "industrial zone" theory is viewed as one origin of the concept, which refers to the co-located relationship of economic actors within a cluster.

Empirical evidence reveals that technology transaction and transfer depend not only on geographical proximity, but also on cognitive, social, cultural, and institutional proximities related to geography (Agrawal *et al.*, 2008; Broekel & Boschma, 2012; Sun & Liu, 2016). However, as research progressed, scholars began to notice the paradox of geographical proximity — the phenomenon of closely located firms that rarely interact due to a lack of shared knowledge, experience, or customs (Mckelvey *et al.*, 2003; Doloreux, 2004; Boschma & Frenken, 2010; Maté-Sánchez-Val & Harris, 2018). This highlights the need to consider a multidimensional analysis of proximity that goes beyond geographical closeness. Interactive learning and collective innovation among actors require combinations of different types of proximities. Since the 1990s, scholars such as those from the French School of Proximity and other EU institutions have recognized the necessity of a multidimensional understanding of proximity. Thus, the concept of proximity has evolved to include cognitive, social, cultural, institutional, and other dimensions to better explain the complex interactions and dynamics of economic activities (Figure 2.40; Table 2.10).

To understand the various channels of technology transfer, researchers have developed several proximity concepts based on their distinct research objectives and perspectives. These include geographical, cognitive, social, organizational, institutional, cultural, economic, strategic, personal, relational, and technological proximities (Bradshaw, 2001; Schamp *et al.*, 2004; Sun & Liu, 2016) as shown in Table 2.11. To better understand the complex interactions involved in technology transfer, these different dimensions of proximity have been integrated into a multiple and interactive conceptual framework. This framework includes geographical, cognitive, technological,

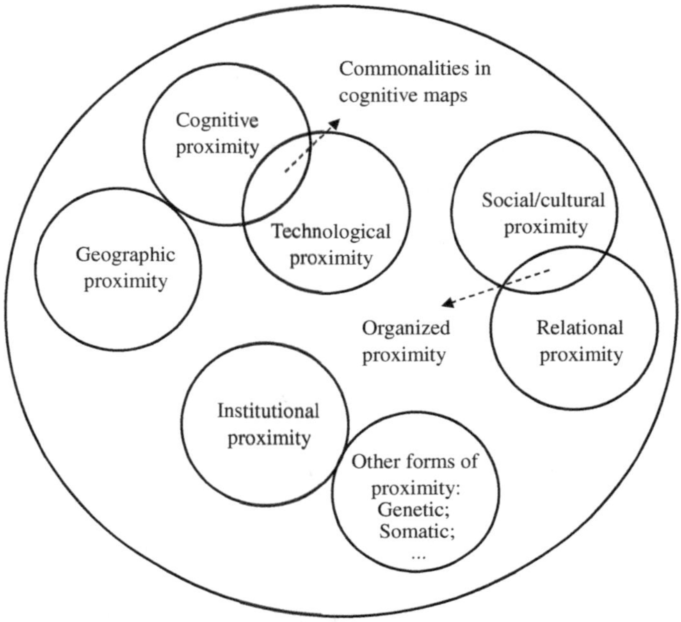

Figure 2.40 Framework of multiple proximities.
Source: Redrawing from Caragliu and Nijkamp (2016).

social or cultural, relational, and institutional dimensions that interact with each other in complementary, alternative, overlapping, or substitutive ways, as depicted in Figure 2.41 (Boschma, 2005; Hansen, 2015; Caragliu & Nijkamp, 2016; Ma & Xu, 2022).

Several proximity concepts have been developed to understand the various channels of technology transfer. Cognitive proximity, for instance, refers to the similarity or closeness of existing knowledge bases between economic actors (Boschma & Lambooy, 1999). Organizational proximity, derived from transaction cost economics, refers to the extent to which relations are shared within or between organizations (Boschma, 2005). Social proximity, originating from research on embeddedness, is defined in terms of socially embedded relationships between actors at the micro-level (Boschma, 2005), emphasizing the importance of trust and reduced opportunism through social ties or relations. Institutional proximity, on the

Table 2.10 Forms of proximity: some features.

Types	Key dimension	Too little proximity	Too much proximity	Possible solutions
Geographical	Distance	No spatial externalities	Lack of geographical openness	Mix of local buzz and extra-local linkages
Cognitive	Knowledge gap	Misunderstanding	Lack of novelty sources	Common knowledge base with diverse but complementary
Organizational	Control	Opportunism	Bureaucracy	Loosely coupled system
Social	Trust based on social relations	Opportunism	No economic rationale	Mixture of embedded and market relations
Institutional	Trust based on common institutions	Opportunism	Lock-in and inertia	Institutional checks and balances

Source: Redrawing from Boschma (2005).

other hand, is associated with the degree to which actors are influenced, shaped, or constrained by the macro-level institutional environment. This includes informal institutions such as cultural norms and habits, as well as formal institutions such as laws and rules. Finally, geographical proximity refers to the physical or spatial distance between economic actors and the ability to achieve face-to-face communication without excessive costs.

Proximity plays an important role in interactive learning and innovation. However, either too much or too little proximity can be detrimental (Boschma, 2005) (Table 2.11). As research has shown, higher degrees of proximity can enhance a firm's innovative capability up to a certain threshold. Beyond this threshold, however, adverse impacts may arise due to lock-in effects. The relationship between the degree of proximity and innovative performance at the firm level is thus inverted-U shaped (Boschma, 2005). In other

Table 2.11 Proximity mechanisms of technology transfer.

Types	Channels	Indicator	Mechanisms of diffusion	Definitions
Geographical	Face-to-face contacts	Differences in regional stocks	Higher probability of meeting	The distance between economic actors
Social/ cultural	Parallel decision under similar culture	Differences in regional stocks	Lower communication cost	The similarities of two areas in term of their social capital
Technological	Technological spillovers	Differences in regional stocks	Higher probability of cross-fertilization	The similar specialization in manufacturing industries
Cognitive	Non-verbal communication	Differences in regional stocks	Sharing of a common communication code	A complementary set of skills and a common knowledge base
Relational	Information networks	Flows between pairs of regions	Lower transaction costs	The capability of regions to learn through cooperation

Source: Redrawing from Caragliu and Nijkamp (2016).

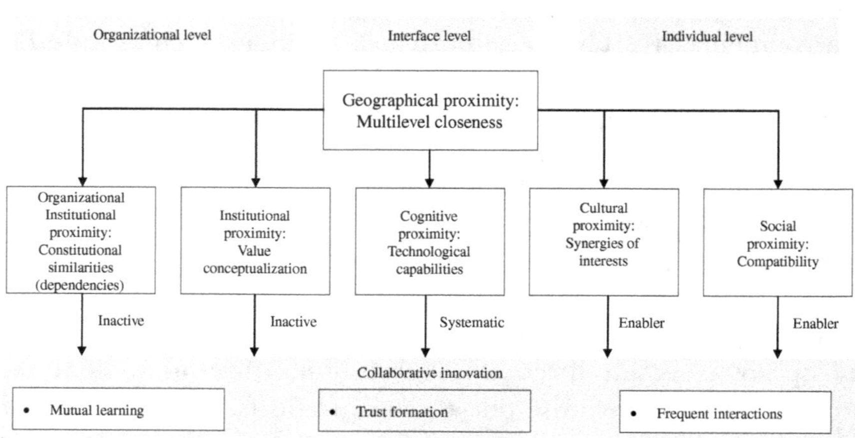

Figure 2.41 Interactions among multiple proximities.

Source: Redrawing from Mahdad (2020).

words, while proximity can facilitate knowledge creation and transfer, excessive proximity can lead to limited exploration of new ideas and perspectives, resulting in lower levels of innovation.

2.1.4.3 *Technological relatedness*

There is a heated debate on knowledge externalities of agglomeration which operate within a specific or same industry (Marshall–Arrow–Romer externalities or localization economies) or between variety of sectors (Jacobs's externalities or urbanization economies) (Glaeser *et al.*, 1992; Feldman & Audretsch, 1999; Frenken *et al.*, 2007; Beaudry & Schiffauerova, 2009; Pierre & Leppälä, 2011). Therefore, technological relatedness and (un)related variety are applied to test the debate (Frenken *et al.*, 2007; Boschma & Frenken, 2009; Neffke *et al.*, 2012).

Similar to the notion of cognitive proximity (Breschi *et al.*, 2003 Boschma & Frenken, 2009), technological relatedness is often understood in terms of both similarity (related knowledge) and complementarity (unrelated knowledge) (Breschi *et al.*, 2003; Tanriverdi & Venkatraman, 2005; Makri *et al.*, 2010), which is defined as the similarity of knowledge base between products based on their cooccurrence frequency (Nelson & Winter, 1982; Teece *et al.*, 1997; and the complementarities referring to the necessity of combination of different technologies or products (Broekel & Brachert, 2015; Boschma, 2017). Complementarities make firms technologically related upstream and downstream (vertically), whereas similarity makes firms technologically related horizontally. From an absorptive capability perspective, higher levels of technology similarity enable easier understanding and application of new knowledge (Cohen & Levinthal, 1990; Makri *et al.*, 2010). Conversely, from the perspective of radical innovation, greater novelty value of created new knowledge is attributed to higher levels of technology complementarity, that is "the more different the combined knowledge items" (Sapienza *et al.*, 2004).

Recent empirical work in economic geography has explored the importance of technological relatedness as a significant

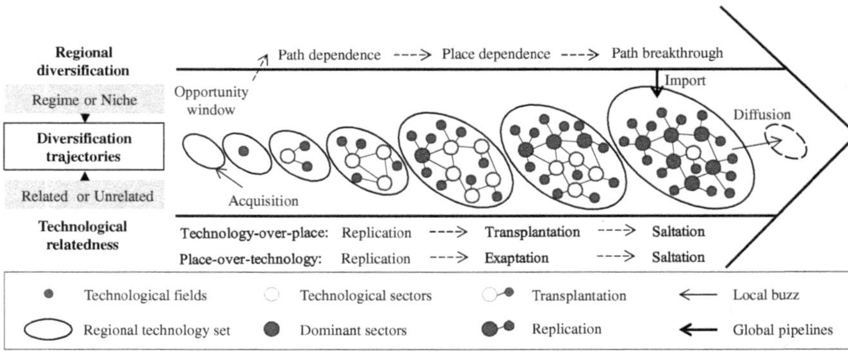

Figure 2.42 Relationship between technology relatedness and regional diversification.

determinant of regional diversification (Bottazzi & Peri, 2003; Owen-Smith & Powell, 2004; Moreno *et al.*, 2005; Gittelman, 2007; Neffke *et al.*, 2011; Zhou & Li, 2012; Boschma *et al.*, 2013; Miguélez *et al.*, 2013; Castaldi *et al.*, 2015; Boschma & Capone, 2016; Grillitsch *et al.*, 2018; Miguelez & Moreno, 2018; Antonietti & Montresor, 2021), which is due to learning processes and knowledge linkages (Breschi *et al.*, 2003) or knowledge spillovers (Boschma & Frenken, 2009; Grillitsch & Nilsson, 2015) (Figure 2.42). There are several mechanisms that play the role in the process of regional diversification or branching, counting firm diversification (reducing uncertainties and switching costs), spinoffs or startups (leading to spatial concentration of economic activity), labor mobility and social networks (fostering localized knowledge spillover) (Neffke, 2009).

At a macro-level perspective, there are several mechanisms through which industries may be technically related (Boschma, 1999), including *producer–user relationships*, *production-system interdependency*, *technological complementarity*, and *technical interdependencies between industries* (Boschma & Frenken, 2009).

Based on the principle of economic competition, the technological relatedness may be measured by the three proxies including industry-specific aspects, technology-based linkages, and firm-specific aspects (Piscitello, 2000; Whittle & Kogler, 2019). Industry-specific aspects are proxied by vertical integration and market similarity, which reflect the hierarchical relatedness of industry classifications. Technology-based linkages are proxied by an input–output measure of R&D flows, which measures the cooccurrence relatedness of technology classes. Finally, firm-specific aspects are proxied by affinity or proximity in the underlying technological competencies, reflecting resource-based relatedness across a firm's technology regime. These proxies provide a comprehensive understanding of technological relatedness and can help firms identify potential partners for collaboration, competition, or acquisition.

In the context of regional diversification, the degree of technological relatedness between industries is further assessed in different ways such as product relatedness (i.e., product co-classification analysis, classification analysis and cooccurrence analysis) (Breschi *et al.*, 2003; Frenken *et al.*, 2007; Hidalgo *et al.*, 2007; Leydesdorff, 2008; Neffke *et al.*, 2011, 2012; Boschma *et al.*, 2012), technological relatedness (i.e., patent classification analysis, co-classification analysis and citation matrix analysis) (Porter, 1998; Rigby, 2015), skill relatedness (i.e., the intensity of human capital flows) (Neffke & Henning, 2013), and input–output relatedness (i.e., the similarities of supplier–buyer relationships) (Essleztbichler, 2015).

2.2 Framework of Technology Transfer Network

Technology transfer refers to a networking system of technology flows among different territorial units and industrial fields. It involves the movement of technology from one site to another, and is recognized as a relational set that includes sociocultural relations (i.e., technology exchanges between transfer actors), spatial relations (i.e., technological flows between origins and destinations), and network relations (i.e., a networked organization consisting of nodes and channels). The process of technology transfer is complex and dynamic, involving multiple actors and interactions across various dimensions of proximity. As such, it requires careful consideration of factors such as the nature of the technology being transferred, the characteristics of the sending and receiving entities, and the broader contextual factors that shape the transfer process.

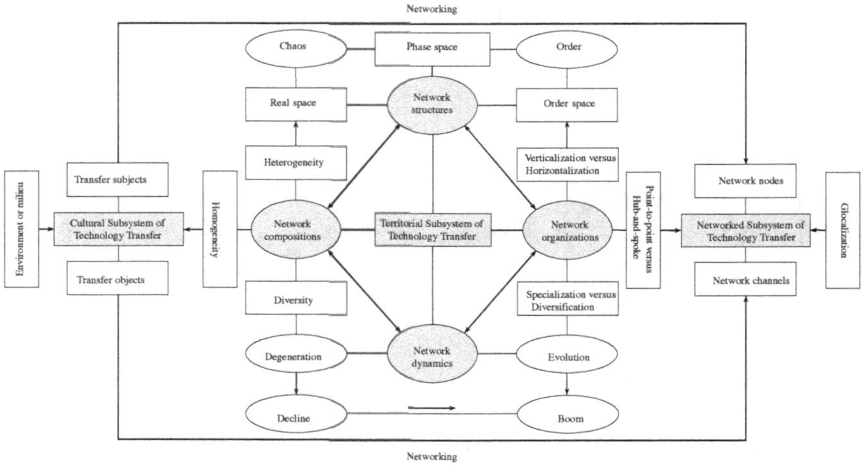

Figure 2.43 Systematic "holy trinity" of TTN.
Source: Own illustration base on Liu (2017).

Accordingly, it consists of the three subsystems (the "holy trinity") with complex and coupling interactions: cultural subsystem (related to human–environment relations between technology transfer activities and their innovative milieu), territorial subsystem (related to interterritorial relations between origins and destinations of technology transfer), and networked subsystem (related to topological relations between nodes and channels of technology transfer) (Figure 2.43).

As a sociocultural system, the process of technology transfer is involved with transfer subjects (transferor and transferee, or sender and recipient), transfer objects (knowledge, skills, technology and intellectual property, business processes, designs and facilities, etc.), and ways of transfer-media (transactions, communications, absorption and spillovers, etc.) and channels (internal and external linkages, or local buzz and global pipelines, or informal and formal vehicle) (Figure 2.44). As cross-border flows of transfer objects (including series of characteristics such as scale, strength, scope, speed and spillover) between subjects or places, the process is determined mainly by its transfer environments and innovation milieu (Bozeman, 2000; Duan, 2020) (Figure 2.45).

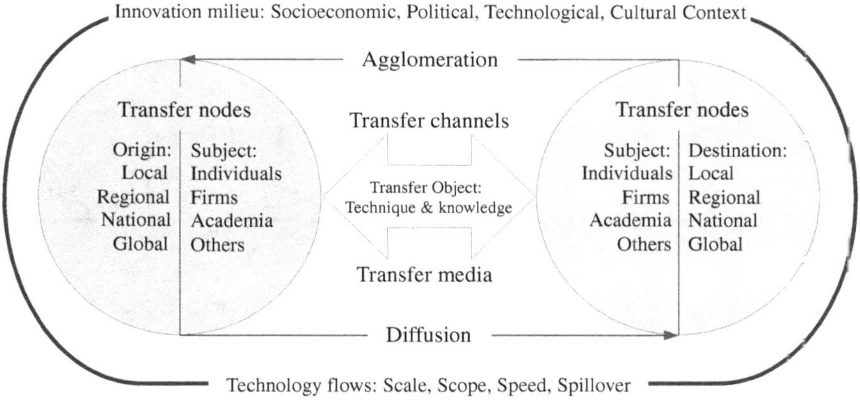

Figure 2.44 Networked subsystem structure of technology transfer.

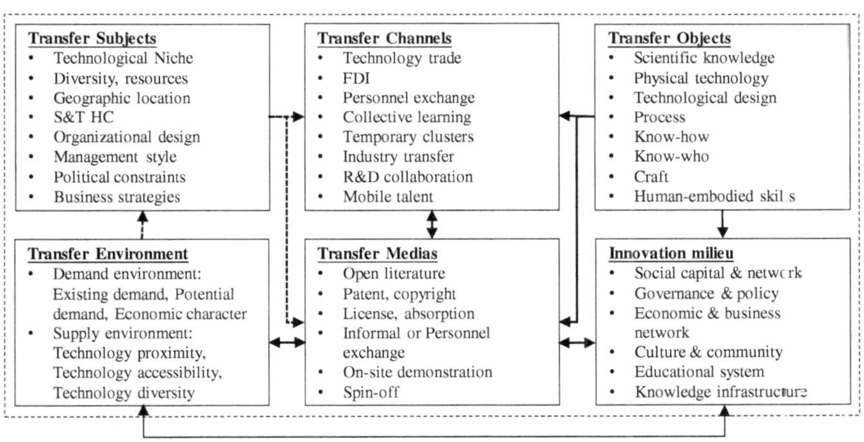

Figure 2.45 Cultural subsystem structure of technology transfer.

Source: Own illustration based on Bozeman (2000).

2.2.1 *Technology transfer system*

The system of technology transfer was extensively defined as a process related to technology life cycle and a network consisting transfer nodes (transfer subjects and their location) and linkages (channels and medias).

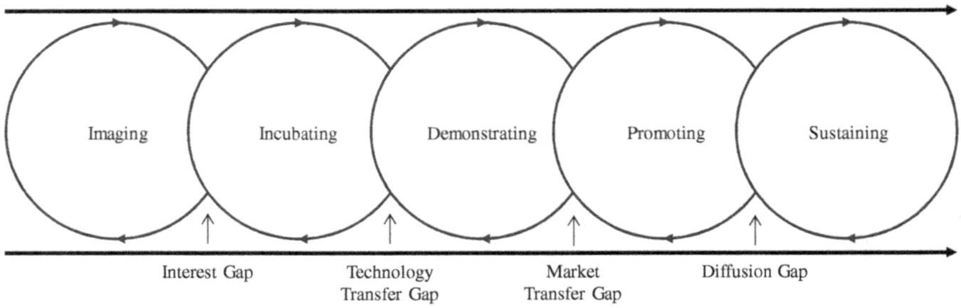

Figure 2.46 Commercialization model of technology transfer process.
Source: Redrawing from Jolly (1997).

2.2.1.1 *Technology transfer as a process*

The first important character for technology transfer is continuity over time (Capello, 1999). The process of technology transfer is often described as interactive, involving ongoing exchange between entities over an extended period (Gibson & Smilor, 1991; Bozeman, 2000; da Silva *et al.*, 2022). This process involves the broader technology commercialization process, which includes imaging, incubating, demonstrating, promoting, and sustaining (Figure 2.46) (Jolly, 1997).

By projecting it in a flow-system model, it is further distinguished into six phrases: technology innovation (communicating ideas of management practice or product innovation), technology confirmation (conducting research on technology innovation), targeting technology consumers (concerning needers benefit from the technology), technology marketing (analyzing demographic profile of anticipated consumers), technology application (identifying consumers' behavior pattern), and technology evaluation (assessing socioeconomic and environmental benefits), whose dynamic is driven by need-resources gaps between transfer actors (Penny, 1992).

In a similar vein, Gibson & Smilor (1991) proposed a technology transfer model consisting four-level stages: technology creation

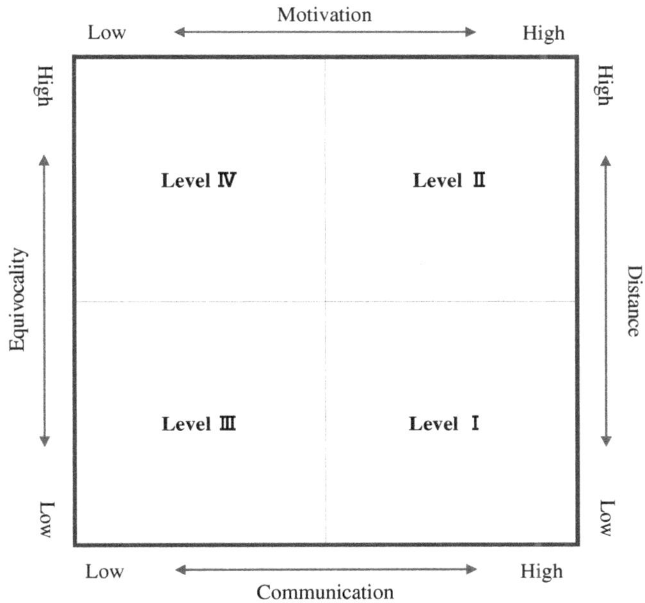

Figure 2.47 Determinants of technology transfer process.
Source: Redrawing from Gibson and Smilor (1991).

(Level I), technology sharing (Level II), technology implementation (Level III), and technology utilization (Level IV) (Gibson & Smilor, 1991). These stages are determined by four key factors such as communication interactivity (the degree to which a medium is able to efficiently and accurately convey task-relevant information), distance (the degree of both geographical and cultural proximity), technological equivocality (the degree of technology concreteness and transfer), and motivation (the recognition of knowledge importance and technology transfer activities) (Figure 2.47) (Gibson & Smilor, 1991; Sung, 2009).

The process is well observed as an STI model of knowledge (science, technology, and innovation), including a process from knowledge generation (science) to knowledge spillover (technology), then to knowledge application (innovation) (Figure 2.48).

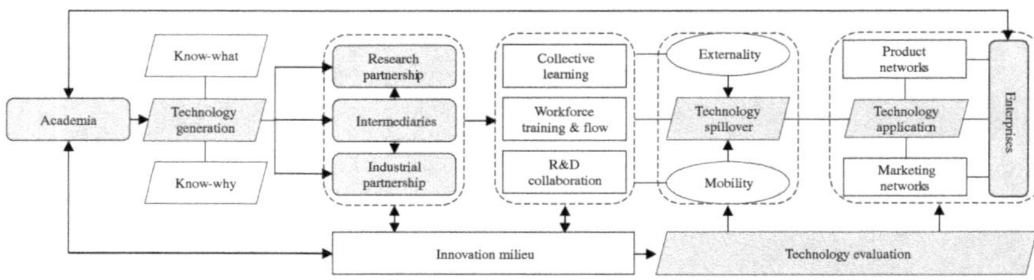

Figure 2.48 STI model of technology transfer process.
Source: Own illustration based on Lyu (2016).

2.2.1.2 *Technology transfer as a network*

It is now widely acknowledged that networks play a critical role in not just technology transfer process but also in ensuring its effectiveness (Jones & Tang, 1996; Singh, 2005; Graf, 2006; Broekel *et al.*, 2014; Coenen *et al.*, 2014; Brennecke & Rank, 2017; Seo & Sonn, 2019). Networks are essentially organizational configurations that serve two essential functions: coordination and transmission (Lambooy, 2004), both vital for generating and transferring technology. Utilizing proximity dimensions, stakeholders can organize technology transfers through dynamic network arrangements (Bessant & Rush, 1995; Coenen *et al.*, 2014; Ma *et al.*, 2020, 2022). These networks can be categorized as horizontal, where technology is transferred within the same geographic unit, industrial sector, or innovative institution, or vertical, where technology is transferred along supply chains from intermediate suppliers to producers, or from foreign-based firms to domestic suppliers (Grosse, 1996; Noh & Lee, 2019). Due to their structure and dynamics, networks facilitate international, intersectoral, and interinstitutional technology transfers with ease.

In spatial terms, technology transfer can be understood as a territorially self-organized network of technology flows. This network comprises "nodes" (subjects) and "connections" (channels or media) through which technology is transferred from one place (technology origins) to another (technology destinations), or from in-house laboratories (R&D generation) to the marketplace for industrialization (Lambooy, 2004). This territorial network is divided

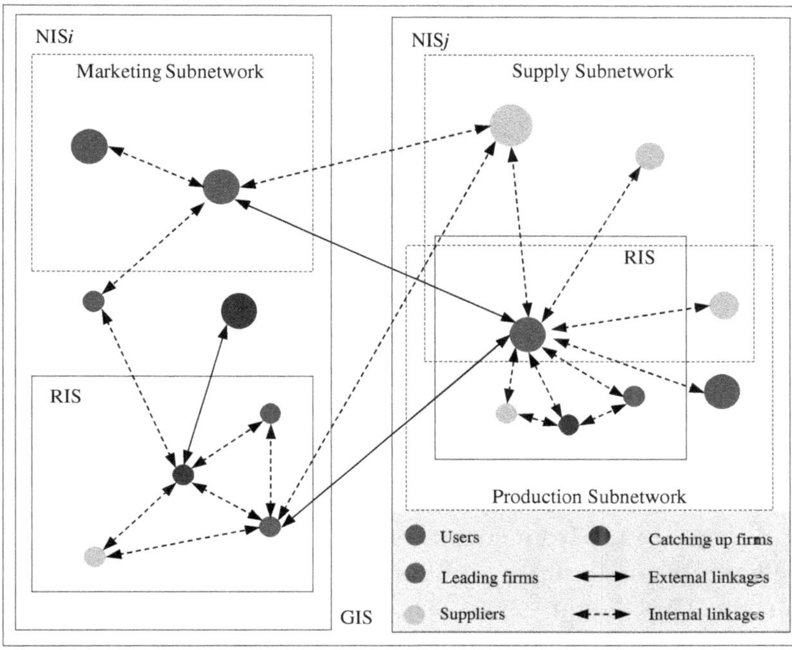

Figure 2.49 DUI model of technology transfer network.
Source: Own illustration based on Lyu (2016).

into two main scales: trans-local networks that connect different places, and local networks that operate within a single place, such as a city. Local networks represent basic and non-basic technology transfer activities within a specific location (Duan, 2020).

As a hybrid network, the TTN is proxied by a DUI model consisting of subnetworks dominated by leading firms. These subnetworks include the supply subnetwork, product subnetwork, and marketing subnetwork. The TTN operates under global–local interactions among regional innovation systems, national innovation systems, and the global innovation system (as illustrated in Figure 2.49).

2.2.2 *Network subjects*

Technology transfer may not be available from a single supplier, but rather through a combination of emitters (Bessant & Rush, 1995).

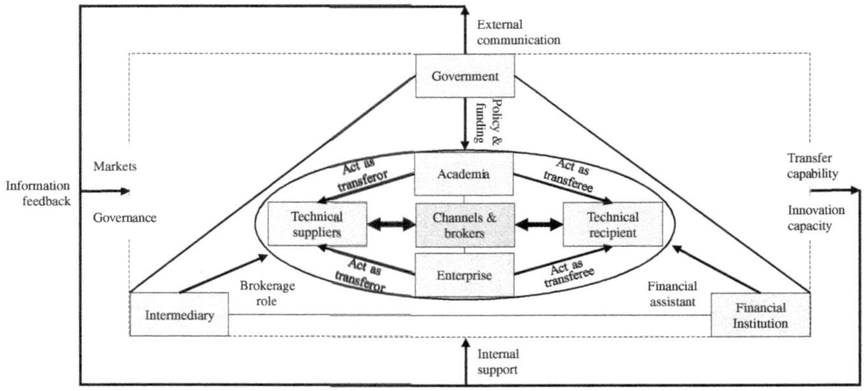

Figure 2.50 The interaction of transfer actors in TTN.

The principal subjects or agents involved in the technology transfer process have shifted from governments to academia and enterprises, as well as intermediaries with different roles and strong interactions within the TTN (Figure 2.50). These entities can be classified from different perspectives, such as institutional, legal, or individual entities; international, national, regional and local collaborators; commercial and non-commercial; state (or public, governmental) and private organizations; companies, universities, research institutions, intermediary institutions, financial institutions, and governmental organizations. Individuals involved in technology transfer may include students, scientists, researchers, programmers, entrepreneurs, and others (Ortega, 2011; Bliznets *et al.*, 2018; Cunningham & O'Reilly, 2018; Hensengerth, 2018; da Silva *et al.*, 2022).

Although a large number of agents are involved in TTN, only certain stakeholders or brokers can play a vital role in the network by bridging the gap between one source and another recipient through intermediaries (Figure 2.51) (Noh & Lee, 2019; Subramanian *et al.*, 2022). These network brokers, which may include small- and medium-sized enterprises (SMEs), transnational enterprises (TNEs), governmental funding organizations, technology transfer offices of universities and research institutes, and

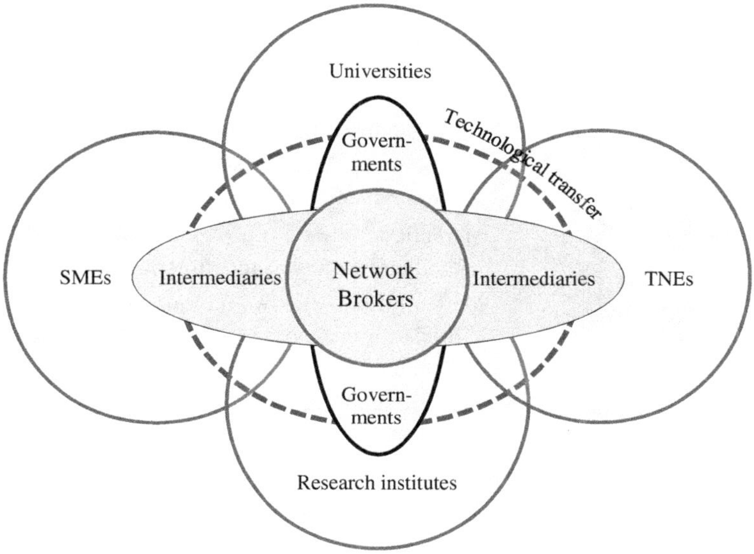

Figure 2.51 Network agents of TTN.

Source: Own illustration based on Engel (2015).

specific intermediaries acting as "middle men", have been shown to be critical in facilitating technology transfer across the TTN (Roxas *et al.*, 2011; Villani *et al.*, 2017; Massa *et al.*, 2022; Palaco *et al.*, 2022; Subramanian *et al.*, 2022).

2.2.2.1 *Brokers*

Institutional as well as individual intermediaries can act as network brokers to bridge gaps in TTNs, which are affiliated with government institutions such as national labs, engineering and technology centers, incubators, science parks; academia institutions such as technology transfer offices, "star scientists" or "principal investigators"; industry players such as large-sized or leading firms, and venture capital firms; and places such as hub regions of the TTN, also known as "brokerage regions", "technoparks", and "knowledge

Table 2.12 Brokerage roles in the inter-regional TTN.

Types	Structure	Descriptions
Coordinator		Mediation between regions of one nation where the mediator is also co-located within a group
Consultant	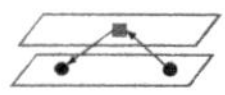	Mediation between regions of one nation where the mediator is not collocated within the nation
Gatekeeper	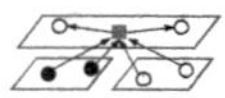	Mediation between two regions where the mediator regulates technology flow to its nation
Representative	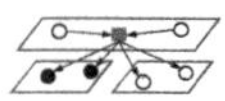	Mediation between two nations where the mediator regulates technology flow from its nation
Liaison	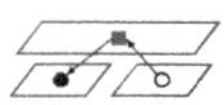	Mediation between two nations where the mediator does not belong to either nation

Note: ▱ represents a nation; ■ represents a brokerage; ○ represents a region collocated with the broker; ● & ○ represent regions outside of the broker region.
Source: Redrawing from Seo (2019).

hubs" (Lee *et al.*, 2003; Bathelt *et al.*, 2011; Kidwell, 2013; Baglieri & Lorenzoni, 2014; Mayer & Cowell, 2016; Olcay & Bulu, 2016; Chiambaretto *et al.*, 2019; Seo, 2019; Taira, 2020). These intermediaries play a crucial role in bridging various gaps, including structural, cognitive, and cultural differences within the TTN.

Based on their structural position within the TTN, network brokers can be classified into five brokerage roles: coordinator, consultant, gatekeeper, representative, and liaison, as outlined in Table 2.12 and illustrated in Figure 2.52 (Gould & Fernandez, 1989; Burt, 1992, 2005; Lazaric *et al.*, 2008; Meyer, 2010; Giuliani, 2011; Seo, 2019). These brokers include a range of entities, such as "star scientists", financial institutions or venture capital firms, science parks and clusters, research organizations, incubators or accelerators, collaborative research centers, technology licensing offices, and technology transfer offices (Groysberg *et al.*, 2008; Kauffeld-Monz & Fritsch, 2013; Villani *et al.*, 2017; da Silva *et al.*, 2018; Massa *et al.*, 2022; Palaco *et al.*, 2022) (Table 2.13).

These network brokerages play a unique role in six areas: conflict resolution, spreading knowledge, connecting idea fragments,

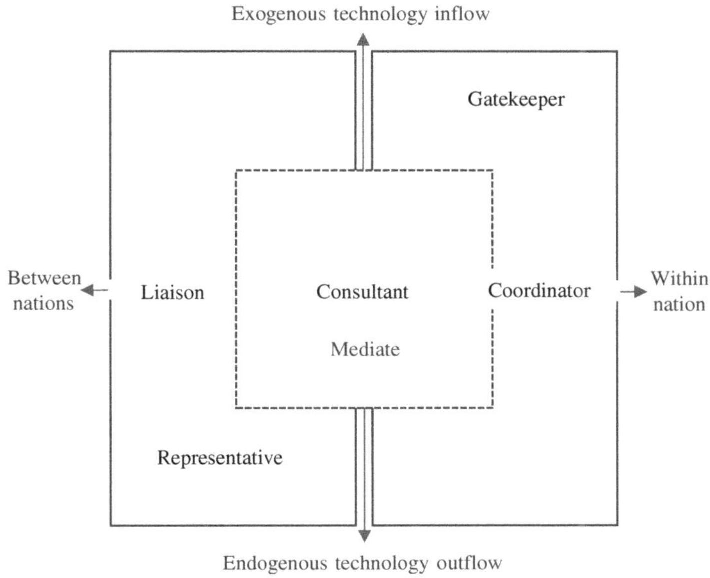

Figure 2.52 Brokerage roles in TTN.
Source: Redrawing from Seo (2019); Gould and Fernandez (1989).

connecting problems to solutions, expanding the network, and strengthening the network, through five mechanisms to support these functions: endorsement, mediation, events, antennas, and digital support systems (Massa *et al.*, 2022).

In TTNs, the effectiveness of network brokers is strongly related to their structural positions in the network, which may be assessed by the degree of nodal centralities and the strength of external ties based on structural hole theory (Burt, 1992).

2.2.2.2 *Governments*

Governments, including governmental organizations, also play a crucial role in the structural dynamics of TTNs, particularly in emerging countries such as China. They establish and enforce policies, standards, laws, regulations, and guidelines, as well as make investments in various scenarios, which significantly influence the

Table 2.13 Types of the international TTN brokers.

Types of intermediaries	Country and its types	Intermediaries example
University industry liaison	Global (developed emerging)	Global Innovation Lab
Technology transfer/ licensing/ commercialization offices	Mexico (emerging)	Technology Transfer Offices network
	Netherlands (developed)	"Digicampus"
Industrial organizations	South Korea (developed)	Korean Integrated Contract Manufacturing Service
	Sweden (developed)	SAFER
	Austria (developed)	Advantage Austria
Science parks & clusters	South Korea (developed)	Daedeok Innopolis
	Colombia (emerging)	"Ruta N"
	France (developed)	"I-Care" cluster
Incubators, accelerators, funding or venture capital firms	Chile (emerging)	CORFO (for its acronym in Spanish)
	Netherlands (developed)	Amsterdam's startup-in-residence programme (SiR)
	Finland (developed)	Finish Innovation Fund (SITRA)
	Mexico (emerging)	FINNOVA
Collaborative R&D centers	Netherlands (developed)	Innovation Network Rural Areas & Agricultural System

Source: Redrawing from Palaco *et al.* (2022).

movement of certain technologies within the TTN (Wang *et al.*, 2003; Ma & Wang, 2008; da Silva *et al.*, 2022).

These policy sets play a crucial role in facilitating the TTN, as they promote and protect technology commercialization (Bellais & Guichard, 2006), minimize impacts on specific areas or sectors (King & Nowack, 2003), and strengthen technology transfer systems (Cunningham *et al.*, 2019). Regulations for public technology management, legal laws of patenting and licensing, and protections of intellectual property are evident examples of policy sets that strengthen TTNs. Governments also ensure institutional funding for R&D and the commercialization of technology through research promotion, government-centered R&D investments, cost reductions for technology transfer, and other initiatives (Audretsch *et al.*, 2016; Noh & Lee, 2019).

Policymaking facilitates the continuous evolution of patterns of technology flows within the TTN by regulating and governing the transfer actors involved. From a policymaking perspective, the governance of governmental organizations in TTNs has been distinguished by different power relationships and modes. These include Mode 1-Hierarchical command-and-control governance (Westphalian or "vertical" structure, led by nation-states through justificatory strategies), Mode 2-Horizontal policy-making governance ("role-based" structure led by diverse transfer actors through soft law, negotiation, competition, agreement, etc.), and Mode 3-Decentralised network governance ("fluid" structure led by network brokers through networked and network-making powers) (Figure 2.53). Within Mode 1, the state or government is sovereign and legitimate in commanding and controlling transfer actors through vertical power relationships. In contrast, the government in Mode 2 takes on a role of "steering" rather than "rowing", performed by non-majoritarian institutions such as central banks, national labs (or science centers), and technology transfer platforms. However, in the context of Mode 3, the government has more limited powers and has been replaced by emerging brokers that command other agents through network-making and networked powers (Zwitter & Hazenberg, 2020).

2.2.2.3 *Academia*

As providers of specialized knowledge and labor, academia (or universities and research institutes) is globally networked actors and plays a central role in academia–industry TTN through their professional intermediaries such as technology transfer offices (TTO), academia incubators, and collaborative R&D centers (Villani

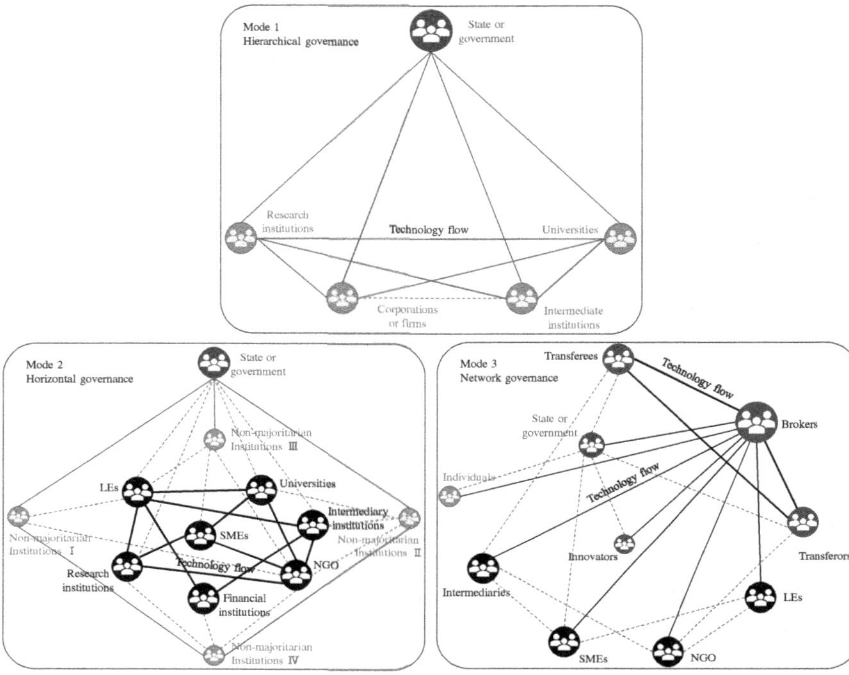

Figure 2.53 Modes of governance in TTN.

Source: Own illustration based on Zwitter and Hazenberg (2020).

et al., 2017), along with university spin-offs or start-ups (Bathelt *et al.*, 2011).

TTO is viewed as one of the most important intermediaries (Brescia *et al.*, 2014; Geoghegan *et al.*, 2015; Olcay & Bulu, 2016), and further grouped into four main types: classical TTO (serves and affiliates one university or research institute), autonomous TTO (higher autonomy degree from academia's administration), discipline-integrating technology transfer alliance (TTA) (serves the technology transfer activities of several universities or research institutes), and discipline-specialize TTA (serves several universities or research institutes) (Schoen *et al.*, 2014), based on its governance structures.

The organizational structure of TTOs is increasingly recognized as an important determinant of TTN. Its dimensions most widely discussed by some scholars include specialization, centralization, standardization, formalization, and configuration. Based on its specialization and configuration, the structure of TTOs is characterized into four typologies: (1) *degree of discipline specialization* (TTO is taken on the departmental level and focused on one specific scientific discipline); (2) *degree of task specialization* (TTO is to act as an intermediary between academia and industry through research funding and services, intellectual property (IP) management, and spin-out services); (3) *level of autonomy* (TTO is distinguished as dependent and independent types; and increase its autonomy through traditional academia's structure, non-profit research foundation, and for-profit venture creation); and (4) *degree of exclusivity* (TTO serves more than one academia and has exclusive relations with others) (Schoen *et al.*, 2014) (Table 2.14).

TTN is also a collaborative innovation community centered on top-ranked universities and research institutions (Brescia *et al.*, 2014) (i.e., world-class universities and national labs in China), which realizes strong alliance and complementary advantages through innovative clusters and industry–university–research collaboration networks including various cooperation between different universities, between universities and R&D institutions, between universities and innovative corporations, along with universities and innovation platforms (Yao, 2019).

2.2.2.4 *Enterprises*

Enterprises, particularly multinational enterprises (MNEs) and leading firms or "lead geese" (Taira, 2020), are indispensable actors and the primary drivers behind TTN and hubs, similar to academic institutions (Lee *et al.*, 2003; Siler *et al.*, 2003; Poon & Sajarattanochote, 2010). During the glocalization process of a cluster, these international-oriented leading firms may play a dominant role in maintaining the cluster's resilience by managing a balanced web of trust-based "pipelines" and "buzz". This balance between detachment and embeddedness is illustrated in Figure 2.54.

MNEs use various methods to transfer and share their technology with host countries, including active or passive, informal or

Table 2.14 Structural dimensions of TTO in TTN.

Types	Descriptions		Values
Degree of discipline specialization	Disciplines	All disciplines	Discipline integrated
		One discipline	Discipline specialized
Degree of task specialization	Technology transfer activities	Research funding services, IP management, and spin-out services	Fully integrated
		IP management and spin-out services	Forward integrated
		Research funding services, and IP management	Backward integrated
		IP management	IP specialized
Level of autonomy	Dependency on academia administration	Independency of academia administration	Independent
		Dependency of academia administration	Dependent
		Serves exclusively one academia	Exclusive
Degree of exclusivity	One serves several academia	Serves several academia	Non-exclusive

Source: Redrawing from Schoen *et al.* (2014).

formal approaches. These can range from copying technology used by local firms as a demonstration effect, creating backward and forward linkages that promote learning-by-doing, bringing skilled employees through training or movement (talent pool effect), and collaborating on research and development, foreign investment, trade, joint ventures, licensing arrangements or agreements, and technology alliances, among others. Regardless of the method

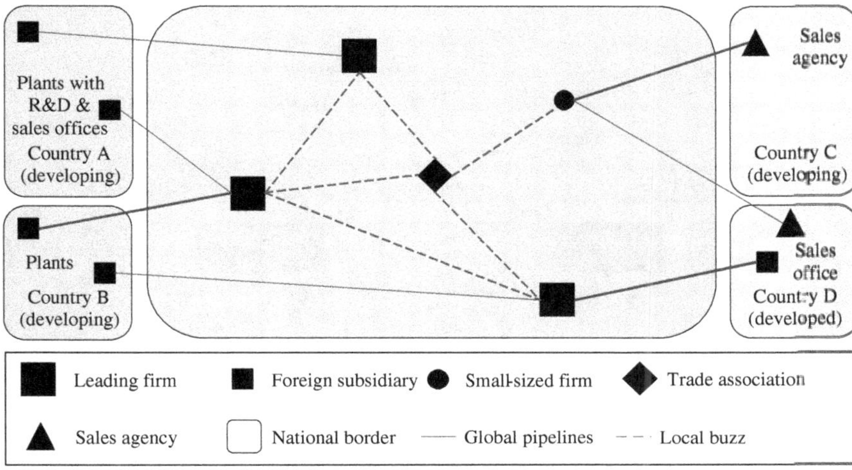

Figure 2.54 Glocalization of leading firm in TTN.

Source: Redrawing from Taira (2020).

employed, this process incurs four types of costs: transaction costs (such as searching, negotiation, and monitoring), costs of opportunistic behavior, provision costs (such as learning-by-doing), and dissipation costs (which refer to the loss of firm-specific advantages) (Eden *et al.*, 1997).

Except for MNEs, the network brokerage roles of small- and medium-sized enterprise (SMEs) are extensively discussed (Chew & Yeung, 2001; Inkinen & Suorsa, 2010). Despite their smaller size, SMEs produce more innovations than large enterprises (Chew & Yeung, 2001). Unlike MNEs, SMEs rely more on informal methods of technology transfer such as joint ventures, technology alliances, and subcontracting and tend to focus on local niche markets rather than foreign ones (Eden *et al.*, 1997). Through subcontracting linkages (including turnkey contracting, consignment contracting, and hybrid contracting), SMEs provide expertise, localized knowledge, and soft technology to MNEs, which is called the "reverse transfer" (Figure 2.55) (Perry & Tan, 1998; Chew & Yeung, 2003).

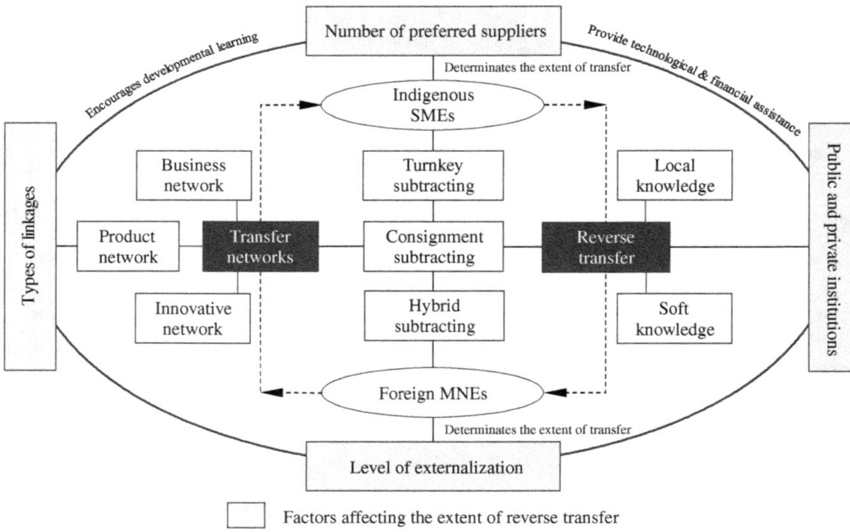

Figure 2.55 Reverse transfer through subcontracting linkages in TTN.
Source: Own illustration based on Chew and Yeung (2001).

2.2.2.5 *Intermediaries*

Various kinds of intermediaries have been identified to facilitate technology transfer, including engineering and technology centers (or collaborative research centers), incubators, technology transfer offices, networking or connecting associations, as well as financial and professional service providers (Inkinen & Suorsa, 2010; Armanios *et al.*, 2017; Clayton *et al.*, 2018; Subramanian *et al.*, 2022). Their primary purpose is to accelerate the flow of knowledge transfer. These intermediaries have been described in different ways in relation to their role in technology transfer, such as innovation agents (Bidault & Fischer, 1994), bridgers (Bessant & Rush, 1995), facilitators (Subramanian *et al.*, 2022), brokers (Albors *et al.*, 2005), and bridging institutions (Sapsed *et al.*, 2007).

Technology intermediaries play a crucial role by acting as a bridge or intermediary in technology transfer between transferors and transferees (Zhang & Li, 2017; Subramanian *et al.*, 2022). Their agents not only support the flow of technology from source to

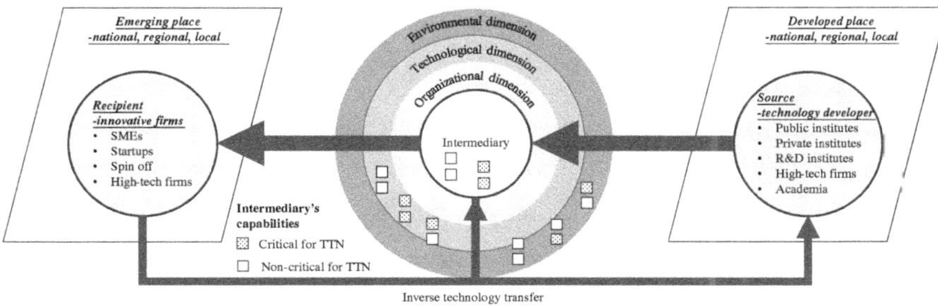

Figure 2.56 Brokerage organization of intermediaries in TTN.
Source: Own illustration based on Palaco *et al.* (2021).

recipient but also store recombinant technology, offer value-added consultation services, and act as co-creators of knowledge, attaining "reverse knowledge benefits" by leveraging the transfer process (Hargadon & Sutton, 1997; Roxas *et al.*, 2011; Fini *et al.*, 2018; Subramanian *et al.*, 2022). In essence, technology intermediaries play a combinational role that is instrumental in cross-pollinating technology between different parties (Figure 2.56).

2.2.3 *Network objects*

Technology is the key and core object of technology transfer process and network. This term can be defined in various ways and categorized according to its intended purpose (Wahab *et al.*, 2012). Current research on technology transfer has linked technology with knowledge or information (Dunning, 1994). Most researchers agree that knowledge is the essential element underlying technology transfer (Hall & Johnson, 1970; Grosse, 1996; Wahab *et al.*, 2012).

Some of the types of knowledge are classified extensively as science versus technology (Gardner, 1994), tacit (or uncodified) versus explicit (or codified) (Polanyi, 1962, 1966; Cooke *et al.* 1997), know-what and know-why versus know-how and know-who (OECD, 1996), STI mode (science-technology-innovation, producing know-why) versus DUI mode (doing-using-interaction, producing know-how and know-who) (Jensen *et al.*, 2007), together with knowledge produced in a disciple-specialized mode (mode 1) versus one produced in a transdisciplinary-networked mode (mode 2) (De la Mothe & Foray, 2001).

2.2.3.1 *Science and technology*

Based on its use purpose, knowledge was classified generally into the two categories: science and technology (Gardner, 1994). They are different but correlated. Scientific knowledge (or science) is generated mainly by academia (or scientist) to understand the natural world, while technological knowledge (or technology) is considered as a transferable asset (Noh & Lee, 2019) mainly developed by enterprises to change the natural world. They can be regarded as a decomposition of knowledge into "know-why" (explanation of specific phenomena and patterns) and "know-how" (development of marketable and industrial research activities) (Karlsson & Johnson, 2005; Ma & Xu, 2022).

2.2.3.2 *Hardware and software technologies*

In knowledge management, technology was originally identified as product-embodied knowledge as follows: (1) *hardware knowledge*, named as capital-embedded technology, including capital-embodied capability, capital equipment and machinery, capital materials and suppliers; (2) *software knowledge*, named as information-embedded technology, includes explicit knowledge (codified knowledge embodied in programs, designs, operations, manuals, blueprints, systems and procedures, etc.) and implicit knowledge (uncodified knowledge embodied in firm-specific skills, experiences, and cultures, etc.) (Wong, 1991), and also consists of physical component (such as products, tooling, equipment, blueprints, processes) and informational component (know-how in management, marketing, production, quality control, reliability, skilled labor, and functional areas) (Kumar *et al.*, 1999). Similarly, it is defined as a cultural system encompassing: (1) the basic knowledge subsystem; (2) the technical support subsystem (software); and (3) the capital-embodied technology subsystem (hardware).

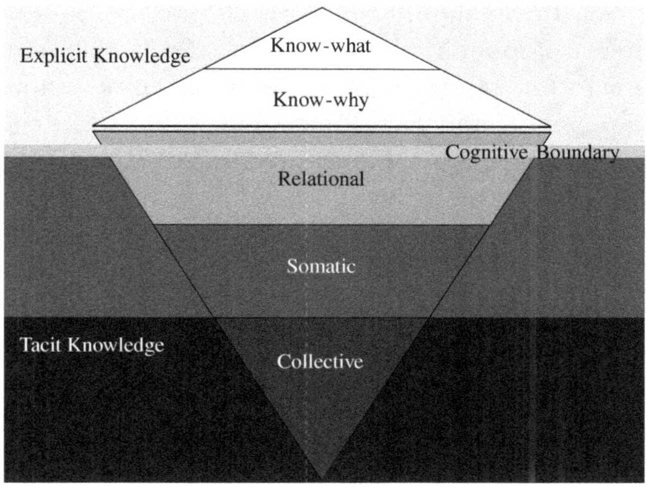

Figure 2.57 The iceberg metaphor of knowledge.

2.2.3.3 *Explicit and tacit knowledges*

Explicit knowledge refers to formal and systematic knowledge that is codified in language, such as know-what and know-why knowledge (Lundvall & Johnson, 1994; OECD, 1996). Implicit or tacit knowledge, on the other hand, includes know-how and know-who knowledge and is often difficult to transfer across large spatial distances as it is "sticky" (Liu *et al.*, 2013). This type of knowledge includes relational, somatic, and collective tacit knowledge (Collins, 2012), which has led to the creation of industrial districts and clusters with co-locality for innovation (Markusen, 1996; Nadvi & Schmitz, 1999). The relationship between explicit and tacit knowledge can be visualized using the iceberg metaphor (Figure 2.57).

2.2.3.4 *Incremental, modular, and architectural technologies*

Technology also varies in terms of the capabilities required by a company to implement a particular type of knowledge (Ernst, 2009). It is helpful to differentiate between four types of technology: incremental technology (dominant component design and architecture with no significant scientific inputs), modular technology (new component technology driven by labor division), architectural technology (changes to the architecture without modifying its components), and radical technology (both new component technology and changes in architectural design) (Ernst, 2009).

2.2.3.5 *Analytical, synthetic, and symbolic technologies*

To understand innovation in different industries, technology can also be categorized into three knowledge bases: analytical, synthetic, and symbolic (Asheim *et al.*, 2007b). Analytical knowledge pertains to industrial settings where knowledge creation primarily relies on deductive formal models, rational processes, and codified science (Martin & Moodysson, 2011). It is similar to science-based knowledge that can take place across distant locations and relies more on globally spread information (Liu *et al.*, 2013). Synthetic knowledge, on the other hand, refers to industrial settings where innovation occurs mainly through the application of existing knowledge or new combinations of knowledge. This type of learning process is akin to engineering-based learning and involves collaborations with organizations in closer geographic proximity, as seen in national or regional innovation systems (Asheim & Coenen, 2005; Asheim *et al.*, 2007b). Finally, symbolic knowledge is artistic or creativity based and highly context-specific, strongly tied to habits and norms, and relies more on face-to-face interaction in locally trust-based networks (Martin & Moodysson, 2011) (Figure 2.34).

2.2.4 *Network channels*

As the informal or informal means, network channels represented by linkages (Coccia, 2008) are key pipelines of technology flows from senders (home nodes) to receivers (host nodes) (Figure 2.58) (Young & Lan, 1997), existing in different forms: (1) formal transfer (legalized through contracts) versus informal transfer (not

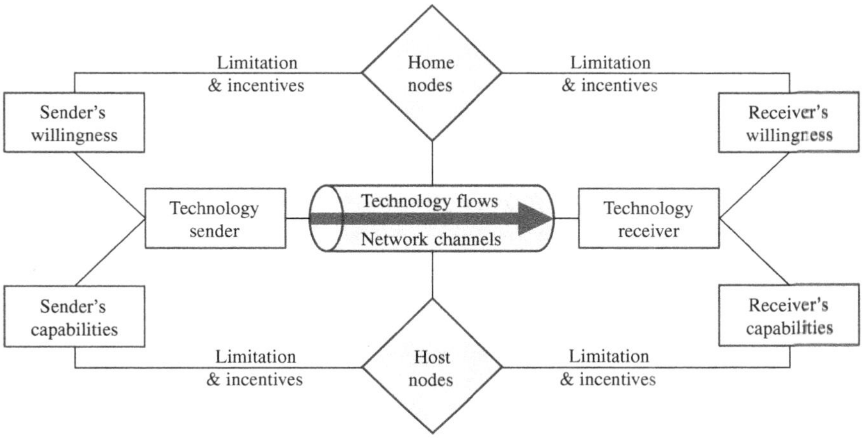

Figure 2.58 Basic model of TTN.

Source: Redrawing from Small (1985); Young and Lan (1997).

legalized by signing a contract, dependent on person-based contacts) (Taira, 2020; Vega-Gomez & Miranda-Gonzalez, 2021); (2) science-based regimes (scientific collaboration, patent licensing and academic spin-offs) versus development-based regimes (R&D collaborations, participation in professional networks, and inflow of high-skilled labors) (Gilsing *et al.*, 2011); (3) institutional interactions (commercial exchanges, contracts, and partnership relations) versus unintentional interactions (Audretsch & Feldman, 1996a; Coccia, 2008). There exists some complementary linkages between them, as informal contacts improve quality of a formal transfer and as formal contracts are accompanied by informal relationships of mutual technology exchange (Grimpe & Hussinger, 2013).

In terms of the degree of formalization, technology transfers can be organized into two categories: internal (or local) buzz, which involves more informal transfer and exchange created by the co-location of transfer actors, and external (or global) pipelines, which pertain to trans-local or extra-regional technology flows through formal and structural means (Figure 2.38). The focus is on spatial interaction between the originations and destinations of technology flows, ranging from local linkages to trans-local linkages.

2.2.4.1 *Global pipelines*

In contrast to local buzz, global pipelines of technology transfer are not limited to a region or cluster. They can be considered as channels for external linkages and exchanges (Bathelt *et al.*, 2004; Bathelt, 2007), which involve non-local interactions (Esposito & Rigby, 2019). Such interactions may take place at a trans-local or cross-cluster level within a city, an extra-regional or interurban level within a nation, or even at a transnational level in the global economy.

The pursuit of external technologies through global pipelines can alleviate the limitations of local buzz resulting from the "death of entropy" (Camagni, 1991). The motivations of transfer actors for seeking external technologies typically fall into four main categories: to commercialize research results, acquire knowledge, access complementary financing for projects, and gain access to other resources within the productive sector (D'Este & Perkmann, 2011).

Global pipelines of technology transfer follow three main pathways: (1) *transfer through procurement*, whereby firms acquire new technologies from collaboration with research laboratories; (2) *transfer through joint development projects*, in which technologies transfer to industry; and (3) *transfer through patents*, which is the most direct and codified way to transfer technology (Amaldi, 2000).

Specifically, it is extensively evidenced that global pipelines include diverse means: foreign technology transfer (which is the primary channel for state-owned enterprises), foreign direct investment, foreign subsidiaries and employees, importing or exporting trade, global value chain participation, inter-industry R&D spillover, and domestic technology transfer (which is the indigenous channel for foreign-funded enterprises) (Young & Lan, 1997; Veugelers & Cassiman, 2004; Santacreu-Vasut & Teshima, 2016; Deng & Lu, 2021; Qian *et al.*, 2022). Temporary clusters, such as professional gatherings, trade fairs, exhibitions, conventions, festivals,

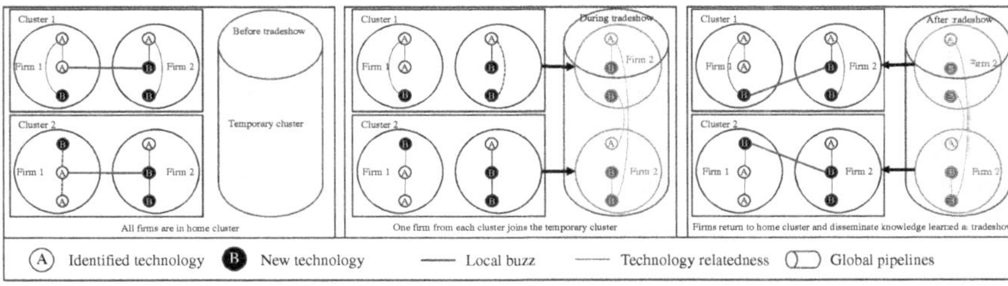

Figure 2.59 Recombination of global pipelines through tradeshow.
Source: Redrawing from Esposito and Rigby (2019); Morrison *et al.* (2013).

congresses, and conferences, also play a role in technology transfer (Bathelt & Schuldt, 2008; Comunian, 2017). By participating in these clusters, firms can share knowledge and learn from other participants, further facilitating technology transfer and spillover (Esposito & Rigby, 2019), as depicted in Figure 2.59.

2.2.4.2 *Local buzz*

Local buzz refers to deliberate face-to-face interactions, as well as unintended "rubbing elbows" or the impact of "being there" (Gertler & Levitte, 2005) that facilitate the establishment of corporate boundaries (Brown & Duguid, 2002). Local buzz is enabled by a specific information culture and a robust social structure within a cluster or local level (Maskell & Malmberg, 1999a, 1999b; Bathelt *et al.*, 2002).

Owing to global online community and temporary external linkages (Torre, 2008; Tamoschus, 2012), buzz is not only generated in the local context affected by spatial proximity but through a virtual relationship or "*gossip*" interaction supported by internet-based technologies (i.e., mobile internet and blogs as "*virtual pipelines*"), as well as short-term interview supported by temporary clusters (such as trade fairs, exhibitions, conventions, congresses, and teleconferences), and evolves into the "*global buzz*" (Tamoschus, 2012) or "*virtual buzz*" (Bathelt & Schuldt, 2008; Jones *et al.*, 2010; Tamoschus, 2012) or "*temporary buzz*" (Maskell *et al.*, 2006) (Figure 2.60). As a result, face-to-face communication is no longer a necessary precondition and geographical proximity can give

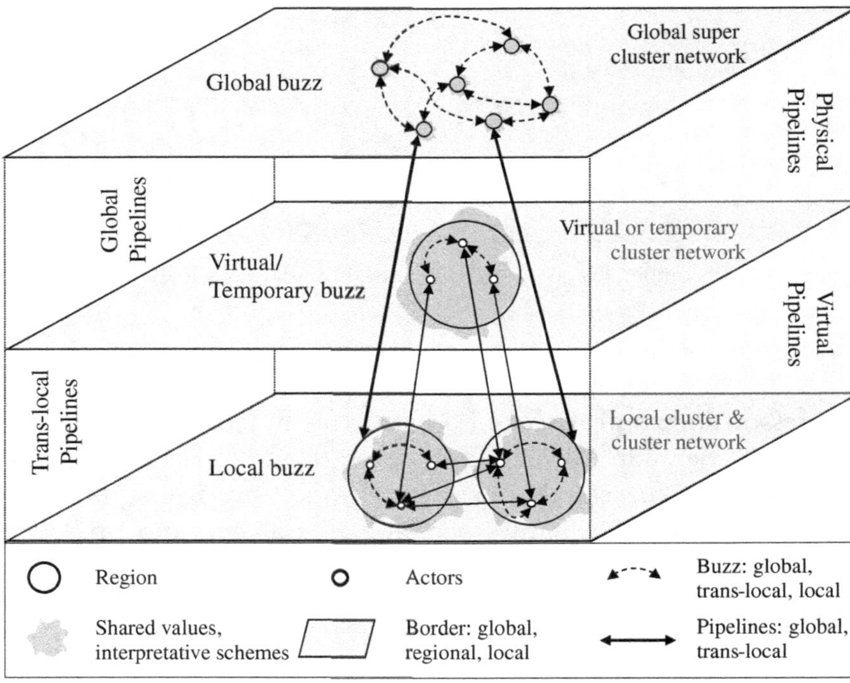

Figure 2.60 Different buzzs of cluster glocalization.

way to relational proximity (Kaufmann *et al.*, 2003), and disperse collaborations can further create buzz elsewhere rather than "being there" (Jones *et al.*, 2010) or "without being there" (Gertler, 2008).

Numerous studies have shown that innovation performance in a cluster is closely related to the strength of local buzz (Esposito & Rigby, 2019). In a cluster with strong local buzz (cluster 1), firm borders are open and recombination of ties occurs between firms as well as within firms, allowing all firms to access the knowledge stocks of their local partners for exploration and enabling technology flows. In contrast, firms in a weaker-tie cluster (cluster 2) tend to be relatively closed and locally focused, with more within-firm recombinant ties than external ties (Figure 2.61).

Local buzz has a direct impact on local innovation performance (Camagni, 1991) and is also related to an inverted U-shaped curve when it comes to innovation capability (Cao *et al.*, 2022). A stronger

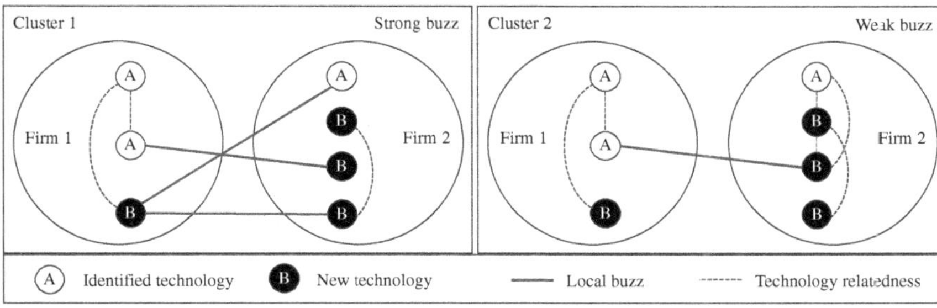

Figure 2.61 Recombination of local buzz within cluster.
Source: Redrawing from Esposito and Rigby (2019).

local buzz leads to more active technology transfer activities (Storper & Venables, 2004). However, greater local interaction may not always benefit innovation due to the potential for technological lock-in effects (Morrison *et al.*, 2013; Esposito & Rigby, 2019).

In addition to occurring at the local cluster level, buzz can also occur within a city where transfer actors are in close proximity and managed by a unified government (Liu & Chen, 2009). Transfer actors within a city tend to share more uniform values and rules, and have higher chances of face-to-face communication due to a common urban milieu (Rossignolo, 1999; Maennig & Ölschläger, 2011). Therefore, cities are often considered at the "local" scale (Maillat, 1998a; Ratti *et al.*, 1999), and technology transfer and diffusion within cities are referred to as "local buzz".

2.2.4.3 *Glocal pipelines*

The buzz-pipelines model has been subject to some challenges. Buzz and informal linkages are found at various spatial scales, as well as formal partnerships and pipelines at both local and national levels (Trippl *et al.*, 2009; Esposito & Rigby, 2019). Knowledge may be person-based as well as locality-based (Taira, 2020), and it can be transferred globally through pipelines beyond the local buzz in a cluster (Bathelt *et al.*, 2004).

Glocal pipelines are a relational mechanism that facilitate the transfer, absorption, and communication of both explicit

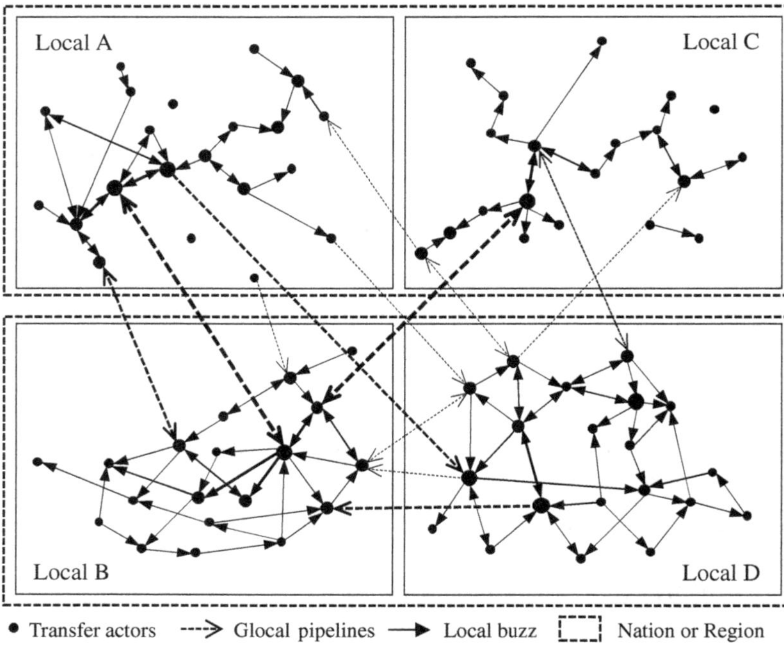

Figure 2.62 Recombination of glocal pipelines in glocalization.
Source: Own illustration based on Si *et al.* (2016).

knowledge (such as production processes, equipment, patents, technical standards, and technology licenses) and tacit knowledge (such as market information and innovation capacity) between individuals, corporations, and places. These pipelines can be formal or informal, physical or virtual in nature, and can serve to enhance the innovation capabilities of those involved. This conceptualization is prevalent in academic literature (Tamoschus, 2012; Si *et al.*, 2016; Mao & He, 2019b; Cao *et al.*, 2022) as illustrated in Figure 2.62.

In addition, glocal pipelines exhibit a multiscale interdependency (Zhang *et al.*, 2013) and can be established both within and across national borders (Taira, 2020; Wang *et al.*, 2023b) (Figure 2.54). At the global level, transnational pipelines are prevalent, where technology transfer occurs between cities located in different countries. In contrast, trans-local pipelines emerge within a nation or region, connecting cities within the same geographic area (Liu & Yan, 2022). These pipelines interact bidirectionally, and their density,

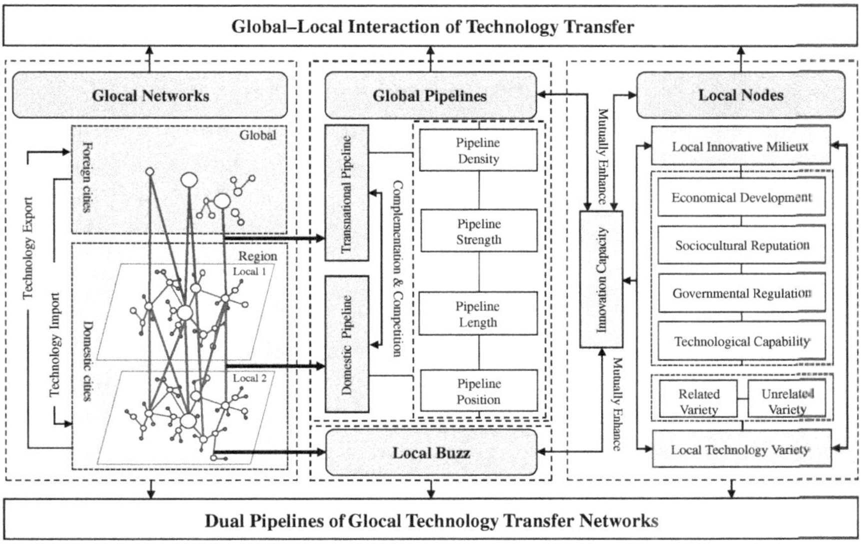

Figure 2.63 Dual-pipeline framework of "global–local" interaction in TTN.
Source: Own illustration based on Liu and Yan (2022).

strength, length, and network position form a complex mechanism of "competition and complementation", as shown in Figure 2.63. Local indices, such as technology variety and innovative milieu, not only influence local innovation capability but also contribute to the development of a local buzz.

The concept of technology variety can be further divided into related variety, which refers to the diversity of local technologies and associated technologies within the same knowledge base, and unrelated variety, which pertains to the diversity of local technologies in different knowledge bases (Frenken *et al.*, 2007; Miguelez & Moreno, 2018). Generally speaking, cities with high related variety exhibit greater technological complementarity, which translates into a lower cost of technology transfers and spillovers. Conversely, cities with high unrelated variety manifest larger technological variability, making it more challenging for technology learning and transferring, thereby hindering the local innovation performance of a city (Frenken *et al.*, 2007). However, once learned, distinct unrelated technologies have the potential to facilitate innovation diffusion and technology transfer

(Boschma, 2017). In turn, an increase in local innovation capacity can contribute to the diversification of local technologies, resulting in higher indices of related and unrelated technology variety.

Similarly, the local innovation milieu also demonstrates a positive relationship with innovation capacity (Feldman, 1999; Görg & Strobl, 2007; Amiti *et al.*, 2012; Dong *et al.*, 2014). The local innovation milieu can be classified into four competitiveness factors: economic development (economic scale and structure), sociocultural reputation (social responsibility, cultural representation, intellectual property awareness, entrepreneurship atmosphere, and social tolerance), technological capability (knowledge infrastructure, R&D collaboration, venture investments, knowledge generation and commercialization), governmental regulation (policies and laws related to local innovation), and external openness (strong global and local business network) (Aula & Harmaakorpi, 2008) (Figure 2.63).

2.2.5 *Network structures*

As a complex system, a network is spatially disaggregated into three equivariant structures (including spatial configurations, temporal dynamics, and hierarchical organizations) by spatial subdivision approach, which is the same as the tripartite division of spaces in physics (namely, real or physical space, phase space, order space, respectively) (Figure 2.64).

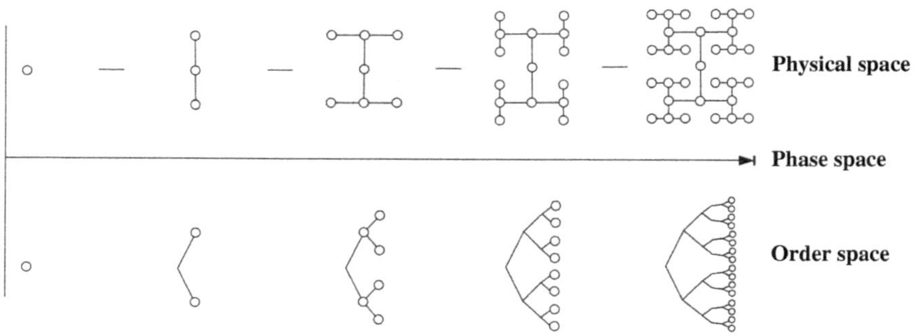

Figure 2.64 Spatial disaggregation: Configurations, dynamics, and hierarchies.
Source: Redrawing from Batty and Longley (1994); Liu (2017).

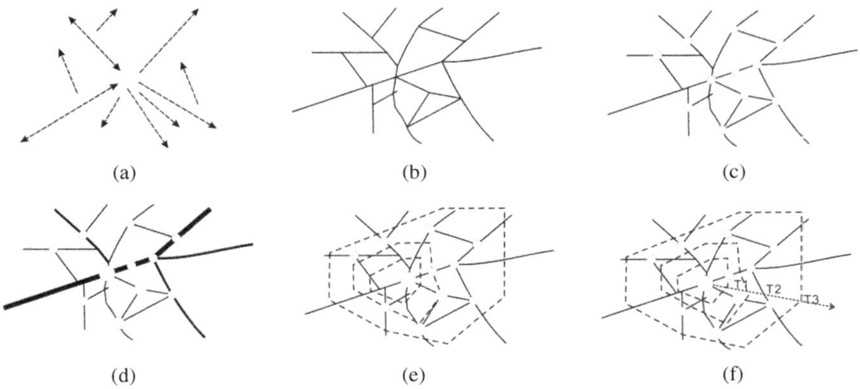

Figure 2.65 "Ternary spaces" of TTN. (a) Flows of technology; (b) Channels of flows; (c) Nodes of transfer actors; (d) Hierarchies of network; (e) Territorializations of network and (f) Diffusions of network.

Source: Redrawing from Liu (2017).

Regarding its spatial organization, the network of technology transfers is perceived as a space of flows that comprises six elements (Castells, 1996): technology flows, connection channels (pipelines and buzz), transfer nodes (individuals, institutions, and places), hierarchical differences, territorializations, and diffusions of the network. These elements are collectively referred to as "*ternary spaces*" (Figure 2.65). Among them, nodes, channels, and flows are linked to the "*real space*", hierarchies represent the "*order space*", while territorializations and diffusions of technology are classified as the "*phase space*".

2.2.5.1 *Real space: Spatial and topological structure*

Physical structure of TTN is depicted as spatial and topological. From a glocalization perspective, TTN includes four spatial forms: (a) global TTN (technology acquisition through international technology transfer); (b) glocal TTN (technology transfer integrated with external linkages (global pipelines) and internal linkages (local buzz)); (c) detached nodes (isolated and peripheral position of

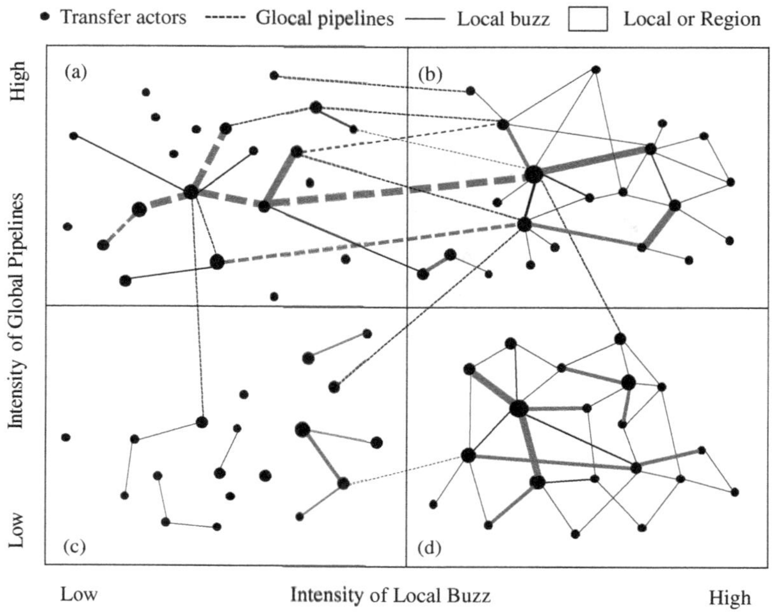

Figure 2.66 Spatial types of TTN based on its glocalization. (a) Global; (b) glocal; (c) detached nodes and (d) local.

TTN); and (d) local TTN (dependent on internal technology transfer with lock-in effect and homogenization risk) (Figure 2.66).

As mentioned previously, TTN is characterized as a topological system comprising nodes (transfer sites, i.e., actors or locations), edges (transfer channels), and flows (intensity of technology transfer) (Lambooy, 2004). Based on the arrangement of nodes and edges, local area networks of technology transfer can be abstracted into several spatial patterns: linear (bus topology), tree (a combination of linear and star topologies), mesh (fishing net topology), and star (hub-and-spoke topology), resulting from point-to-point and hub-and-spoke mechanisms (Figure 2.67).

According to network motif, the topological structure of TTN is classified into five situations: independent application (technology transfer within a same subject), collaborative application (technology transfer between different subjects through patent

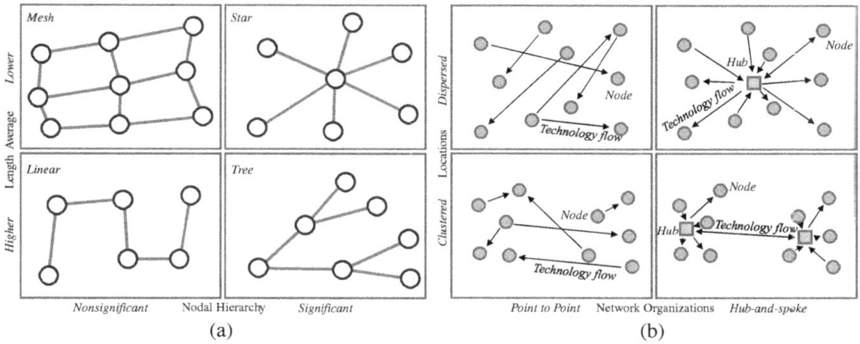

Figure 2.67 Spatial patterns of TTN based on its organizational mechanism. (a) Main patterns and (b) Organizational mechanism.

Source: Redrawing from Rodrigue *et al.* (2020).

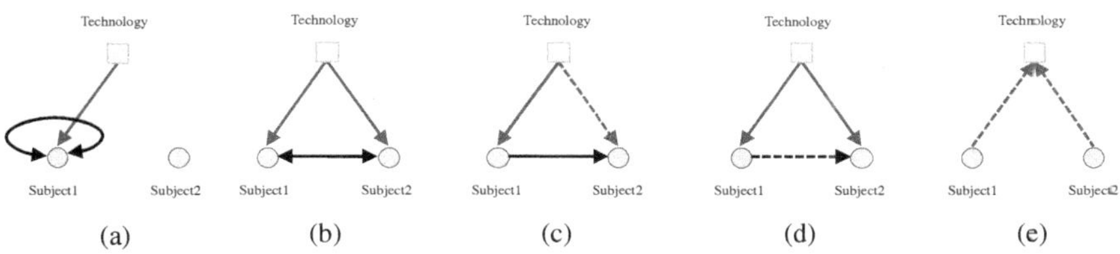

Figure 2.68 Topological structure of TTN based on its motif. (a) Independent application; (b) collaborative application; (c) technology transaction; (d) technology licensing and (e) technology co-citation.

Source: Redrawing from Lin *et al.* (2020).

collaborations), technology transaction (technology transfer between subjects through patent right transaction), technology licensing (technology transfer between subjects through patent licensing), technology co-citation (technology transfer between subjects through patent co-citation) (Figure 2.68).

TTN is characterized by strong and weak ties based on the closeness and intensity of its edges (Granovetter, 1973), whether they are adjacent or non-adjacent to a focal node in a triad network. Consequently, it can be further classified into six types of network configurations (Moon *et al.*, 2022), as depicted in Figure 2.69. In TTN, strong ties of technology transfer may offer homogeneous but

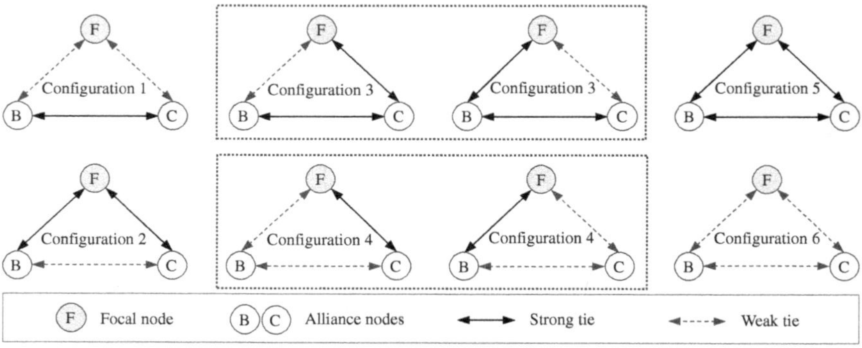

Figure 2.69 Topological structure of TTN based on the strong and weak ties.
Source: Redrawing from Moon *et al.* (2022).

redundant technology, while weak ties can provide novel, diverse, and non-redundant technology (Moon *et al.*, 2022).

2.2.5.2 *Order space: Hierarchical and organizational structure*

In TTN, there are fewer dominant agents that are hard to be replaced with an already involved agent or new entrant, which results in the emergence of hierarchical network rather than peer-to-peer network (Seo & Sonn, 2019). The hierarchy not only creates a vertical connection from focal nodes to peripheral ones embedding in core–peripheral organization but also a horizontal linkage between hub nodes by hub-and-spoke organization (Figure 2.70).

In the first context, the nodal degree distribution of TTN exhibits a small-world characteristic whereby a strong preference for core nodes resulting in the "rich-club effect". This means that a smaller number of core nodes dominate a larger amount of technology transfer nodes, activities and flows due to their shorter path length and higher clustering coefficient. On the other hand, the second organization typically displays a scale-free network with a power law distribution of nodal connectivity. This network is characterized by a majority of connections concentrated on less than 20% hubs

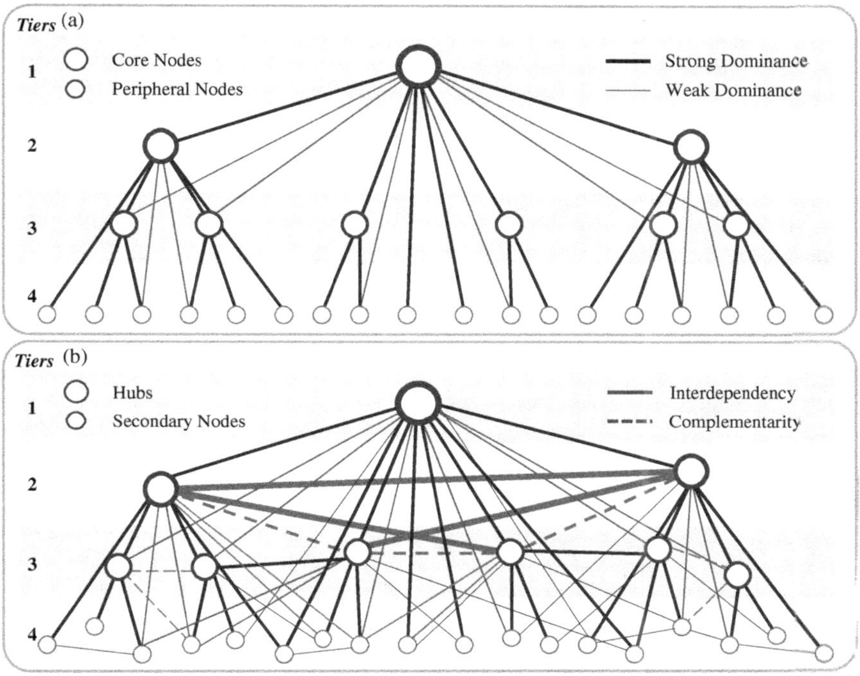

Figure 2.70 Network organizations with hierarchies. (a) Core-peripheral organization with hierarchy and (b) hub-and-spoke network organization with hierarchy.
Source: Redrawing from Rodrigue *et al.* (2020).

following the "Pareto distribution", with strong horizontal ties between these hubs (including secondary hubs) demonstrating interdependency and complementarity (Figure 2.70).

In terms of its internal organization, TTN is considered as a network of networks or multilevel network disaggregated into different levels from motifs, modules, to subgraphs, which is divided into three hierarchies: ego-network level, community level, and whole network level (Sytch & Tatarynowicz, 2014; Wang & Lu, 2021) (Figure 2.71).

Furthermore, the hierarchical structure of TTN is also evident across different spatial scales. Within the framework of glocalization, TTN creates a glocal network that encompasses trans-border

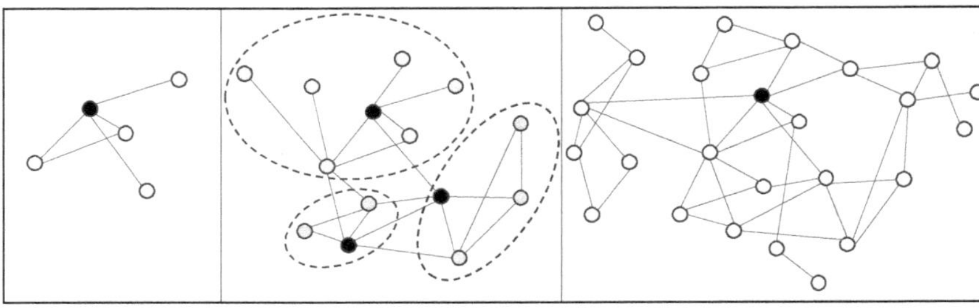

(a) (b) (c)

Figure 2.71 Network hierarchies of TTN. (a) Ego-network level; (b) community level and (c) whole network level.

Source: Redrawing from Syth and Tatarynowicz (2014).

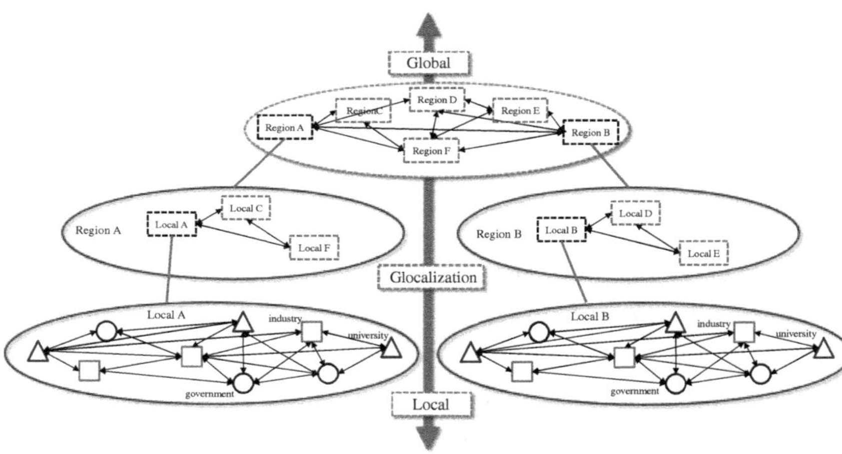

Figure 2.72 Scalar hierarchies of TTN.

technology flows by integrating local innovation clusters with the global innovation network (Figure 2.72). Consequently, it can be classified into various scales such as local TTN (mainly characterized by symbolic knowledge transfer), regional TTN (predominantly driven by synthetic knowledge transfer), national TTN, and global TTN (dominated by analytical knowledge transfer) as shown in Figure 2.34.

2.2.5.3 *Phase space: Temporal and self-organizational structure*

Similar to a territorial network, TTN can be defined as a dynamic and long-term relationship network among transfer actors or places that is influenced by various context-specific peculiarities such as integration, cooperation, individuation, involvement, and so on (Cannarella & Piccioni, 2008).

This dynamic process can be illustrated by two variables: firstly, the rapid growth of nodes over time due to the preferential attachment mechanism. An increase in the number of nodes signifies a rise in the number of actors participating in various technology transfer activities. Secondly, the gradual creation of edges caused by the network's pathologies. The presence of links between nodes indicates the establishment of interactions such as cooperation or competition among transfer actors within the network.

TTN is considered to be a cumulative self-organizing process, whereby the properties of transferred actors and technologies undergo irreversible changes (Lambooy, 2004). This selective process is strongly influenced by its embeddedness in the environmental context, particularly in the market and institutional environment. Additionally, the path and duration of TTN's development and application are important aspects of this process (Lambooy, 2004). Similar to the technology life cycle, the path creation of TNN can be seen as a self-organizing life-cycle process consisting of five stages: the initial stage, growth stage, mature stage, and the renaissance or decline stage (Figure 2.73).

Throughout these stages, the complexity of TNN emerges as a dynamic networking process that involves varying degrees of intensity ranging from low to high levels, organization from disorderly to orderly, scale from local to glocal, and connectivity from simple to complex. This process is not static or linear but instead offers "many roads to success" and "many actors to exit", which are highly dependent on endogenous (mostly specific factors) and exogenous (mostly industry-driven factors) determinants such as the abundance and agglomeration of transfer actors and activities, the connectivity and diversity of nodes and channels, the

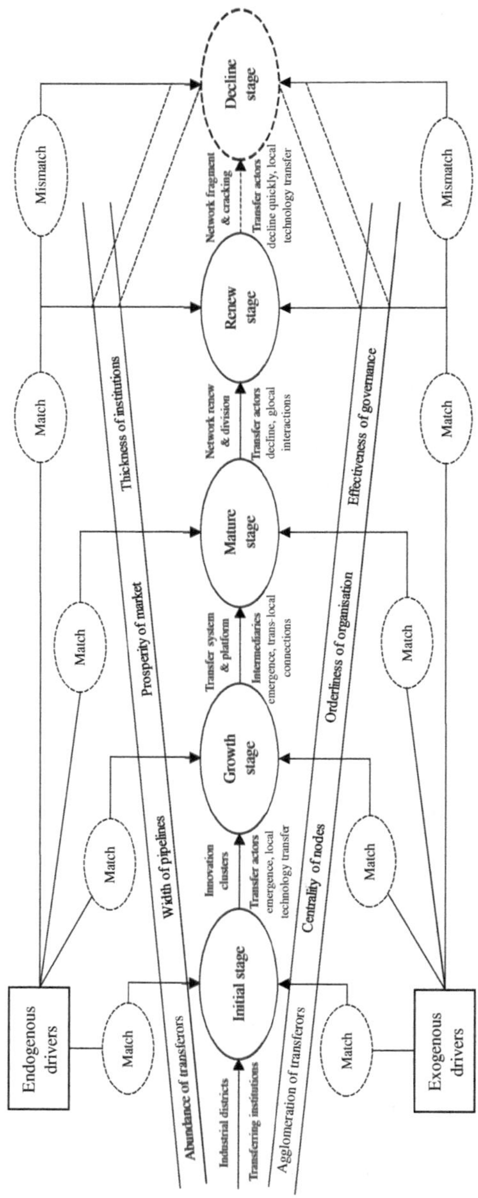

Figure 2.73 Dynamic structure of TTN.

Source: Own illustration based on Liu (2008).

innovative milieu including market and institutional environment, and the effectiveness of organizations and governance structures as shown in Figure 2.73 (Maskell & Malmberg, 2007; Dyba *et al.*, 2020).

2.2.6 *Network dynamics*

The formation and evolution of TTN is a process of networking flows (or movement) of new technology among organizations or places, which follow the evolutionary rules such as life-cycle law, path dependence-and-creation law, and structural catastrophe law, etc.

2.2.6.1 *Life-circle law*

As mentioned earlier, TNN undergoes a life-cycle process similar to that of a living organism (Dyba *et al.*, 2020). This process is divided into four stages, namely emergence (initial phase), expansion (growth phase), equilibration (mature phase), and exhaustion (decline phase), or extension (renewal phase) of nodes and links, which are collectively referred to as the "5E" stages. This process can be depicted as an inverted U-shaped curve, which is influenced by two dynamics. Firstly, the transmission of technology from in-house generation, through commercialization between academia and industries, to diffusion, then to spillover or lock-in of technology, along with diversification of its transfer actors and technologies. Secondly, the networking of technology transfer from science-oriented and local loose-up networks driven by academia, through academia-to-industry network led by small- and medium-sized enterprises, to leading firm-oriented and trans-local cluster network, then growing up to glocal network led by leading enterprises, or declining to local networks, along with the ferment-and-incremental dynamics of its nodes, channels, and communities (Table 2.15; Figures 2.74 and 2.75).

The process of technology transfer is a dynamic networking process that involves a progression from lower to higher levels, a

Table 2.15 Life-cycle characteristics of TTN.

Phase	Transfer technology	Transfer actors	Transfer means	External linkages	Innovation value chains	Networking speed
Initial	Lower diversity	Academia	Scientific collaboration network	Low	Technology generation	Slower
Growth	Higher diversity	Small- & medium-sized enterprises	Academia-industry collaboration network	Higher	Technology commercialization	Very Fast
Mature	Very high diversity	Leading enterprises	Trans-local TTN	High	Technology diffusion	Saturation
Renewal	Higher diversity	Leading enterprises	Glocal TTN	Higher	New technology emergence	Faster
Decline	Lower diversity	Academia	Local R&D network	Low	Technology lock-in	Slower

Source: Own illustration based on Zhang (2018).

transition from disorder to order, a movement from simple to complex, and a shift from cohesion to openness. This process is largely determined by the life cycle of industry or technology (Audretsch & Feldman, 1996b; Klepper, 2010). In the initial stage of this process, academic institutions such as universities and research institutes play a dominant role in technology sourcing. The interactions during this stage are informal, and technology spin-offs occur as a result of geographic proximity (Dyba *et al.*, 2020). At this stage, the capacity for transfer is still limited, and primarily focuses on the accumulation of existing technologies and tacit and heterogeneous knowledge exchanges. The process is highly dependent on face-to-face interaction due to the role of geographic proximity and the lack of cognitive reputation. As a result, technology transfer intensity is lower and unstable, leading to a looser and more local network of transfer activities and actors that are relatively well-distributed throughout the location (Figure 2.75).

During the growth stage, enterprises and intermediaries become increasingly involved in technology transfer, forming an open network that is integrated with innovation cluster networks. In these clusters, technology flows are concentrated, primarily within leading firms who benefit from greater collective learning and technology sharing. The influx of external technologies into these clusters expands local technology pools (Dyba *et al.*, 2020) and facilitates growth into transfer centers or core regions of the network, further widening the gap between the core and peripheral nodes. The entire network is organized in a core–peripheral structure driven by the cumulative or self-reinforcing mechanism.

As the technology matures, various flows of formal and informal transfers grow in number and density, forming a dense network including large amount of nodes with hierarchy and stable linkages including local buzz and global pipelines. Due to their well-established technology pools, a few core nodes with majority of glocal paths act as technology hubs or gatekeepers of the network (Dyba *et al.*, 2020) under the effect of "place dependence". As a result, it is organized as a hub-and-spoke and glocal network.

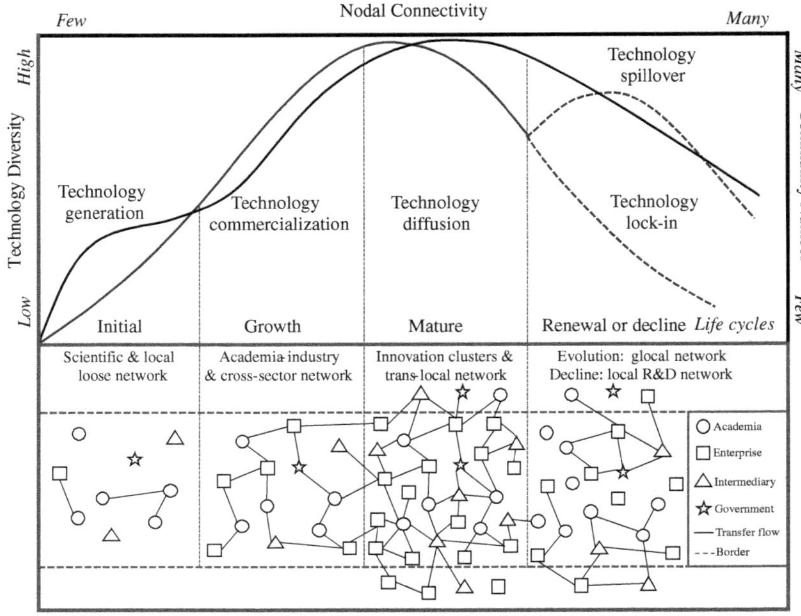

Figure 2.74 Life-cycle law of TTN.

Source: Own illustration based on Huang *et al.* (2022).

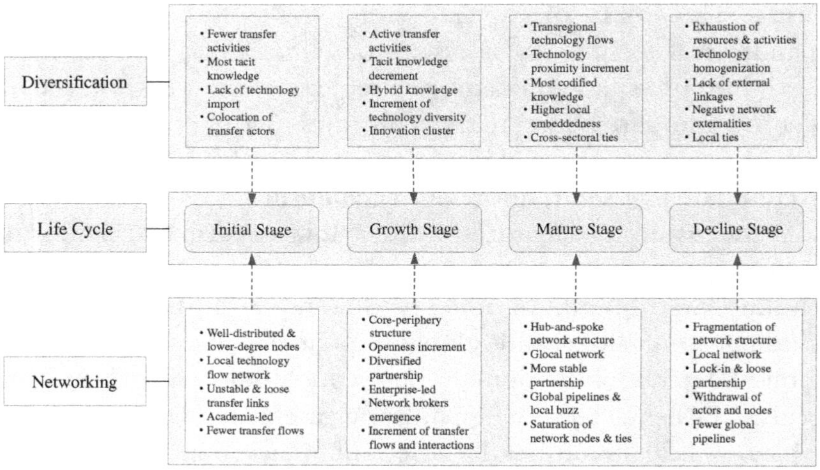

Figure 2.75 Diversification and networking process of TTN.

Source: Own illustration based on Zhou (2008).

During the decline stage, growth in new paths will slow down, with most of them being expansions or extensions of prior nodes. The number of nodes, channels, and communities will also sharply decrease due to mergers between different sub-technologies (Huang *et al.*, 2022). At the same time, having too narrow knowledge bases and focusing too much on specific markets and technologies can hinder the network's adaptability (Giuliani, 2005), and ultimately lead to the closure of external pipelines and a lock-in of local paths in the network (Figure 2.75).

2.2.6.2 *Path dependence-creation law*

The notion of path dependence originates from research on diffusion of technology (David, 1985; Arthur, 1989). It plays a major role in affecting technology transfer process, by offering increasing returns (Arthur, 1989).

Path dependence is a double-edged sword for technology transfer networking progress. Similar to the specification of clusters, it can reinforce networking through positive feedback mechanisms such as *coordination effects, complementarity effects, learning effects* (Bathelt & Boggs, 2003), and *adaptive expectations* (Sydow *et al.*, 2010). However, path dependence is not the same as increasing returns and may go through periods of constant or even decreasing returns (Dobusch & Schüßler, 2012) leading to a homogenization of technology and a lock-in of the TTN.

Path dependence is a mechanism that can cause a network to become locked into a specific trajectory based on past decisions and events, whether they be geographic, technological, or cognitive in nature (Dobusch & Schüßler, 2012). This can lead newcomers in TTN to connect with the original nodes that have high connectivity along existing channels, creating what is known as "place dependence" (Martin *et al.*, 2006) or "spatial structural inertia" (Narula, 2002; Sydow *et al.*, 2010). These phenomena are consistent with "retention" mechanisms like "preferred attachment" and the "Matthew effect" (Glückler, 2007), which suggest that nodes with high connectivity will continue to attract more connections over time.

An important challenge for firms is how to break away from an established trajectory and create a new path (referred to as "path creation" by Garud & Karnøe, 2001) in order to overcome network lock-ins. The process of path creation involves attracting new entrants into the network through the sudden emergence of external bridges or global pipelines, which stands in contrast to the gradual and cumulative nature of path dependence.

The path dependence and creation of the network are thus strongly correlated to three categories of critical parameters: (1) the externalities of location decision, and the varieties of technology relatedness; (2) the diffusion of technologies, and the overlap between technology phases; (3) the properties of network structure (Figure 2.76) (Suire & Vicente, 2014). Based on their combination, TTN follows different pathways with path dependence and creation along the market phases of composite technologies, as shown in Figure 2.77.

Starting from a beta-test market, TTN is organized in a very scattered structure. Due to the role of geographical proximity and "audience effect", locational cascades of technologies occur along with strong technological homophily and rather low technology interactions (Suire & Vicente, 2014). Path dependence and creation strategies can help TTN transition into a core–peripheral structure with disassortativity during early growth. New entrants preferentially

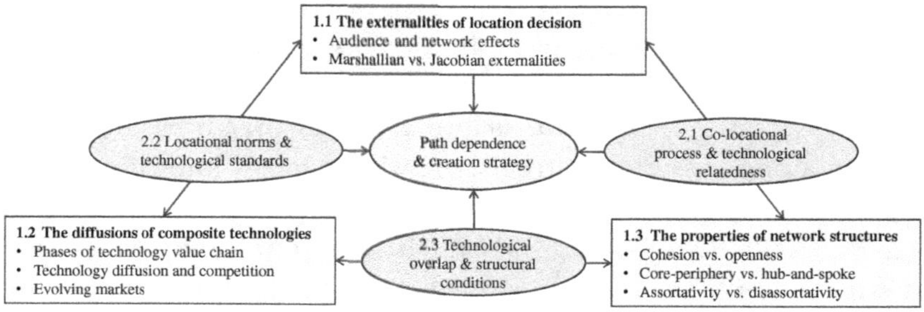

Figure 2.76 Critical factors of path dependence and creation of TTN.

Source: Own illustration based on Suire and Vicente (2014).

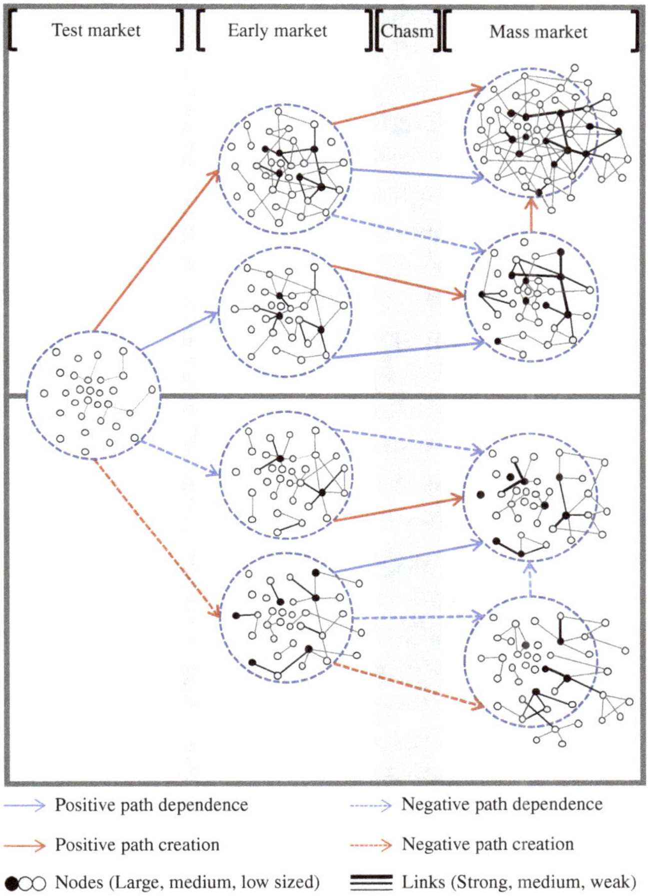

Figure 2.77 Path dependence and creation of TTN.

Source: Own illustration based on Suire and Vicente (2014).

connect with core initial nodes, increasing network cohesiveness, while diverse pathways must be followed to cross the chasm between early and mass markets (Mazurkiewicz *et al.*, 2022). TTN's appropriate structure features a few hubs and mainly spokes with global pipelines/bridges and local buzz/cohesiveness. Hubs serve as geographical gatekeepers between local structures and external organizations (Cowan *et al.*, 2004).

During the process, there are two typical effects of path dependence creation: positive and negative (Roe, 1996; Martin & Sunley, 2006, 2007). These effects result in two related situations within the network: dominant versus declining (Suire & Vicente, 2014). The positive effect leads to the expansion of TTN from a looser network to a complex hub-and-spoke network with scale-free and small-world properties. On the other hand, the negative effect leads to a declining network that fails to cross the chasm. When coupled together, TTN exhibits complex networking dynamics with both gradual and abrupt changes in its structures. It follows diverse pathways toward maturity, as illustrated in Figure 2.78.

2.2.6.3 *Structural catastrophe law*

As the process of technology transfer is getting specialized and fragmented due to open innovation, the overall interaction is evolving from a sparse and simple network to a complex one, with increasing actors and objectives involved, which needs to be reevaluated.

First of all, the structure of actors involved in TTN has evolved from triple helix to quadruple helix, and the linkage form evolved from point-to-point to hub-and-spoke organization.

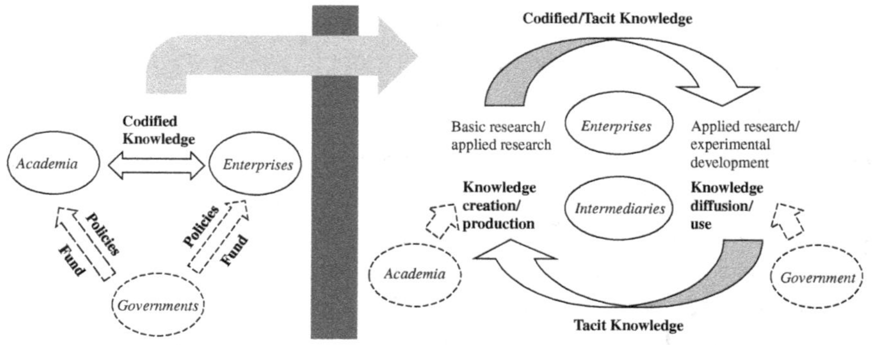

Figure 2.78 Structural dynamics of actors in TTN.

During the early stages of developing a TTN, the network is typically relatively simple and not highly connected. At this stage, technology transfer primarily occurs between universities and enterprises. The triple helix model, as described by Etzkowitz and Leydesdorff (2000), is a powerful way to understand the interconnectedness of university–industry–government relationships that drive knowledge generation and transfer. In the context of technology transfer, this model emphasizes the importance of interaction between universities and industry to promote knowledge and technology commercialization (Siegel *et al.*, 2003). Universities can benefit from transferring their knowledge to enterprises in various ways, such as acquiring financial support, enhancing academic reputation, increasing income, and improving research in applied areas. On the other hand, enterprises can benefit from commercializing and applying university technology to reap economic benefits, improve productivity, reduce costs, bridge organizational gaps, and enhance their reputation (Lee, 2000; Perkmann & Walsh, 2008; Ankrah *et al.*, 2013; Franco & Haase, 2015). On this model, the government plays a crucial role in proposing policies that raise demand and offering funds that provide supply for both universities and enterprises in related R&D activities (Ankrah *et al.*, 2013). Therefore, the original structure of actors in the TTN can be seen as a triple helix with universities and enterprises as the primary actors and the government as a subsidiary actor. Furthermore, the technology transfer process usually occurs directly among universities, enterprises, and the government, creating a point-to-point linkage form.

As various social and intermediary actors are being promoted by open innovation, the technology transfer mechanism has been faced with large conflict (Bruneel *et al.*, 2010). Carayannis and Campbell (2010) then improved triple helix to quadruple helix model which added the "media-based and culture-based public" that includes creativity and value by intermediaries, customers, suppliers, and media. During the technology transfer process, the entry of intermediaries such as technology transfer office, incubator, collaborative research center and social capital acting as the integrator of resource can have both financial value such as knowledge capitalization (Al-Tabbaa & Ankrah, 2016) as well as non-financial value such as improving knowledge absorptive capacity (De Silva *et al.*, 2018) and decreasing multidimensional proximity between actors (Villani *et al.*, 2017). At the same time, with the rise of knowledge economy, multinational enterprises have become the primary actor in TTN (Chaminade & Vang, 2008) which set research department directly and interact with intermediaries for knowledge commercialization. Therefore, the linkage formed here becomes the hub-and-spoke with some authoritative multinational enterprises and intermediaries acting as the hub to integrate and transfer knowledge and technology. The quadruple helix can better illustrate this kind of interaction as the enterprises and intermediaries become the primary while universities and government subsidiary.

Instead of only targeting on codified knowledge, the importance of both codified knowledge and tacit knowledge has become increasingly highlighted during technology transfer process. A large number of enterprises, universities, and research institutes used to highly rest on codified knowledge transfer to protect their innovation ideas (Graham *et al.*, 2009) when the general network was sparse. The codified knowledge can be transferred with less friction with the help of trade regimes, markets for intellectual property rights, and improvements in communication media (Bathelt *et al.*, 2004). However, when the network is getting dense and technology getting complex, individuals who acquire knowledge and technology usually need additional learning through explanations, adaptions, modifications, demonstrations, and testing, that is, the tacit part of that knowledge (Maurseth & Svensson, 2020). The tacit knowledge can be transferred only through the direct interaction of the actors (Maurseth & Svensson, 2020). Therefore, the objective of technology transfer now involves not only the codified knowledge that ensures the intellectual property rights but also the tacit knowledge including the private knowledge about the nature of inventions.

Secondly, the TTN has gone beyond local scale to break geographical borders and has formed the structure of global–local interaction. A large amount of literature has demonstrated that over-introduction of local elements may lead to information redundancy which is exposed to the risk of lock-in (Asheim & Isaksen, 2002; Bathelt *et al.*, 2004; Miguelez & Moreno, 2013; Trippl *et al.*, 2018). In this way, integrating internal knowledge and external knowledge, that is, the interaction between global and local can achieve the path to a breakthrough to enhance city's innovation capacity (Moodysson, 2008; Trippl *et al.*, 2009; Cao *et al.*, 2022). Under the framework proposed by Bathelt *et al.* (2004), the interactions within local actors are through internal channels or "local buzz" while the interactions cross-regionally are through external channels dubbed as "global pipelines". In TTN, the structure is therefore evolved from local transfer network to a glocal (global–local) transfer network involving both local knowledge learning and external knowledge communication, which is in accordance with the basic concept of global–local interaction.

Thirdly, as the fragmentation and differentiation at all levels become the corollary of globalization, the network success is strongly dependent on the governance framework where the regional and local economies are embedded (Swyngedouw, 2004). Thus, variety of governance is needed to manage the complex relationship such as "virtual enterprises" and "extended enterprises" with clearly defined rules, common business process, and shared information infrastructures (Ernst, 2009). Governance mechanisms for TTN can be divided into internal governance of local TTN and coordination among various TTN, which include both formal laws and regulations (e.g., intellectual property protection laws), and informal community governance (e.g., trade associations and internal company governance regulations). Furthermore, the coordination mechanism between local TTN consists of two aspects, one is the cooperation mechanism between local innovation networks at the same level, and the other is between a local TTN and a higher level one (e.g., global TTN).

2.2.7 *Network effects*

Currently, there has been an increasing number of empirical works related to network effects such as technology transfer, on innovation capability, performance, and productivity (Siler *et al.*, 2003; Guan *et al.*, 2006; Min *et al.*, 2020). Beyond all doubt, technology transfer and networking are found to be significant in increasing innovativeness of a firm or region (Love & Roper, 1999; Coccia, 2008).

The network effect comprises both positive and negative aspects of network externalities or spillovers (Coccia, 2008; Meeteren *et al.*, 2016; Huang *et al.*, 2020; Li & Yayavaram, 2021). These effects are mainly driven by various network elements, such as the size of transfer subjects or members and diversity of transfer objects or technologies (Ahuja, 2000; Lazer & Friedman, 2007; Min *et al.*, 2020), as well as network structures, including feasibility, centrality, connectivity, structural holes, position and centralization, among others. Network conducts, such as opportunistic behavior, reputation signaling, and perceptions of trust, also play a crucial role in shaping the network effect (Afuah, 2013; Opper & Nee, 2015; Hur & Park, 2016; Gui *et al.*, 2018b; Moon *et al.*, 2022). Finally, changes in network membership, number of ties, and structure of ties can greatly impact the network effect (Li & Yayavaram, 2021).

The impact of network effects (or externalities and spillovers) on TTN can be analyzed through the contacts between nodes involved in transfer activities (Lambooy, 2004). These internodal contacts can have a positive effect on local innovation capacity, with technological structures and network structures acting as moderators (Choi *et al.*, 2010). Specifically, these two types of structures can help enhance the positive effects of internodal contacts, leading to stronger innovation capacity in the local environment.

2.2.7.1 *Effect of technology structures*

Research has shown that technological structures can have a spillover effect on local innovation performance, particularly through the relatedness of technology. Importing external technologies can strengthen the local knowledge base and contribute to the generation of new local knowledge and technology, which is essential for sustaining long-term innovation in a region. However, not all technologies are equally effective in this regard. In fact, spillover benefits on local innovation are greatest for imported technologies that are highly related to but distinct from local technologies (Boschma & Iammarino, 2009; Boschma *et al.*, 2012; Hartog *et al.*, 2012). This is because such technologies can be more easily understood and learned by local innovation actors while also providing some differentiation from local knowledge, thereby boosting innovative efficiency.

When external technologies that are homogenous with the local technology base are imported, their impact on local innovation capacity is limited. While it may be easier to comprehend and assimilate similar or identical technologies, they do not bring new ideas or knowledge to the local level, hindering the emergence of fresh thinking. Additionally, similar technologies often pose the risk of "path lock-in" (Boschma, 2005; Martin, 2010). As a result, local innovators may find it challenging to be inspired by these technologies and create new ones. The absence of incentives to innovate can lead to lower innovation performance and slower growth in innovation capacity.

Alternatively, a diverse range of external technology categories can positively influence the growth of local innovation capacities. Diverse technologies have the potential to promote various technology fields within a geographical site and enhance its comprehensive innovation capacity (Rigby & Essletzbichler, 1997; Neffke *et al.*, 2011). This, in turn, can encourage greater willingness among enterprises and other transfer actors to engage in R&D activities, leading to increased investments in R&D and technology transfers, and forming a positive feedback effect (Meng *et al.*, 2004).

2.2.7.2 *Effect of network structures*

Local innovation activities are deeply embedded in social networks, which play a vital role in enabling the diffusion and exchange of knowledge and technology (Granovetter, 1985). To measure and analyze the structural properties of TTNs from a social network perspective, various topological indicators such as nodal centralities, strength of ties, structural holes, and network heterogeneity are commonly used (Ahuja, 2000; Hur & Park, 2016; Li & Yayavaram, 2021; Moon *et al.*, 2022; Piazza *et al.*, 2019).

Nodal centralities refer to various measures that capture the different aspects of a node's importance in a TTN, such as degree centrality (the breadth of nodal external links), weighted degree centrality (the intensity of external links), closeness centrality (the proximity of external links), and betweenness centrality (the transshipment role in external links) (Gilsing *et al.*, 2008; Gui *et al.*, 2018a, 2018b). Nodes with greater centralities tend to have broader and deeper technological exchange with others, stronger capacity to control and deploy technological elements, and access to more external technologies, all of which can facilitate innovation capability to a greater extent. However, this nodal importance has an obvious threshold effect, depending on its absorptive capability (Moon *et al.*, 2022). Positive innovation effects only occur when nodal degree passes its technological threshold point.

The strength of nodal ties in TTN can be classified into strong and weak ties based on the combination of closeness, intensity, and intimacy (Granovetter, 1973). However, there is an ongoing debate about the relative advantages of strong and weak ties (Moon *et al.*, 2022). Some studies suggest that local innovation performance is likely to be superior in places with strong ties of technology transfer, as these ties typically offer deeper commitment and trust due to greater closeness and intensity (Guan *et al.*, 2006). On the other hand, other scholars have found that weak ties generate superior innovation results because they provide access to diverse and novel knowledge (Chiambaretto *et al.*, 2019).

The structural hole of a network refers to the fact that some nodes within the network are only directly connected to certain other nodes (Burt, 1992), resulting in a network structure that resembles a cave. The node occupying the structural hole position in the network is the only one directly connected to all other nodes, thereby defining its power within the network (Zhang & Li, 2011). Cities that occupy such structural holes have an advantage in terms of accessing and controlling a wide range of technologies. The more structural holes a city occupies, the greater its presence on the critical path of technology flows, and the greater its ability to access external technologies, knowledge, and control of their transfer (Vasudeva *et al.*, 2013; Sun *et al.*, 2014).

Network heterogeneity refers to the variation in the types of nodes and resources held by each node within a TTN (Reagans & McEvily, 2003). While a node's position in the network may offer opportunities to access technological resources, its capacity to maximize these opportunities for innovative development depends on its own characteristics. Individual cognitive distances and attribute differences result in distinct innovation performance among various nodes (Zhang & Lang, 2013). The greater the network heterogeneity, the wider the variation among network nodes in terms of knowledge, technology, resources, and information. This variation widens the pipelines for acquiring knowledge and information, and the presence of diverse but complementary nodes is more conducive to generating innovation (Berends & Reymen, 2011). Additionally, greater network heterogeneity also means greater differences in the type, quality, and quantity of accessible technologies between nodes. Integrating complementary and divergent technological elements can enhance knowledge spillover effects and ultimately boost the technological innovation capacity of urban nodes.

2.2.8 *Network determinants*

Networking of technology transfer is self-organized and influenced by endogenous and exogenous determinant factors including multiple proximities (geographical, cognitive, social, institutional, and organizational proximities), buzz-and-pipeline interactions (local face-to-face interviews and trans-local technology interactions), structural properties of technologies, structural properties of network, and innovative milieux (economic, sociopolitical, and cultural contexts), along with transport infrastructure (speeding up and networking of transportation and communication infrastructures) (Figure 2.79).

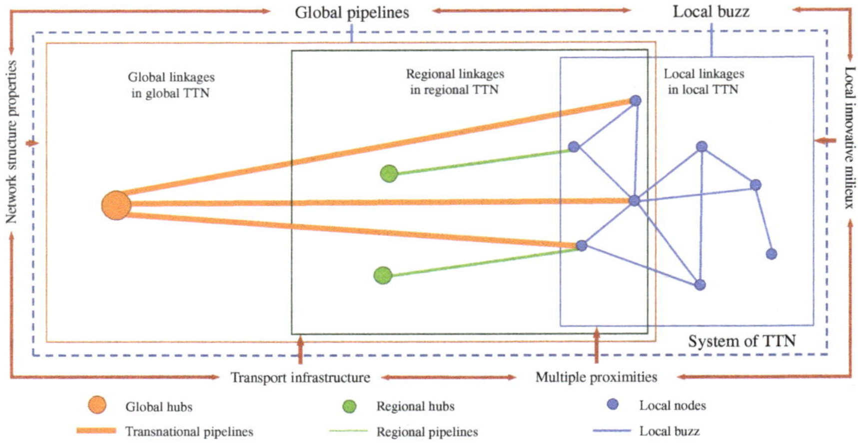

Figure 2.79 Multiple evolutionary mechanisms of TTN.

Among them, multiple proximities are mechanisms related to dynamics on the network. Buzz-and-pipelines interactions are viewed as dynamic mechanism within the network, while innovative milieux and transport infrastructures are acting as dynamic mechanism outside the network; these groups of factors present two-side effects (positively or negatively) on the networking of technology transfer. As a negative factor, barriers are usually attached with great importance to obtain high effectiveness of TTN.

2.2.8.1 *Multiple proximities*

Different types of proximities are viewed as drivers of knowledge transfer network (Balland, 2012; Broekel & Boschma, 2012; Broekel *et al.*, 2014; Ter Wal, 2014). Effective and efficient network of technology transfer depend strongly on some degree of similarity in frames of reference of the transfer actors (Villani *et al.*, 2017) (Figure 2.80).

It is extensively evidenced that geographical proximity contributes to technology transfer, due to low transaction and communication costs, great access to resources, and high mutual trust among

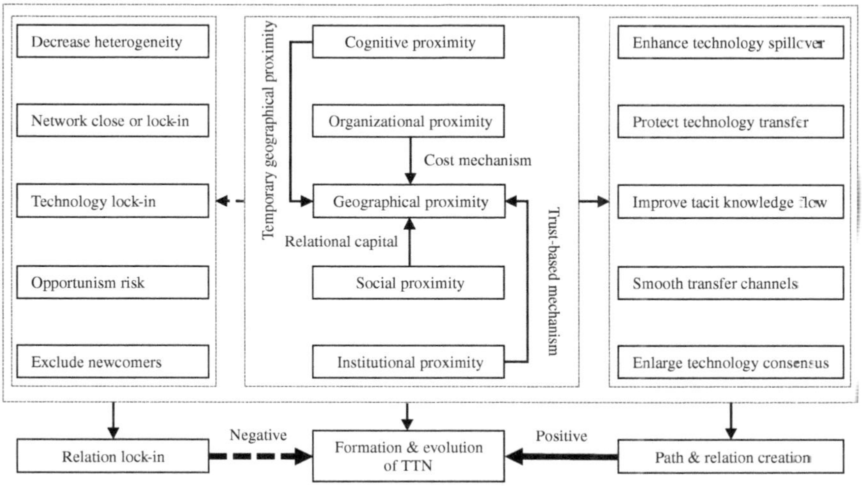

Figure 2.80 Multiple proximities mechanism of TTN evolution.
Source: Redrawing from Wang (2018).

neighboring areas (Oerlemans *et al.*, 2001; Ivarsson & Alvstam, 2009; Ma & Xu, 2022). The extent to which geographical proximity matters for technology transfers may depend on the type of the knowledge bases (Martin & Moodysson, 2013).

The influences on technology transfer are primarily implemented in two ways. First, the vast majority of technologies are deeply rooted in local contexts and are highly tacit, "sticky", and path-dependent (Asheim & Isaksen, 2002). As a result, the dissemination of tacit knowledge frequently requires face-to-face interactions and access to knowledge (Massard & Mehier, 2009). Second, geographical proximity allows actors to establish trust based on shared local milieu or embeddedness (Granovetter, 1985). This relationship of trust enables both parties to develop more open and meaningful communication, which facilitates the transfer of technology between actors.

In addition, the importance of non-spatial proximities has been emphasized, including cognitive (or technological), organizational, and social dimensions (Nooteboom *et al.*, 2000b; Boschma, 2005; Huber, 2011; Balland, 2012; D'Este *et al.*, 2013; Cassi & Plunket,

2014). Cognitive proximity, sometimes referred to as technological proximity (Ma & Xu, 2022), is essential for efficient information exchange between technology transfer stakeholders (Boschma & Ter Wal, 2007; Brennecke & Rank, 2017; Lian *et al.*, 2021). Firstly, it facilitates effective communication between stakeholders, resulting in low-cost knowledge spillover and the development of TTN. Secondly, it is a critical factor in the transfer of tacit knowledge, thereby fostering collaborative innovation. However, excessive cognitive proximity can reduce actor heterogeneity and lead to technological lock-in, having a negative impact on technological cooperation and transfer.

Social proximity, which refers to the relational embeddedness of actors in terms of partnerships, kinships, and friendships (Boschma, 2005), can strengthen the path-dependent tendency of TTN formation and evolution. Existing socially embedded relationships and shared experiences between technology transfer actors increase mutual trust, decrease the difficulty of transferring and absorbing tacit knowledge, and facilitate the exchange of complex or sensitive knowledge. Consequently, technology transfer is more likely to occur between actors who share social proximity. However, excessive social proximity can lead to an underestimation of the risk of opportunism and lock technology transfer actors into established social networks that tend to be closed and discourage technology flow.

Institutional proximity refers to the level of proximity between the political, economic, and social systems of actors involved in technology transfer. At the macro level, institutional proximity encompasses the institutional frameworks of countries and regions, including legal texts, labor relations, business practices, accounting rules, and training systems. Formal institutions, such as laws, and informal institutions, such as cultural practices, work together to coordinate technology transfer activities. At the micro level, institutional proximity refers to the system of business norms and procedures that influence inter-institutional technology transfer.

Organizational proximity refers to the level of strategic interdependence between two or more organizations, which can reduce uncertainty about potential future partners. Greater organizational proximity between technology transfer actors can prevent unintended technology spillovers, reduce transfer costs, and facilitate technology exchange and transfer.

In general, the importance of various proximities differs not only in different phases of technology transfer networking, but also between different types. Institutional and social proximities are relatively important in the early phase of TTN, while geographical and cognitive proximities are essential in subsequent phases (Broekel *et al.*, 2021). Geographical and technological proximity of agents are considered to be the main factors in knowledge transfer, as both have an impact on the learning capabilities of firms (Sutton, 1998).

2.2.8.2 *Buzz and pipelines*

As a network bridge, "Buzz" and "Pipelines" not only serve as an important channel for the flow of technology but also drive the networking of technology transfer. From a glocal perspective, the glocalization of the TTN is integratively led by local buzz and global pipelines, dominated by knowledge communities: local (or inter-organizational) technology alliances and trans-local (or international) business coalitions, respectively (Moodysson, 2008) (Figure 2.81).

Buzz refers to the local interaction between actors with geographical proximity, formed by daily face-to-face interviews and communications. This can greatly facilitate the transfer of implicit technologies. Pipelines refer to a more formalized interaction between actors at all scalar levels, even long distances, which fosters the exchange of heterogeneous technologies (Zhang *et al.*, 2016). Their impact on networking of technology transfer is implemented along three lines: the connectivity of exchange (geographical or physical proximity versus relational proximity, local embeddedness versus globally configured embeddedness, communities of practice versus epistemic communities), the content of exchange (tacit versus codified knowledge, and different combinations), and the form of exchange (face-to-face versus distanced contacts) (Moodysson, 2008).

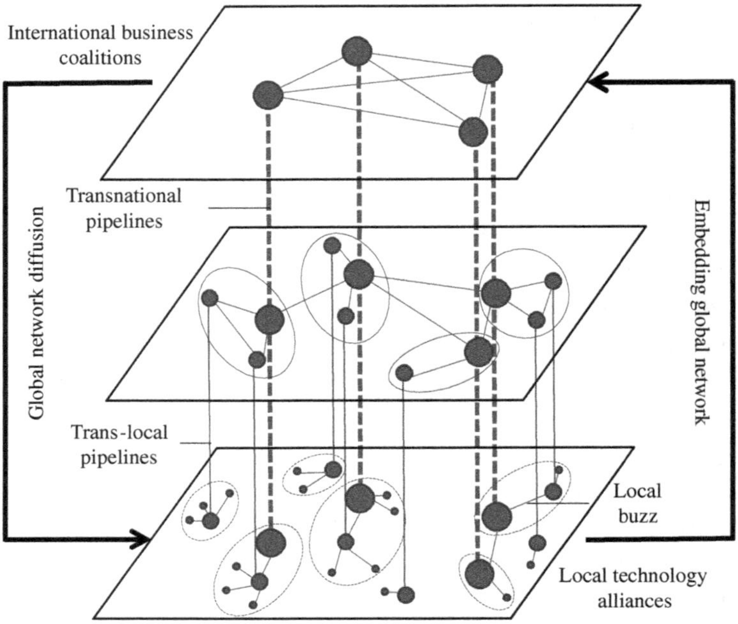

Figure 2.81 Buzz-and-pipelines effect on TTN.

Both pipelines and buzz are equally important as channel mechanisms for spatial technology transfer, and their effects complement each other (Zhang *et al.*, 2016). The stronger the buzz, the greater the demand for pipeline construction, and then more efficient cross-regional pipelines can promote the buzz to self-renew and self-reinforce. As a result of integrated internal and external linkages, new knowledge and technologies are developed and spread.

However, excessive buzz or pipeline can be detrimental to technology transfer and its network (Esposito & Rigby, 2019). Overly strong buzz can lead to information overload, forced trust, and passive interaction, inhibiting the initiative of transfer actors and reducing transfer effectiveness. Meanwhile, an over-reliance on pipelines can create significant technology gaps between local actors and external networks, making it difficult for local actors to absorb external technology and reducing the flexibility of local transfer networks.

2.2.8.3 *Innovative milieu*

Innovative milieu, as proposed by GREMI, is a set or network of informally social relationships in a limited geographical area (Camagni, 1991, 1995; Maillat, 1991, 1998a, 1998b) embedded in a specific innovation environment and consistent with network embeddedness (Oerlemans *et al.*, 2001; Hsueh *et al.*, 2010). It is regarded as something localized and intangible that enables innovation to occur in certain regions but not in others (Crevoisier, 2004), emphasizing the significance of a local trustworthy atmosphere (Tura & Harmaakorpi, 2005) and collective learning (Capello, 1999) as sources of innovation.

Its definitions vary, but many of theorists use "network" as their principal organizational metaphor (Tripple & Bergman, 2014). Social capital is a central factor in promoting innovativeness and competitiveness within the network (Aula & Harmaakorpi, 2006). Social capital cannot be traded but is built only through synergetic and collective interaction and learning processes (Camagni, 1991; Aula & Harmaakorpi, 2006). Based on these principles, three conceptual sets of elements define innovative milieus: effective relationships between transfer actors within a region (forming a common cultural and political basis); informal social contacts (enhancing learning processes and mutual trust); and image & sense of belonging (resulting in homogenization and coordination of transfer actors) (Fromhold-Eisebith, 2004). Based on the principles of interaction and learning, innovative milieux are depicted as dynamics of spatial systems ranging from simple geographical proximity, through industrial districts, to milieu innovateur (Capello, 1999) (as shown in Figure 2.82). Furthermore, they can be categorized into four types: technopole (know-how learning), pole (exchange of information), industrial district (exchange of knowledge), and milieu innovateur (creative milieux) (Lambooy, 2004; Amara *et al.*, 2005) (Figure 2.83).

In the social context, technology transfer and diffusion of an organization or a place are networked and networking. They depend not only on their own efforts but more importantly on their innovative milieu, which can be classified as economic context, cultural context, and sociopolitical context (Crevoisier & Maillat, 1991;

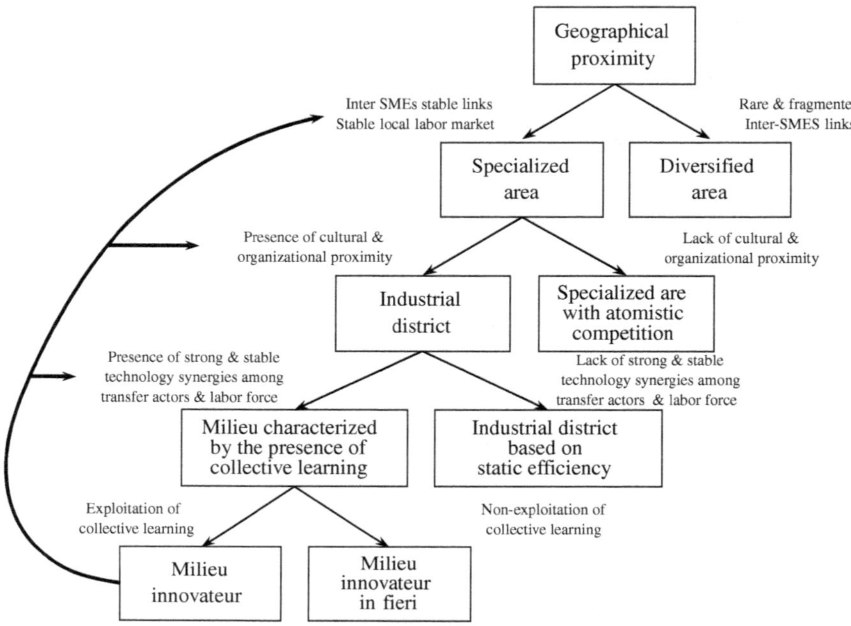

Figure 2.82 Territorial dynamics of innovative milieu.

Source: Redrawing from Capello (1999).

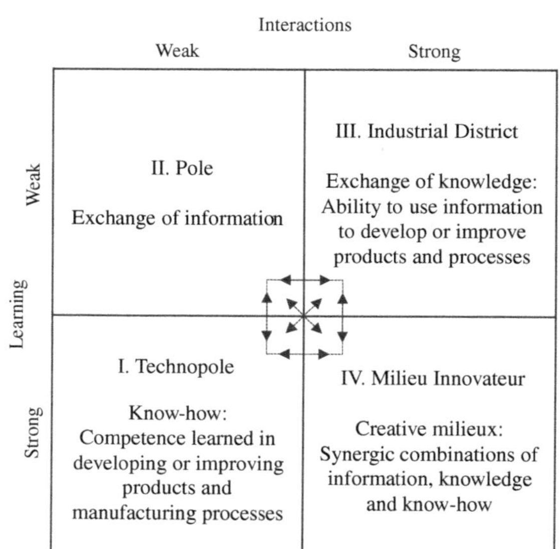

Figure 2.83 Territorial types of innovative milieu.

Source: Redrawing from Amara *et al.* (2005).

Pudelko & Mendenhall, 2009) or divided into economic embeddedness, cultural embeddedness, political embeddedness (Hsueh *et al.*, 2010), similar to local socioeconomic filter conditions measured by human capital, educational level, economic structure, social exclusion and institutional efficiency (Bilbao-Osorio & Rodríguez-Pose, 2004; Rodríguez-Pose & Comptour, 2012; D'Agostino & Scarlato, 2015). There is a positive effect of innovative milieux on the TTN, which occurs through three paths: organizational (laws and rules of transfer), technological (generation and diffusion of technology), and territorial paradigm (proximity and complementarity between transfer actors) (Crevoisier, 2004) (Figure 2.84).

2.2.8.4 *Structural properties*

The structural properties of TTN are identified as two broad research designs: the characteristics of technology structure versus the characteristics of network structure (Lian *et al.*, 2021).

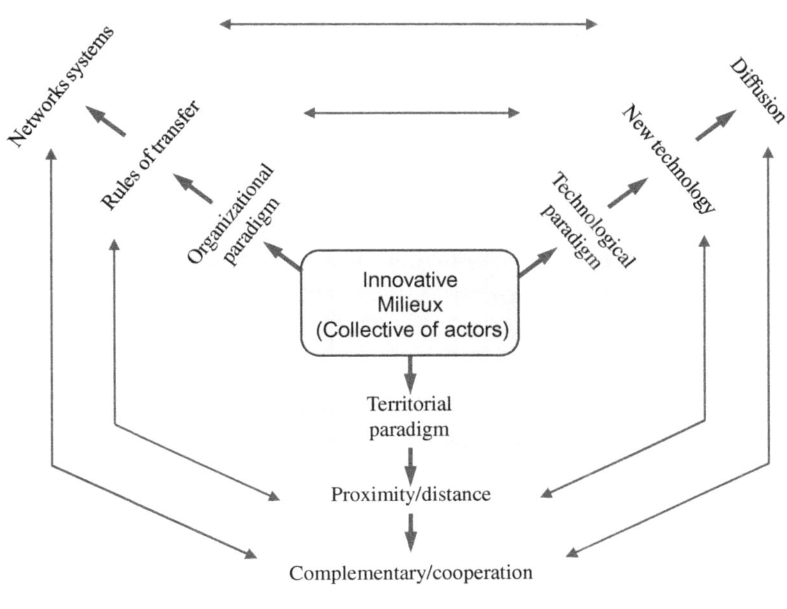

Figure 2.84 Effect of innovative milieu on TTN.

Source: Redrawing from Crevoisier (2004).

In the first category, technological properties are discretely measured by two proxies in a macroscopic view: diversity and uniqueness of technology stocks (Lian *et al.*, 2021). Technology diversity is understood as the variety of knowledge elements an entity holds (Lian *et al.*, 2021). It is considered one of the most important factors in improving technology transfer, thanks to the recombination of diverse knowledge (Liu *et al.*, 2015). However, high diversity can result in more time and money being spent identifying new technologies (Sampson, 2007). Technological uniqueness refers to strengths in areas of technology that other organizations do not have (Brennecke & Rank, 2017). Evidence shows that the more unique the experience or knowledge of a unit compared with its peers, the stronger its outflows of knowledge (Schulz, 2001). However, technologies with unique strengths may not attract great attention from transferors because of a lack of importance or value (Yayavaram & Ahuja, 2008).

Regarding the second item, extensive research has documented the role of networks in facilitating technology emergence, transfer, and diffusion in social network science. The most critical factor is network embeddedness (Chen *et al.*, 2019), which originated from Grannovetter's "embeddedness" concept (1985, 1992) and includes different divisions: relational aspect (dyadic relations of innovators) (Sun & Grimes, 2017) versus structural aspect (structures of relations network) (Oerlemans *et al.*, 2001; Chen *et al.*, 2019); organizational aspect (external mobilization of resources) (Häkansson, 1989) versus spatial aspect (Oerlemans *et al.*, 2001). Besides the concept, it is further developed into measurements that include four items: density of network, number of connections, existence of mutual trust, and existence of social norms (Moran, 2005).

Network positions, also known as positional embeddedness by Hsueh *et al.* (2010), are commonly used to measure structural effects (Baum & Dutton, 1996). It is believed that different positions of nodes in a network have varying abilities to access new information and resources (Burt, 1992; Oerlemans *et al.*, 2001). The power of TTN also varies depending on its combinations, which are reflected by two indicators: structural holes (representing opportunities) and nodal centralities (representing potential) (Lian *et al.*, 2021).

2.2.8.5 *Transport infrastructures*

As evidenced in existing studies, innovative capabilities and interregional linkages shape access to and transfer of technology through, which may stimulate technology transfer and its networking (Xiao *et al.*, 2022). Obviously, transport infrastructure plays a leading role in heading for the process through its networking and high-speeding (Luo *et al.*, 2022; Guan *et al.*, 2021; Xiao *et al.*, 2022).

The high-speed networking of transport infrastructure, such as high-speed railways (HSR) (Gao & Zheng, 2020; Xiao *et al.*, 2022) and airlines (Guan *et al.*, 2021), enables space-time compression while strengthening locational advantages and enhancing interregional connectivity (referred to as the "*direct effect*" set by Xiao *et al.* (2022)). This reshapes interregional linkages (including human mobility, FDI flows, collective learning, etc.) and multiple proximities (such as geographical, institutional, organizational, technological, industrial, and sociocultural proximities) (*the first set of indirect effects*). It further improves internal capabilities of creation and diffusion through the knowledge spillover effect (*the second set of indirect effects*), culminating in the promotion of interregional technology transfer and its networking. Figure 2.85 illustrates how internal capabilities and external linkages co-shape the technology transfer

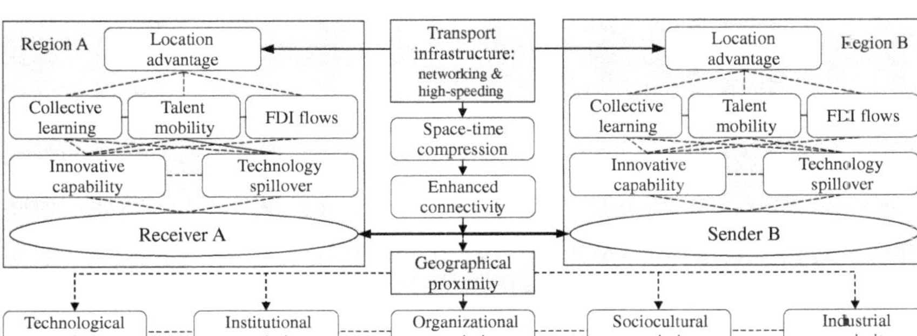

Figure 2.85 Effect of transport infrastructure on TTN.

Source: Own illustration based on Xiao *et al.* (2022).

process and network through the high-speed networking of transport infrastructure.

2.2.8.6 *Network barriers*

Despite policymakers efforts to activate technology transfer, much technology remains in universities and R&D centers, due to negative effects from various influencing factors. These obstacles, referred to as barriers, hinder the effectiveness and efficiency of TTN (Mazurkiewicz *et al.*, 2022), particularly in technology developing countries.

Multiple proximities are classified as cultural (communication difficulties and cultural proximity), geographical (geographic location and geographical proximity), organizational (organizational restrictions and proximity), personal (cognitive proximity and capabilities of actors) (Jung, 1980; Lian *et al.*, 2021), with a particular focus on their negative effects.

From an ecosystem perspective, the barriers to technology transfer are also divided into four interactive groups: technological (e.g., lack of technological infrastructure, resource constraints, inadequate technology, uncertainty of technology), economic (e.g., lack of financial resources), institutional (organizational rigidity, lack of skilled labor, managerial support, communication or cooperation trust, and risk controlment), and political (e.g., lack of regulations, rewards, interest, and incentives) (Smith, 1998; Gnyawali & Srivastava, 2013). This generally aligns with the viewpoint of the innovative milieu school.

Systematically, they are further integrated into three categories, with respect to: (1) the content of transfer objects (the consistency and complexity of what will be transferred); (2) the context of transfer activities (physical or virtual environment where TTN takes place); and (3) the characteristics of transfer subjects (the capabilities and interests of who are responsible for sharing and absorbing technology) (Duan *et al.*, 2010). These barriers are further induced

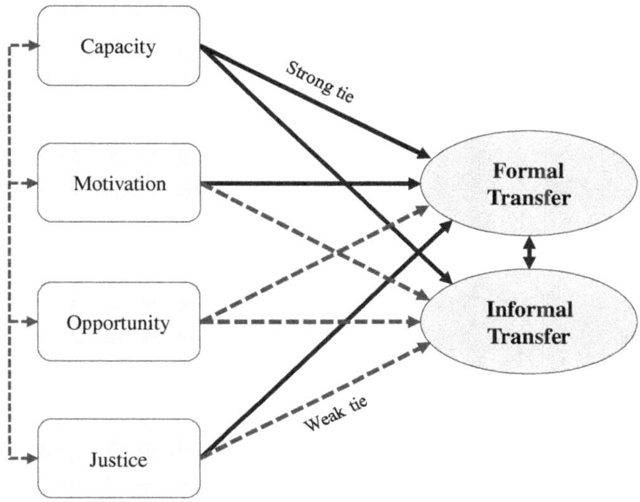

Figure 2.86 Subjects' barriers of TTN.
Source: Redrawing from Vega-Gomez *et al.* (2021).

at three levels: technological level (technology stocks of transfer actors), ecosystem level (law and policy of technology transfer), and individual-and-organizational level (individual capabilities, cultural & institutional differences, and geographical distances) (Lian *et al.*, 2021).

Regarding the transfer subjects, Siegel *et al.* (2003) identified three fundamental obstacles: the existence of informational and cultural barriers, inadequacy of existing rewards to encourage transfer activities, and limited business experience of TTO staff, together with high turnover rates. Similarly, Vega-Gomez *et al.* (2021) proposed the four determinant barriers of academic knowledge transfer: capacity (lack of self-perceived capacity and awareness to transfer), motivation (lack of participating motivation in transfer activities), opportunity (lack of opportunity offered by innovative environment), and justice (lack of fairness established by different organizations) (Figure 2.86).

CHAPTER 3

METHODS OF TECHNOLOGY TRANSFER NETWORK

3.1 Research Flowchart

As mentioned in Section 2.2, despite increasing literature on technology transfer network (TTN), the issue of spatial complexity has not been fully addressed due to its limited approaches and technologies. Specifically, the following questions remain unclear: Where do the complexity and heterogeneity of TTNs in China originate? What are the similarities and differences in their structures and dynamics among different scales? What are the effects of their evolutions on urban innovation capability? What driving forces will dominate their evolutions?

To address these questions, this book explores three critical aspects related to "global–local" interaction: (1) the spatial evolution of TTNs at global, national, and regional scales; (2) the spillover effects of multiscalar TTNs on local innovation capability; (3) and the determinant factors of transnational and national TTNs, which are analyzed quantitatively by integrating social network analysis (SNA), GIS mapping, topological visualization, and spatial econometric models (Figure 3.1).

In detail, this book first constructs a geographical and relational database of intercity technology transfer from 2001 to 2018 by data

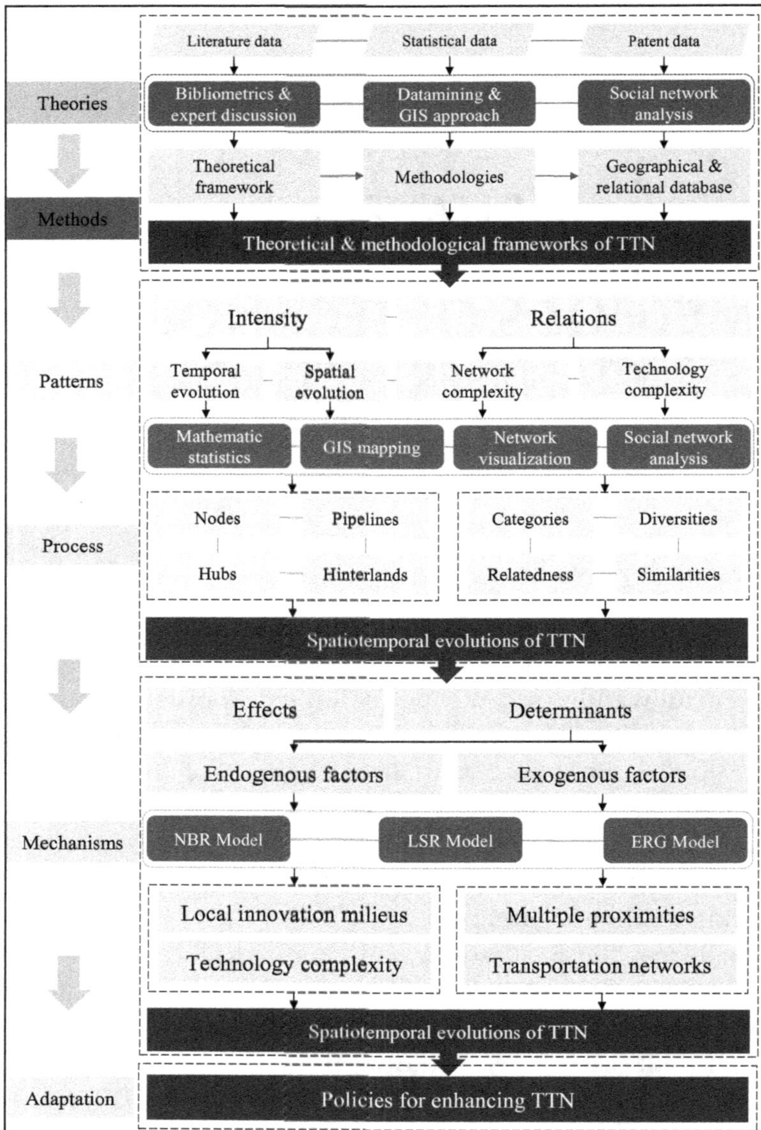

Figure 3.1 Flowchart of research methods.

mining (Pathon software) and GIS (ArcGIS software) techniques. Next, it comparatively examines the spatiotemporal structures and the dynamics of TTNs at three scales. Additionally, geographical and topological heterogeneity of technological diversity and relatedness are visualized by using SNA (Gephi, https//gephi.org/) and technology space approach. Then, the Negative Binomial Regression Model (NBRM) is used to quantitatively analyze the dynamic determinants of the TTNs at the three scales, and the Least Squares Regression Model (LSRM) and Exponential Random Graph Model (ERGM) are used to explore the spillover effects of the TTNs on local innovation capabilities. Finally, targeted policy recommendations are made to effectively enhance technology transfer of some emerging countries based on policy text analysis and China's case study. The precise research framework is outlined in Figure 3.1.

3.2 Network Construction

A large number of natural and social systems are structured in the form of networks (Barrat *et al.*, 2004). Technology transfer is based on a three-mode relational network consisting of knowledge network (relatedness among technological categories), social network (connection among transfer actors) (Brennecke & Rank, 2017), and territorial network (technology flows across places) (Figure 3.2).

The first involves a technology space analysis, which examines the diversities and relatedness of technologies to demonstrate their transferability and availability. Meanwhile, SNA is used to represent the latter two aspects — multilevel and multiscalar mapping of networking properties of technology transfer. This method has been employed in various studies (Grabher, 2006; Krätke, 2010; Ter Wal, 2009; Ter Wal & Boschma, 2009; Yokura *et al.*, 2013). To evaluate the accessibility and territoriality of technology transfer, multiple centrality assessment (MCA), weighted average central rank model (WACR), and weighted hinterworlds division method (WHD) are utilized (as shown in Figure 3.1).

However, studies on the geography of innovation are often limited by indicators related to research collaboration data including co-authored patent and co-cited patent (Chaminade *et al.*, 2015). Regarding data, a very promising line of research is based on the construction of supply- and demand-side relational data from transfers of patent license and ownership (Ma & Xu, 2022), and on the interaction of global and local scales over time.

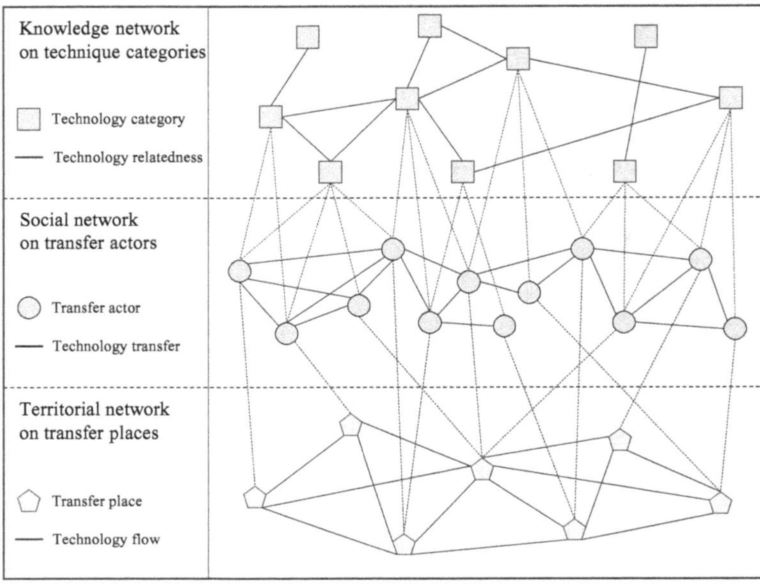

Figure 3.2 Construction of TTN.

Source: Own illustration based on Brennecke and Rank (2017).

Since intercity patent transaction is regarded as a flow with magnitude and direction, a directed-and-weighted network with a single-mode graph $G(v, e, w)$ is constructed based on graph theory (Figure 3.3), where node v represents the city involved in patent transfer process, edge e represents the link of patent transfer between cities, and weight w is the number of technology transfer. The weighted adjacency matrix $M(=(w_{ij}a_{ij})_{n*n})$ at a set of glocal scales including transnational, national, and regional scales is constructed as:

$$M = \begin{bmatrix} 0 & M_{12} & \cdots & M_{1(n-1)} & M_{1n} \\ M_{21} & 0 & \cdots & M_{2(n-1)} & M_{2n} \\ \cdots & \cdots & 0 & \cdots & \cdots \\ M_{(n-1)1} & M_{(n-1)2} & \cdots & 0 & M_{(n-1)n} \\ M_{n1} & M_{n2} & \cdots & M_{n(n-1)} & 0 \end{bmatrix} \tag{3.1}$$

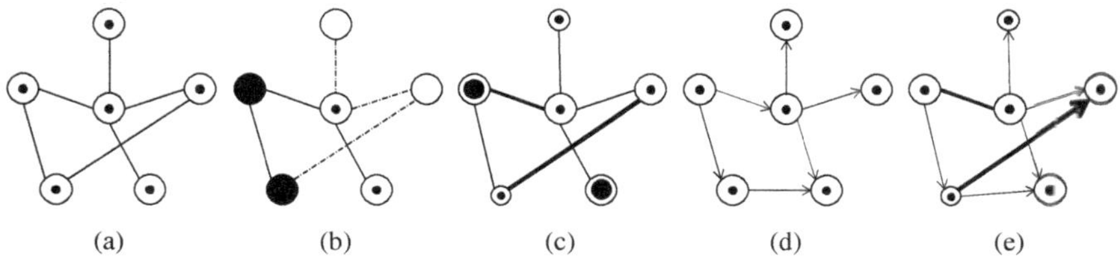

Figure 3.3 Graph representations of complex networks. (a) Nondirectional and mono-mode network; (b) Nondirectional and multi-mode network; (c) Nondirectional, weighted and bipartite network; (d) Directional and mono-mode network and (e) Directional and weighted 1-mode network.

The matrix represents the intensity and direction of association of the network, where a_{ij} takes a value in [0, 1] relying on the relationship between cities i and j, and weight w_{ij} represents the patent transfer flows from cities i to j.

As the networking structure and function at different scales are heterogeneous (Cai *et al.*, 2022), we propose to explore glocalization of China's TTN from three different subnetworks: international network (between domestic cities and foreign ones), national network (among domestic cities), and regional network (among cities in some classic regions such as the three most innovative growth poles: the BTH, the YGD, and the GBA urban agglomerations (Liu *et al.*, 2019), along with the innovation declining region — Northeast China) (Liu & Niu, 2019).

3.3 Social Network Analysis

Derived from the graph theory, SNA is a set of specifications and methods for characterizing the web structure and dynamics of social relations or interactions (Wasserman & Faust, 1994; Borgatti *et al.*, 2009). The recent development of SNA has resulted in improved visualization and interpretation of TTN (Li *et al.*, 2019). To better understand the magnitude and direction of technology flow in TTN, new methods have been introduced to order and group cities based on their hierarchical structure and networked organization, such as the statistical complexity of the entire network.

3.3.1 *Multiple centrality assessment*

Key nodes or hubs in a complex network are identified quantitatively by centrality proxies. *Centrality* is the most important and common measurement to reflect the position and power of nodes in networks (Ye *et al.*, 2020). The notion proposed by Freeman (1978) is used to describe how central a node is in a network, or how easily an actor obtains or controls resources from any node pairs (Yang *et al.*, 2021).

MCA is a computer-operated procedure for mapping centralities of nodes in complex networks (Porta *et al.*, 2009), including a set of models such as degree centrality, closeness centrality, betweenness centrality, etc.

3.3.1.1 *Degree centrality*

Degree centrality (C_D) is defined as the number of other cities directly connected with a city to evaluate the importance of the city in a network (Yang *et al.*, 2021), equated as:

$$C_D(i) = \sum_{j=1}^{N} a_{ij} \tag{3.2}$$

where a_{ij} represents the adjacency matrix of intercity patent transfer. Assign 1 if there exists a transferred patent between cities, otherwise 0. In the directional network, nodal degree can also be divided into in-degree and out-degree (Cai *et al.*, 2022). In-degree represents the number of cities transferring patents to this city, while out-degree denotes the number of cities importing patents from this city. Both of them are used to illustrate the size of technology transfer pipelines in each city.

3.3.1.2 *Weight degree centrality*

Weight degree centrality (C_{WD}) reflects the sum of patent transactions in a certain city, also including weighted in-degree and weighted out-degree, defined as:

$$C_{WD}(i) = \sum_{j=1}^{N} X_{ij} \tag{3.3}$$

where x_{ij} represents the number of transferred patents that city i exports (imports) to (from) city j, and indicates the intensity of interurban technology transfer.

3.3.1.3 *Closeness centrality*

Closeness centrality (C_C) means the ratio of the number of other cities to the sum of the shortest paths from this city to all other cities, and is calculated in the following form:

$$C_C(i) = \frac{N-1}{\sum_{j=1;j\neq i}^{N} d_{ij}} \tag{3.4}$$

where d_{ij} denotes the number of shortest paths between cities i and j, and N represents the number of cities that have evolved in technology transfer. To some extent, it demonstrates the average Euclidean distance of technology transfer pipelines between a certain city and others.

3.3.1.4 *Betweenness centrality*

Betweenness centrality (C_B) denotes the frequency with which a certain city appears on the shortest path connecting other cities in a network, and is given as:

$$C_B(i) = \sum_{j=1;k=1;j\neq k\neq i}^{N} \frac{jk(i)}{N_{jk}} \tag{3.5}$$

Here, N_{jk} is the total number of shortest paths between cities j and k, while N_{jk} (i) represents the number of these paths passing through city i. In TTN, it measures the potential ability of a certain city with respect to its role as a "gatekeeper" or "bridger" of technology flows between cities, indicating the extent of technology brokerage opportunities for a city (Yang *et al.*, 2021). The larger betweenness the city has, the greater control ability this city has in TTN.

3.3.2 *Weighted ego network analysis*

The weighted ego network analysis (WENA), described by Zhang and Li (2011), provides a more reliable means for characterizing structural hierarchies of real-life networks such as maritime network (Liu *et al.*, 2018b, 2018c) and street network (Zhang & Li, 2011). In WENA, the links between pairs of transfer actors need to be directed and weighted. In addition, structural holes theory, first proposed by Burt (1992), is viewed as one of the most efficient ways to define the positional status of each node with two main measurements: constraint and aggregate constraint (Burt, 1992).

3.3.2.1 *Strength of link*

Strength considers not only the number of a node of neighbors but also the weights of the ties between them (Barrat *et al.*, 2004). According to its adjacence to a focal node (called "ego"), the tie is classified as a two-way interaction: strong and weak (Granovetter, 1973; Borgatti *et al.*, 2009). The strength of weak ties, exhibiting less closeness and intensity (Granovetter, 1973), is regarded as a critical driver of radical innovation (Moon *et al.*, 2022), owning to providing novel and diverse knowledge (Burt, 1992; Levin & Cross, 2004; Chiambaretto *et al.*, 2019). While strong tie with higher levels of closeness and intensity is beneficial to incremental innovation because of deeper knowledge and greater trust (Ryssel *et al.*, 2004; Lee & Jonhson, 2010). But these findings still present an extensive debate on the relationship between tie strength and innovation outcome (Moon *et al.*, 2022), which throws light on this book.

In ego-network analysis, the strength of each link (p_{ij}) from a city i to all its neighbor j is defined as the reciprocal of the weighted degree (wk) of the city i. Mathematically,

$$p_{ij} = \frac{1}{wk_i}(j \in N_i) \tag{3.6}$$

where N_i is the set of adjacent nodes of node i. For link strength deficiencies, this book revised the link weight (w_{ij}) from city i to its neighbor j as the ratio of city j to the sum of all neighboring degrees of city i.

$$W_{ij} = \frac{k_j}{\sum_{s=1}^{s=m} k_s}(j \in N_i) \tag{3.7}$$

where m is the number of nodes in the N_i set, k_j is the degree of city j, and s and k_s are the sum of the degrees and degrees of all neighboring cities of city i, respectively.

3.3.2.2 *Constraint of link*

The constraint from city i to city j (C_{ij}) is calculated as the square of the sum of the direct and indirect link strengths from city i to city j (Burt, 1992):

$$C_{ij} = (w_{ij} + \sum w'_{ij})^2 = (w_{ij} + \sum_q w_{iq} w_{qj}) \quad (j \in N_i, q \in N_i, q \neq i, q \neq j) \tag{3.8}$$

where w_{ij} is the intensity of direct connection from city i to its neighbor j. While w'_{ij} indicates that i is not directly connected to j but city i is directly connected to two connected cities (city j and city q). In TTN, the larger the C_{ij}, the smaller the opportunity for the city i to transfer technology (Zhang & Li, 2011).

3.3.2.3 *Weighted average centrality rank*

In order to characterize nodal hierarchies, the weighted average centrality rank (WACR) is developed to measure nodal opportunity in the technology flow transmission, as follows:

$$\text{WACR}_i = \frac{k_i}{\sum_{j=1}^{m} C_{ij}} = \frac{k_i}{\sum_{j=1}^{m} (w_{ij} + \sum w_{iq} w_{qj})^2} \quad (j \in N_i, \quad q \in N_i, \quad j \neq s) \tag{3.9}$$

where $\sum_{j=1}^{m} C_{ij}$ denotes aggregate constraint of a node j, and is then normalized in order to make comparisons across years. From Equation (3.9), WACR combines the concept of degree centrality and clustering coefficient in traditional complex network analysis (Zhang & Li, 2011), and provides a more effective classification of structural hierarchies. In TTN, the larger the WACR value of city i, the more significant the role of city i in the transfer of flows and the higher the technology transfer size or importance of the city.

3.3.3 *Hub hinterworlds division*

Despite some improvements given by WACR, the hierarchical classification system is still subjected to judgment based on percentage (Zhang & Li, 2011). Therefore, some methods are developed to quantitatively display the hierarchies of nodes and further identify the hubs, such as Ht-index and dominance flow analysis.

3.3.3.1 *Head/tail breaks index*

Hierarchical scaling structure is prevalent in the majority of geographic features (Jiang & Yin, 2014). This unbalanced scaling hierarchy can be captured by a new classification scheme-the head/tail breaks approach for data with heavy-tailed distributions (Jiang, 2013; Jiang & Yin, 2014). The so-called method *Ht-index* is a technique for recursively deriving inherent classes or hierarchies and showing patterns of far more smaller things than larger ones (Jiang, 2015).

By calculating the mean, all data values are split into two parts: the head (the above-mean portion) and the tail (the below-mean portion). This procedure is continued until the head values are no longer heavy-tailed (Jiang, 2013). After a specific number of rounds, the intervals for the various hierarchies and the levels of technology transfer hubs are obtained (Figure 3.4). This book applies a uniform categorization criterion to all generated WACR values using the Ht-index in order to facilitate longitudinal comparisons. In TTN, the higher the Ht-index, the more obvious the power–law distribution of the whole network, in which a minority of transfer hubs dominates a majority of technology flows.

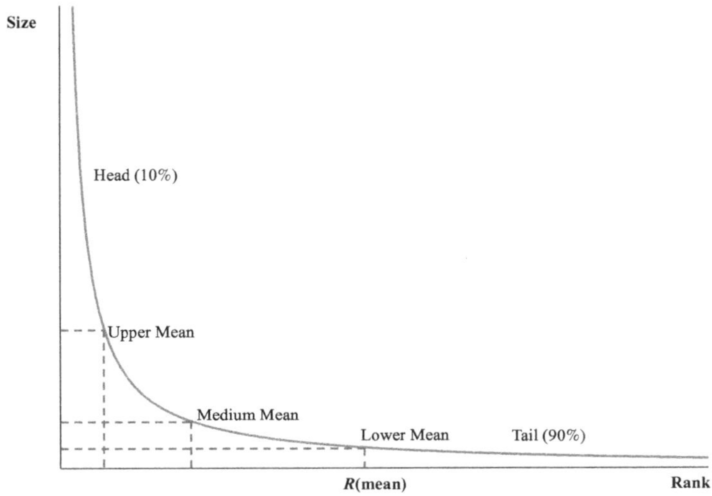

Figure 3.4 The head/tail breaks rule.
Source: Own illustration based on Jiang (2013).

3.3.3.2 *Dominant flow relations*

The Ht-index classification is more suitable for detecting the urban hierarchies of city's magnitude, while it is hard to underline the external direction of their dominance in TTN. Thus, a "largest flow" method, proposed by Nystuen and Dacey (1961) and named as "N–D method" (Grubesic *et al.*, 2009) or "dominance flow analysis" (Gui *et al.*, 2019b), can be utilized to clarify the "nodal regions" — the pivotal hubs and their dominant directions within a subnetwork (Ducruet & Notteboom, 2012; Ducruet, 2017).

Within myriad relations existing between cities, the largest flow will be the ones outlining the skeleton of spatial organization of technology transfer system. The "dominant relations" display the hierarchical position of a city in a network system (Nystuen & Dacey, 1961). It is identified as the maximum element by simple inspection of a matrix of flows between all pairs. According to aggregation of "largest flow", central city (dominate center) is recognized into three properties such as *independent* (the largest flow to a smaller city), *transitivity* (implies that if a city *a* is subordinate

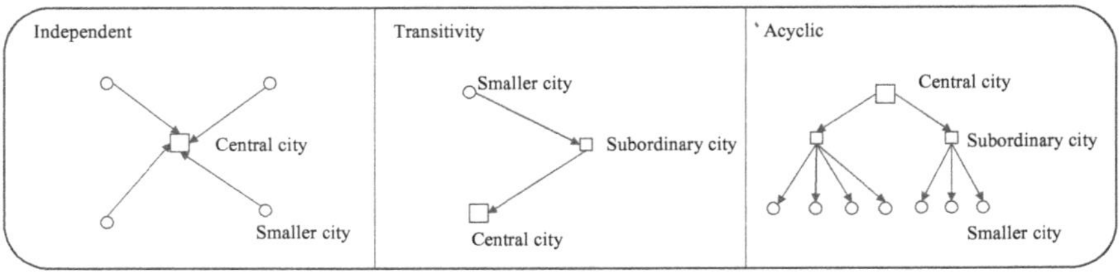

Figure 3.5 Three organizational properties of dominant relations.

Source: Own illustration based on Nystuen and Dacey (1961).

to city *b* and *b* is subordinate to city *c*, then *a* is subordinate to *c*), and *acyclic* (stipulates that a city is not a subordinate of any of its subordinates) (Nystuen & Dacey, 1961), which is illustrated in Figure 3.5.

In this book, the dominance flow analysis is applied to the TTN based on the following rules: (1) the size of cities is evaluated by the WACR; (2) the largest flow is identified based on the total strength of patent transfer between two cities; (3) the dominant center of technology transfer has its largest technology flow to a smaller city; and (4) the subdominant center has its largest flow to a larger city.

3.3.3.3 *Hinterworlds of hubs*

Driven by space of flows (Castells, 1996), there exists a larger amount of non-local linkages across the boundaries of cities in TTN (Lu *et al.*, 2020). Based on central flow theory, Taylor (2001) proposed the concept of "hinterworld", and further constructed the "interlocking network" to identify hub's hinterworld with trans-local and external relations (Taylor & Derudder, 2006; Taylor *et al.*, 2010).

According to the method of GaWC (Globalization and World Cities Research Network), "relative city-dyad connectivity" is usually used to delineate the hinterworld of technology transfer hubs (Taylor & Derudder, 2006; Li & Zhou, 2016), defined as:

$$C_{pq} = a + bN_q(+R_{pq})\ (q = 1,2,3,\ldots,n, p\neq q) \tag{3.10}$$

where C_{pq} represents the connectivity of technology transfer between city q and other city p (p = 1, 2,...), and N_q denotes the total connectivity scores of all other cities. C_{pq} is linearly regressed against N_q, and the connectivity of city q is analyzed using the standardized residual R_{pq}. If the residual R_{pq} is greater than 1, then city p is over-linked with city q. If the residual R_{pq} is less than 1, then city p is under-linked with city q. Where the over-linked cities are defined as the core hinterworld of the hub p, and the under-linked cities are defined as its peripheral hinterworld (Taylor, 2001; Li & Zhou, 2016).

3.3.3.4 *Weighted hinterworld size*

The hinterworld size of technology transfer hubs, defined as geographical scope of weighted patent transfers (Wang *et al.*, 2023a), is formulized by:

$$\mathrm{HS}_i = \left(\overline{a_{ij}} \times \overline{d_{ij}}\right)^2 \tag{3.11}$$

where HS_i denotes the weighted hinterworld size of technology transfer hub i, and $\overline{a_{ij}}$ and $\overline{d_{ij}}$ define the average volume of patent transfer and the average Euclidean distance between cities i and j, respectively.

3.3.4 *Network properties*

In network science, the structural properties of a real network include small-world, scale-free, community, and assortativity/disassortativity, which are measured based on some topological indicators, such as degree distribution, clustering coefficient, path length, degree correlations, etc. (Wasserman & Faust, 1994; Borgatti *et al.*, 2009).

3.3.4.1 *Small-world property*

Small-world property is defined as a characteristic with far greater average path length and clustering coefficient than a random graph with equal size, exhibiting a large component with a relatively short path (Desmarchelier & Zhang, 2018). It implies a small number of transfer brokers and fast circulation of technology flows in TTNs.

3.3.4.2 *Scale-free property*

Scale-free property is detected as a power–law behavior in statistical degree distributions, which are characterized by the abundance of hubs with a very large number of connections compared to the average degree value (Desmarchelier & Zhang, 2018).

3.3.4.3 *Community*

Community is defined as a subnetwork containing nodes that are more densely tied to each other than to the rest of the network (Newman & Girvan, 2004). Within a community, technology is more easily accessible to its members (Ye *et al.*, 2020).

3.3.4.4 *Assortativity*

Assortativity is often operationalized as a correlation between two nodes and refers to the preference for a network's nodes to attach to others that are similar in some way (Newman, 2002). In TTN, it shows assortative mixing when nodes with great technology flow frequently link with others with a high degree. Alternatively, the network exhibits a disassortative mixing if nodes with high transfer connections have neighbors with low connections (Ferraro & Iovanella, 2016, 2017; Li *et al.*, 2015).

3.4 Technology Space Proxies

According to a related approach (Hidalgo *et al.*, 2007; Kim *et al.*, 2021), we connect the IPC codes of transferred patent and cities of transfer actors by building a code-city bipartite network in which the weight of the link is the number of IPC codes possessed by the city.

The network exhibits technology relatedness between transferred technologies, which is referred to as "technology space" or "knowledge space". This concept is similar to "produce space" and has been studied by researchers such as Breschi *et al.* (2003), Hidalgo *et al.* (2007), Kogler *et al.* (2013, 2017), and Rigby (2015). In this space, each node represents a technological category presented as a tree, while each link represents the proximity between technologies as a forest. Proximity is determined based on the minimum value of pairwise conditional probability that two technologies have a comparative advantage together within the same city, as measured by technology diversity and relatedness (Kim *et al.*, 2021).

3.4.1 *Technology diversity*

The degree of diversity, specifically technology variety, is considered a source of regional knowledge spillovers and a local knowledge portfolio protecting a region from external shocks (Frenken *et al.*, 2007). It is classified as either related variety, which refers to the diversity within sectors, or unrelated variety, which pertains to the diversity between sectors. It is evidenced that related variety is expected to be beneficial for Jacobs externalities, while unrelated variety protects a region best against external shocks (Frenken *et al.*, 2007).

3.4.1.1 *Variety*

A methodology concerning entropy measure has been specifically applied in the context of technology variety. It is measured by the entropy of technology subcategories (Frenken *et al.*, 2007), and is given by:

$$V_j = \sum_{j=1}^{J} p_j \log_2 \left(\frac{1}{p_j} \right) \tag{3.12}$$

where p_j indicates the shares of a technology subcategory j under all technology sectors. The subcategories are calculated at the four-digit level of IPC patent codes based on Schmoch's (2008) criteria for technology classification. In TTN, a higher value of variety in a city indicates that the city has more diverse and differentiated technologies, or that the city's original knowledge base is richer and full of network spillovers.

3.4.1.2 *Related variety*

Related variety (RV) of technology refers to the externality degree of interrelated technologies in local knowledge base. It is evidenced by the weighted sum of entropy within each technology sector (Frenken *et al.*, 2007). In TTN, related variety in a technology sector is expected to be beneficial for Jacobs externalities in the form of knowledge spillovers.

The calculation is carried out in three main steps: firstly, let all four-digit subcategories j fall exclusively under a two-digit technology sector S_g, where $g = 1,\ldots,G$. One can derive the shares P_g by summing the four-digit shares p_j within each two-digit class (Frenken *et al.*, 2007):

$$P_g = \sum_{j \in s_g} p_j \tag{3.13}$$

Then the entropy at each four-digit subcategory is defined as a percentage of the sum shares of each two-digit technology sector:

$$H_g = \sum_{j \in s_g} \frac{p_j}{p_g} \log_2 \left(\frac{1}{p_j / p_g} \right) \tag{3.14}$$

Finally, the entropy within each two-digit sector is weighted and summed by:

$$\text{RV} = \sum_{g=1}^{G} P_g H_g \tag{3.15}$$

where Equation (3.15) characterizes Jacobs' externalities as it measures the diversity within each of the two-digit classes. It is expected that transferred technologies arising from a variety are strongly interactive between transfer actors. Therefore, it is used to analyze the spillover effects present in local technology pools.

3.4.1.3 *Unrelated variety*

Unrelated variety (UV) of technology measures the extent to which a city is diversified in widely varying types of technology activities, which is instrumental in technological breakthroughs (Castaldi *et al.*, 2015). It is indicated by the entropy of patents distribution at the two-digit level, and is given by (Frenken *et al.*, 2007):

$$\mathrm{UV} = \sum_{g=1}^{G} P_g \log_2\left(\frac{1}{p_g}\right) \tag{3.16}$$

where unrelated variety characterizes how diversified each city is across unrelated technology domains. It provides the building blocks for breakthrough innovation (Castaldi *et al.*, 2015), although unrelated technologies are difficult to transfer, learn, and absorb. So it can be used to analyze the presence of edge spillovers from the local technology pool.

3.4.2 *Technology relatedness*

Does regional probability of successfully entering a new technology increase when it already has related technologies? To answer this question, a measurement of technology relatedness is used to estimate the proportion of related technologies already existing in a region (Hidalgo *et al.*, 2007; Boschma *et al.*, 2014). The colocation of patents from different technology classes is a reasonable proxy for technological relatedness.

3.4.2.1 *Relatedness*

Relatedness of technology refers to the degree to which external knowledge is related, but not identical, to the local knowledge base. This relationship is often characterized by both similarity and complementarity (Makri *et al.*, 2010). It reflects the difficulty or cost associated with transferring from one technology to another (Balland *et al.*, 2019). To measure relatedness, researchers use the entropy of imported technologies within a four-digit technology subcategory and multiply it by the share of total exported four-digit technologies (Boschma & Iammarino, 2009):

$$R_j = \sum_j E_4^M(j) P_4(j) \tag{3.17}$$

where Equation (3.17) is a standardized measure of the frequency with which two IPC classes appear on the same technology (Balland *et al.*, 2019). $E_4^M(j)$ represents the import entropy within the four-digit classes, but belongs to the same two-digit domain j; and $P_4(j)$ is the relative size of the four-digit export subcategory j in the entire export technology of a city. In TTN, a larger technology relatedness of a city indicates a greater correlational degree between its external technology pipelines and local knowledge base, which implies a lower cost of technology diffusion from one transferor to another transferee.

3.4.2.2 *Similarity*

Similarity, represented by the patent similarity, refers to the IPC standard (Jaff *et al.*, 1986; Liu *et al.*, 2018a). It is a measure of the proximity of external knowledge to existing local knowledge bases. Based on the trade similarity variable (Boschma & Iammarino, 2009), technology similarity is measured by multiplying specific values for import and export technologies in a particular subcategory, and can be calculated by:

$$S_i = \log \sum_j \mathrm{AV}_4(j) P_4(j) \tag{3.18}$$

where $\mathrm{AV}_4(j)$ represents the absolute value of the four-digit subcategory j of export technologies, and $P_4(j)$ is the absolute value of the four-digit import subcategory j in a city. This indicator gets greater value when a city is more specialized in just one and the same sector in both imports and exports, which is not expected to lead to additional radical innovation and regional growth (Boschma & Iammarino, 2009). This book, therefore, anticipates that technological similarity may play a minimal or non-existent role in regional innovation performance.

By measuring the relatedness and similarity of these external technologies to the city's current knowledge base, we can infer how relevant they are to the development of new technologies in that city. This can be a valuable tool for identifying opportunities for innovation and growth, as well as for determining which technologies are most likely to succeed in a given market.

3.5 Network Dynamics Models

In economic geography, four different regression models have been constructed to explain the spatial context of network dynamics in TTN, including gravity models (GMs), quadratic assignment procedure (QAP), exponential random graph models (ERGMs), and stochastic actor-oriented models (SAOMs) (Broekel *et al.*, 2014).

Intercity technology flow has significant heterogeneity and hierarchy driven by different determinant factors. Previous studies have used a gravity model to analyze its network determinants (Scherngell & Hu, 2011). Because the dependent variable is a count with "overdispersion" (Cameron & Trivedi, 1986), it is inappropriate to use an ordinary least squares estimation instead of negative binomial regression models (Hilbe, 2011).

This book implies that the negative binomial regression model (NBRM) can be used to demonstrate the positive effects of various driving forces, such as local innovation capabilities, innovation milieux, and high-speed transports, on multiscalar technology transfer size based on several empirical studies (Cohen & Levinthal, 1990; Glass & Saggi, 1998; Hoekman *et al.*, 2009; Sigler & Martinus, 2017; Nepelski & De Prato, 2018). Additionally, the book also employs ERGMs to further provide a more comprehensive demonstration of the evolutionary mechanisms of TTN. Finally, the spillover effect of the network on local technology variety, relatedness, and similarity is explained using a simpler least squares regression model (LSRM).

3.5.1 *Negative binomial regression model*

Because of the "overdispersion" of patent transfer data, the negative binomial regression model (NBRM) is introduced to explore the factors determining the size of intercity TTNs at international, national, and regional scales.

3.5.1.1 NBRM of transnational TTN

For the international TTN, the model referring to its determinants is defined as:

$$\begin{aligned} \mathrm{IPT}_i = \alpha + \beta_1 pergdp + \beta_2 terti + \beta_3 rd + \beta_4 talent + \beta_5 enterp \\ + \beta_6 college + \beta_7 fdi + \beta_8 exp + \beta_9 imp + \varepsilon \end{aligned} \tag{3.19}$$

where α is a constant term; ε is a random error term; IPT_i represents the total number of international patent transfer of city i; and *pergdp, terti, rd, talent, enterp, college, fdi, exp, imp* represent the GDP per capita, the share of tertiary industry, the amount of governmental investment in R&D, the number of researchers and IT employees, the number of enterprises, the number of colleges, the amount of foreign direct investment, the volume of exported goods, and the volume of imported goods of city i, respectively.

3.5.1.2 NBRM of domestic TTN

In terms of the national TTN, its drivers are shown as follow:

$$\begin{aligned} \mathrm{DPT}_i = \alpha + \beta_1 pergdp + \beta_2 terti + \beta_3 rd + \beta_4 talent + \beta_5 enterp + \\ \beta_6 comp + \beta_7 college + \beta_8 fdi + \varepsilon \end{aligned} \tag{3.20}$$

where α is a constant term; ε is a random error term; DPT_i is the total amount of domestic patent transfer in China; *comp* denotes the number of listed companies; and the meanings of *pergdp, terti, rd, talent, enterp, college, and fdi* are same as Equation (3.19).

3.5.1.3 NBRM of regional TTN

It is extensively evidenced that regional technology transfers are strongly related to interregional distance, technology gap, industrial similarity, foreign investment, etc., and are also dependent on innovative capacity, absorptive capacity, economy size, and administrative hierarchy (Table 3.1).

Table 3.1 Main variables and their interpretations.

Variables	Symbols	Indicator interpretation or formula
Interpreted variable		
Number of RPT	*trans*	$trans{=}Count_{ij}$
Explanatory variables		
Geographical proximity	*geo*	$geo = 1 - ln\left(\frac{d_{ij}}{max\, d_{ij}}\right)$
Industrial similarity	*ins*	$ins = \frac{\sum_{K=1}^{K}[(P_{i.k} - \overline{P}_i)(P_{j.k} - \overline{P}_j)]}{\sqrt{\sum_{K=1}^{K}(P_{i.k} - \overline{P}_i)^2 \sum_{K=1}^{K}(P_{j.k} - \overline{P}_j)^2}}$ $(k = 1,2,3)$ P_{ik} is the output of industry k in city i as a proportion of total output
Economic gap	*ecgap*	$ecgap = 1 - \ln(min\ gdp/max\ gdp)$
Innovative similarity	*ino*	$ino = \frac{min\ p}{max\ p}\left(\frac{min\ p + max\ p}{2}\right)$ p represents the volume of city's patent application
Absorptive capacity	*ac*	Represented by the number of patent transferred to city i
Foreign direct investment	*fdi*	Represented by FDI transferred to city
Control variables		
Subprovincial city	*Key city*	when subprovincial city is assigned a value of 1, other 0

Source: Liu *et al.* (2019).

In order to ensure the confidence of results, a variance inflation factor (VIF) test was performed, and the test results indicated that there was no multicollinearity. The model was established as follows:

$$\text{RPT} = \alpha_0 + \beta_1 geo + \beta_2 ins + \beta_3 ecgap + \beta_4 ino + \beta_5 ac + \beta_6 fdi + \varepsilon_{it} \quad (3.21)$$

where ε_{it} denotes the error term in the usual sense; α_0 and β are the parameters to be estimated. The explanatory variables *fdi* and *ac* are logarithmical transformed, as the logarithmic transformation eliminates heteroskedasticity and does not affect their changing dynamics.

3.5.1.4 *NBRM of aviation effects*

Air transport has a significantly positive effect on technology flows owing to its networking and high speed (Luo *et al.*, 2022; Guan *et al.*, 2021). Accordingly, the NBRM is applied to clarify the effect of a city's accessibility in national aviation network on its technology transfer, in which the number of technology transfer is regarded as the dependent variable and city's aviation accessibility is the core independent variable (Dong *et al.*, 2020; Liu *et al.*, 2018a; Guan *et al.*, 2021). The model includes some control variables such as economic level, human capital, innovation investment, industrial structure, technology similarity, and geographical distance, and calculated as:

$$Y_i^{CX} = \alpha_0 + \beta_1 AF_i + \beta_2 Gdp_i + \beta_3 Edu_i + \beta_4 Rd_i + \beta_5 Ind_i + \beta_6 Hig_i + \beta_7 Hsr_i + \varepsilon_{it} \quad (3.22)$$

where Y_i^{CX} represents the volume of city *i*'s technology transfer in the network; AF_i represents the total number of city *i*'s airline flights with other cities; α_0 is the intercept term of the regression model; the control variables Gdp_i, Edu_i, Rd_i, Ind_i, Hig, and Hsr_i represent the GDP per capita, the volume of researchers and IT employees, the proportion of R&D investment in GDP, the proportion of tertiary industry in GDP, the mileage of highway, and the total number of HSR stations of a city *i*, respectively; β_1, β_2, β_3, β_4, β_5, β_6 and β_7 are the regression coefficient; ε_{it} is the random error term. When conducting the empirical research, we concentrate on the coefficients of β_1.

In addition, the effect of aviation on technology transfer is highly dependent on geographical distance. There exists an inverted U-shaped relationship between aviation accessibility and technology flows at different distance intervals (Guan *et al.*, 2021). Currently, the speed of most commercial aircraft in China is about 750–800 km/h. According to the half-hour division rule, the intervals can be divided into below 600 km, 600–900 km, 1,200–1,500 km, 1,500–1,800 km, and above 1,800 km. Moreover, technology transfer activities are greatly concentrated on Eastern China, noticeably in the three metropolises — Beijing, Shanghai, and Shenzhen. The research scope, therefore, takes Beijing, Shanghai, and Shenzhen as the centers and respectively establish five circles whose radii are 600 km, 900 km, 1,200 km, 1,500 km, and 1,800 km, which aims at discovering the spillover effect from Beijing, Shanghai, and Shenzhen to the rest of Chinese cities. The regression model is established as:

$$F_{ij}^{CX} = \alpha_g + \gamma_1 AF_{ij} + R_{1,2,3,4,5,6} + \gamma_2 Gdp_{ij} + \gamma_3 Edu_{ij} + \gamma_4 Rd_{ij} + \gamma_5 Ind_{ij} + \gamma_6 Tec_{ij} + \gamma_7 Geo_{ij} + \gamma_8 Hiq_{ij} + \gamma_9 Hsr_{ij} + \varepsilon_{ig} \quad (3.23)$$

where F_{ij}^{CX} represents the volume of technology transfer from city i to city j; AF_{ij} represents the total number of flights from city i to city j; Gdp_{ij}, Edu_{ij}, Rd_{ij}, Ind_{ij}, Geo_{ij}, Tec_{ij}, Hig_{ij}, Hsr_{ij} represent the same variables as Equation (3.22); $Geo_{1,2,3,4,5,6}$ represents a threshold variable that can be divided into 6 ranges: below 600 km, 600–900 km, 1,200–1,500 km, 1,500–1,800 km, and above 1,800 km; γ_1, γ_2, γ_3, γ_4, γ_5 , γ_6, γ_7, γ_8 and γ_9 are the coefficients of related variables; ε_{ig} is the random error term.

3.5.2 *Differences-in-differences model*

Due to its high speed and networking, HSR network has played a role in strengthening interregional connectivity of socioeconomic flows (Gao & Zheng, 2020; Xiao *et al.*, 2022). Differences-in-differences (DID) model is applied to explore the impact of exogenous shock on interregional technology flows.

3.5.2.1 *DID method*

The opening of HSR is usually treated as a quasi-natural experiment, in which cities located along HSR are chosen as experimental groups, while cities without HSR are selected as control or comparison groups. Thus, a multiphase DID model is defined as:

$$Y_{it} = \alpha_0 + \beta_1 \mathrm{HSR}_{it} + \delta_t + \partial_i + \gamma X_{it} + \varepsilon_{it} \quad (3.24)$$

where, the dependent variable Y_{it} reflects the number of technology transfer of city i in year t; HSR_{it} means the dummy variable of HSR.

Different cities opened HSR at different periods, so a multiphase dummy variable-opening time was adopted. X_{it} denotes the control variables (i.e., socioeconomic characteristics, industrial characteristics, innovation resources, and topographic characteristics) influencing regional technology transfer size (Table 3.2); γ means the coefficient of control variable; δ_t indicates the time-fixed effect; ∂_i denotes the city-fixed effect; ε_{it} signifies the stochastic error term.

3.5.2.2 *PSM-DID method*

The scores of the treatment group and control group were gained according to the propensity score matching (PSM) method. The model is finally established as follows:

$$Y_{it}^{\text{PSM}} = \alpha_0 + \beta_1 \text{HSR}_{it} + \delta_t + \partial_i + \gamma X_{it} + \varepsilon_{it} \tag{3.25}$$

where Y_{it}^{PSM} represents the regional technology transfer size after matching; other variables have same connotations as that in Equation (3.24). In the model, if the core factor β_1 is positive, it means that the opening of HSR enhances the technology transfer; otherwise, a negative effect will be produced.

3.5.3 *Spatial econometric models*

Since knowledge spillover is spatially intercorrelated, this chapter uses spatial econometric models to analyze agglomeration and network externalities of HSR opening on urban innovation performance.

Firstly, the agglomeration size is calculated to analyze the impact of HSR opening on urban economic agglomeration. Secondly, based on the patent transaction data, the intercity linkage strength is calculated to analyze the impact of HSR opening on interurban innovation cooperation. Finally, the weighted distance matrix is constructed, and the impact of agglomeration externalities and network externalities on urban innovation performance is explored using spatial lag model (SLM) and spatial error model (SEM) as follows:

Table 3.2 Major variables and their measurement methods.

Variables	Index layer	Variable meaning (Y)	Measurement method
Dependent variable	/	Urban innovation performance	Number of patents per 10000 people
Independent variable	/	HSR (HSR)	A multi-phase dummy variable; its value is 1 after a city opened HSR; its value is 0 before the opening.
Control variables	Socioeconomic characteristics	Urban economic level	Logarithm of GDP per capita
		City size	Logarithm of urban population
		Opening degree	Logarithm of foreign investment
		Urban traffic	Logarithm of road area per capita
		City internet level	Logarithm of mobile phone ownership ratio per capita
		Economic agglomeration level	Economic output per unit area
	Innovation resources	Human resources	Logarithm of ten thousand college students
		Research input	Proportion of science and education expenditure in fiscal expenditure
		Knowledge stock	Logarithm of the number of patents lagging two phases
	Geographical proximity	Urban geographical gradient	Mean value of steepness and slowness of urban surface units

$$Agg_{it} = \alpha_0 + \beta_1 \mathrm{HSR}_{it} + pW^{*}X_{it} + \delta_t + \partial_i + \gamma X_{it} + \varepsilon_{it} \tag{3.26}$$

$$\ln Y_{it} = \alpha_0 + \beta_1 \mathrm{HSR}_{it} + \beta_2 Agg_{it} + pW^{*}X_{it} + \delta_t + \partial_i + \gamma X_{it} + \varepsilon_{it} \tag{3.27}$$

$$Deg_{it} = \alpha_0 + \beta_1 \mathrm{HSR}_{it} + pW^{*}X_{it} + \delta_t + \partial_i + \gamma X_{it} + \varepsilon_{it} \tag{3.28}$$

$$\ln Y_{it} = \alpha_0 + \beta_1 \mathrm{HSRi}_t + \beta_2 Deg_{it} + pW^{*}X_{it} + \delta_t + \partial_i + \gamma X_{it} + \varepsilon_{it} \tag{3.29}$$

where Agg_{it} denotes the size of urban economic agglomeration; the degree centrality Deg_{it} of a city i is used to characterize the strength of interurban innovation linkages; $W{*}X_{it}$ represents the spatial lag term. X_{it} denotes the control variables same as Equation (3.25) (Table 3.2).

3.5.4 *Exponential random graph models*

Exponential Random Graph Model (ERGM) is a statistical method designed for relationship-based data that addresses several fitting difficulties (Cai *et al.*, 2022). It is a generative model with series of variables including self-organizing structure, nodal-related attributes, and network exogenous covariates to explain how local node relationships shape the global network structure (Lusher *et al.*, 2013).

3.5.4.1 *Exponential random graph model*

ERGM estimates the probability of ties among nodes and models the choice process of advice tie creation (Brennecke & Rank, 2017), parameterized in the form:

$$Pr_{\vartheta;\eta,g}(Y = y/x) = \frac{exp(\eta(\vartheta)\cdot g(y;x))}{k_{n,g}(\vartheta;x)}, \quad y \in Y \tag{3.30}$$

where $Pr_{\vartheta;\eta,g}(Y = y/x)$ is the probability estimating network similar to the observed network. Y is the random network variable and its corresponding realization is represented by y. $\vartheta \in \Theta \subseteq R^q$ (q is the vector of parameters) is the combining vector of network variables, and mapped by the canonical parameters η: $\Theta \to R^p$. The vector of sufficient statistics g: $y \to R^p$, may also depend on data x, assumed fixed and known; and the standardized constant (in y) $k_{\eta,g}$: $R^q \to R$, relevant with the network variable, ensures that $Pr_{\vartheta;\eta,g}$ sum to 1.

3.5.4.2 *Weighted exponential random graph model*

ERGM is a model that uses binary networks as dependent variables and aims to identify the factors that affect the probability of link formation between nodes. However, it does not account for the varying weights of links in intercity technology transfers, which could be significant (Liu *et al.*, 2017). To address this limitation, a variant of ERGM called weighted ERGM has been proposed, which incorporates edge weights into the modeling framework. This approach has been shown to overcome some of the limitations of traditional models, as noted by Krivitsky (2012), and takes the following form:

$$Pr_{\vartheta;h,\eta,g}(Y=y)=\frac{h(y)}{k_{h,\eta,g}(\vartheta)}\exp\left(\sum_i \vartheta_i \cdot nodeattr_i\right) \tag{3.31}$$

$$Pr_{\vartheta;h,\eta,g}(Y=y)=\frac{h(y)}{k_{h,\eta,g}(\vartheta)}\exp\left(\sum_i \vartheta_i \cdot edgecov_j\right) \tag{3.32}$$

$$Pr_{\vartheta;h,\eta,g}(Y=y)=\frac{h(y)}{k_{h,\eta,g}(\vartheta)}\exp(\vartheta_k \cdot Mutual) \tag{3.33}$$

$$Pr_{\vartheta;h,\eta,g}(Y=y)=\frac{h(y)}{k_{h,\eta,g}(\vartheta)}\exp\begin{pmatrix}\sum_i \vartheta_i \cdot nodeattr_i + \sum_j \vartheta_j \cdot edgecov_j \\ +\vartheta_k \cdot Mutual\end{pmatrix} \tag{3.34}$$

$$Pr_{\vartheta;h,\eta,g}(Y=y)=\frac{h(y)}{k_{h,\eta,g}(\vartheta)}\exp\begin{pmatrix}\sum_i \vartheta_i \cdot nodeattr\_in_i + \sum_i \vartheta_i \cdot nodeattr\_out_i \\ +\sum_j \vartheta_j \cdot edgecov_j + \vartheta_k \cdot Mutual\end{pmatrix} \tag{3.35}$$

$$Pr_{\vartheta;h,\eta,g}(Y=y)=\frac{h(y)}{k_{h,\eta,g}(\vartheta)}\exp\Big(\sum_i \vartheta_i \cdot nodeattr\_in_i + \sum_i \vartheta_i \cdot nodeattr\_out_i + \sum_j \vartheta_j \cdot edgecov_j + \vartheta_k \cdot Mutual_j + \sum_j \vartheta_j \cdot Tec\_in\&out_m\Big) \tag{3.36}$$

In Equation (3.31), the model parameters are identical to those specified earlier. $Pr_{\vartheta;h,\eta,g}(Y=y)$ is the probability estimating network similar to the observed network. $nodeattr_i$ represents the urban innovation capability measured by a series of node attributes (Table 3.3). $edgecov_j$ denotes the interurban proximity of city pairs (Table 3.4). *Mutual* denotes the reciprocity of TTN, while $Tec\_in\&out_m$

Table 3.3 Main variables of nodal attributes and their illustrations.

Dimension	Variables	Indicators	Expression	Structure
Underlying elements	Economic development level	City's GDP per capita (Ten thousand Yuan)	GDP_per_ca	●⟶●
	Economic development level (Transfer-in)	Transfer-in city's GDP per capita (Ten thousand Yuan)	GDP_per_ca_in	○⟶●
	Economic development level (Transfer-out)	Transfer-out city's GDP per capita (Ten thousand Yuan)	GDP_per_ca_out	●⟶○
Innovation input	Innovation fund	City's R&D expenditure/GDP (Scientific expenditure/public budget expenditure)	$R\&D_{GDP}$ (Sci_{Pub})	●⟶●
	Innovation fund (Transfer-in)	Transfer-in city's R&D expenditure/GDP (Scientific expenditure/public budget expenditure)	$R\&D_{GDP}$_in (Sci_{Pub}_in)	○⟶●
	Innovation fund (Transfer-out)	Transfer-out city's R&D expenditure/GDP (Scientific expenditure/public budget expenditure)	$R\&D_{GDP}$_out (Sci_{Pub}_out)	●⟶○
	Innovation human capital	City's R&D employees/Total employees (Number of employees in the research integrated technology industry/number of employed persons)	$R\&D_{Work}$ (Sci_{Work})	●⟶●

Dimension	Variables	Indicators	Expression	Structure
	Innovation human capital (Transfer-in)	Transfer-in city's R&D employees/ Total Employees (Number of employees in the research integrated technology industry/number of employed persons)	$R\&D_{Work}\_in$ (Sci_{Work}_in)	○→●
	Innovation human capital (Transfer-out)	Transfer-out city's R&D employees/ Total Employees (Number of employees in the research integrated technology industry/number of employed persons)	$R\&D_{Work}\_out$ (Sci_{Work}_out)	●→○
Local system	Level of decentralization	City's decentralization index	Dec_Index	●→●
	Level of decentralization (Transfer-in)	Transfer-in city's decentralization index	Dec_Index_in	○→●
	Level of decentralization (Transfer-out)	Transfer-out city's decentralization index	Dec_Index_out	●→○
	Level of marketization	City's level of marketization	Mar_Index	●→●
	Level of marketization (Transfer-in)	Transfer-in city's level of marketization	Mar_Index_in	○→●
	Level of marketization (Transfer-out)	Transfer-out city's level of marketization	Mar_Index_out	●→○
	Administrative grade homogeneity	Administrative grade homogeneity is achieved or not	*Homo*_Adm	●⇢●

(*Continued*)

Table 3.3 (*Continued*)

Dimension	Variables	Indicators	Expression	Structure
Innovative activities	Technology innovation capacity	City's number of new high-tech enterprises (100)	New_ Hitech	●⟶●
	Technology absorption capacity	Transfer-in city's number of new high-tech enterprises (100)	New_ Hitech_in	○⟶●
	Technology export capacity	Transfer-out city's number of new high-tech enterprises (100)	New_ Hitech_ out	●⟶○
	Technology intermediary capacity	City's number of IP service companies	ITS	●⟶●
	Technology intermediary capacity (Transfer-in)	Transfer-in city's number of IP service companies	ITS_in	○⟶●
	Technology intermediary capacity (Transfer-out)	Transfer-out city's number of IP service companies	ITS_out	●⟶○
Technology characters	Technology-related diversity (Transfer-out)	Transfer-out city's technology-related diversity	Rel_Div_ out	●⟶○
	Technology-unrelated diversity (Transfer-out)	Transfer-out city's technology–unrelated diversity	Non_Rel_ Div_out	●⟶○
	Technology relatedness (Transfer-in)	The relatedness between transfer-in city's imported technologies and local technologies	Tech_ Simi_in	○⟶●

represents the preferential link toward technology transfer-in or transfer-out (Table 3.5). The weighted ERGM incorporates *h(y)* as the reference distribution, and given the discrete characteristics of the technology transfer size, the binomial distribution is chosen as the reference distribution in this book.

Table 3.4 Main variables of city pair's dualism and their illustrations.

Variables	Indicators	Expressions	Structures
Geographical proximity	Standardized indices of geographical distance between cities	*Pro*_Geo	
Industrial proximity	Cosine similarity index of patented technology categories between cities	*Pro*_Tec	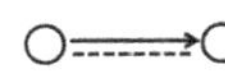
Technological proximity	Cosine similarity index of the number of employees in secondary national economy industries between City	*Pro*_Ins	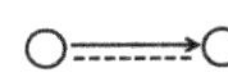
STI potential gap	Difference in patent filings between cities	*Diff*_Inno	
Economic potential gap	Difference in GDP between cities	*Diff*_Eco	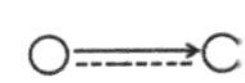

Table 3.5 Main variables of network structure and their illustrations.

Variables	Descriptions	Expressions	Structures
Intercept term	Weighted edge links	Edges	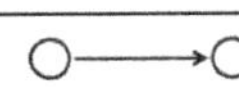
Reciprocity	The two-way link between cities	Mutual	
Aggregation	The preference link toward technology transfer-out	Indegree	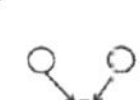
Diffusion	The preference link toward technology transfer-in	Outdegree	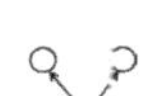
Transmissibility	Transmissive tripartite transfer relationships	Transitive	

ERGM is usually tested by Markov chain Monte Carlo maximum likelihood (MCMCML) estimation (Snijders, 2002; Snijders *et al.*, 2006; Brennecke & Rank, 2017; Cai *et al.*, 2022), and the significance of each parameter is characterized by the *t*-statistic, and the model fit is tested by the Akaike Information Criterion (AIC) and the Bayesian Information Criterion (BIC) to test the fit of the model, with smaller values of both representing a better fit (van der Pol, 2019).

Compared to traditional models such as logistic regression model, NBRM, and QAP, the weighted ERGM can provide a more comprehensive description of the evolutionary mechanisms of TTNs.

3.5.4.3 *Explained variables*

(1) The explained variables referred to as determinants of TTN are mainly conducted as the three following aspects (Tables 3.3–3.5): urban innovation milieux, including such nodal attributes as economic development level, R&D investment, human capital, degree of marketization, degree of decentralization, technology absorptive capacity, technology intermediation capacity, administrative levels, technology (un)related diversity, and technology similarity.

Among them, the economic development level of a city is proxied by GDP per capita. Innovation capital investments are relatively quantified by the ratio of public expenditure on science to local public budget expenditure (Equations (3.37) and (3.38)) and the ratio of the number of staffs in R&D integrated technology industry to the total workforce (Equation (3.39) and (3.40)), respectively.

$$R\&D_{GDP} = R\&D\_Expend/GDP \tag{3.37}$$

$$R\&D_{Work} = R\&D\_Stuff/Work \tag{3.38}$$

$$Sci_{Pub} = Sci\_Expend/Pub\_Expend \tag{3.39}$$

$$Sci_{Work} = Sci\_Stuff/Work \tag{3.40}$$

The degree of decentralization is influenced by expenditure decentralization, income decentralization, and fiscal self-sufficiency, and is defined as:

$$Dec_{Index} = F(Dec_{Exp} + Dec_{Rev} + Finance_{Self}) \tag{3.41}$$

$$Dec_{Exp} = Pub_{Exp}/Provin_{PubExp} \tag{3.42}$$

$$Finance_{Self} = Pub_{Rev}/Pub_{Exp} \tag{3.43}$$

$$Dec_{Rev} = Pub_{Rev}/Provin_{PubRev} \tag{3.44}$$

The marketization index is taken directly from Fan and Wang's method, including the government–market relationship score, non-state economic development score, product market development score, factor market development score, and intermediary organization and legal score (Fan & Wang, 2001).

The hierarchy of administrative levels is considered to be a driving force in TTN. The 297 cities in China are divided into four classes: municipalities directly under the central government valued as 4, provincial capitals valued as 3, subprovincial and separate planning cities valued as 2, prefecture-level cities valued as 1. If two cities have the same rank, the administrative rank congruency is 1; otherwise, its 0.

Technology absorptive capacity is characterized by the number of new high-tech enterprises engaged in technological innovation activities. Technology intermediary capacity is measured by the number of enterprises providing services of intellectual property rights.

The measurements of technology-related diversity, technology-unrelated diversities, along with technology similarity are same as the previous methods, as shown in Section 3.4.

(2) interurban multiple proximities, calculated by connective attributes of each edge such as geographical proximity, technological proximity, industrial proximity, technological innovation potential, and economic development potential.

Geographical proximity is calculated by using the straight geographic distance between cities. Considering that large differences in actual geographic distances between cities may bias the estimation results, the actual distances were standardized (Liu *et al.*, 2018a) and calculated as follows:

$$ProGeo_{ij} = 1 - \ln[d_{ij}/max(d_{ij})] \tag{3.45}$$

where $ProGeo_{ij}$ is the geographic proximity between cities i and j, d_{ij} is the standardized geographical distance between cities i and j.

Technological proximity is defined as the cosine similarity of patent subcategories between two cities (Jaffe, 1986). The formula was calculated as follows:

$$ProTec_{ij} = \sum_{t=1}^{p} T_{ti} \times T_{tj} / \sqrt{\sum_{t=1}^{p} T_{ti}^2 \times \sum_{t=1}^{p} T_{tj}^2} \tag{3.46}$$

where $ProTec_{ij}$ is the technological proximity between cities i and j. T_{ti} and T_{tj} are the number of patents applicated by city i and j in the technology field t (IPC category) in the study period, and p is the sum of the number of technology fields applied by both parties.

Similarly, industry proximity is defined as the cosine similarity of industry structures between cities and calculated by:

$$ProIns_{ij} = \sum_{e=1}^{I} W_{ei} \times W_{ej} / \sqrt{\sum_{e=1}^{I} W_{ei}^2 \times \sum_{e=1}^{I} W_{ej}^2} \tag{3.47}$$

where $ProIns_{ij}$ is the structural proximity of industries between cities i and j. W_{ei} and W_{ej} are the number of employees in cities i and j, and e is the number of staff and I is the number of industries.

Technological innovation potential is represented by the difference in the number of patent applications (Elkan *et al.*, 1996) and the economic development potential is represented by the difference in GDP. They are calculated as follows:

$$DiffInno_{ij} = 1 - \ln[min(Inno_i, Inno_j) / max(Inno_i, Inno_j)] \tag{3.48}$$

$$DiffEco_{ij} = 1 - \ln[\,min(Eco_i,Eco_j)\,/\,max(Eco_i,Eco_j)\,] \tag{3.49}$$

where $DiffInno_{ij}$ and $DiffEco_{ij}$ are the potential difference in technological innovation and economic development level between cities i and j. $Inno_i$ and $Inno_j$ are the number of patent applications in cities i and j during the study period. Eco_i and Eco_j are the GDP of cities i and j, respectively.

(3) and network structure properties, represented by possible endogenous structures such as reciprocity, transmissibility, aggregation, and diffusion.

Reciprocity indicates the tendency to form or exhibit bi-directional linkages between cities in the network. The existence of TTN is corroborated by whether the structure of bi-directional linkages between nodes can significantly facilitate technology transfer linkages or not.

Transferability indicates the tendency for a network to form or exhibit a tripartite transfer relationship between cities, where cities are linked by an intermediary party to generate technology transfer with a third party, and only one-way closed-loop relationships are discussed in this chapter. The existence of TTN is corroborated by the ability of the structure of triadic transmission to significantly facilitate technology transfer linkages whether there is transmissibility or not.

Aggregation and diffusion indicate a tendency to form or exhibit meritocratic links between cities in the network. The presence of aggregation implies that cities that are more active in technology import (higher centrality of entry) tend to generate new technology transfer links, while diffusivity reflects cities that are more active in technology transfer-out (higher centrality of exit) tend to generate new technology transfer links.

3.5.5 *Least squares regression model*

To address the issue of overdispersion, the number of transferred patents is logged to make it more compatible with the least squares regression model. This approach is used in conjunction with the individual and time dual fixed effects model.

3.5.5.1 *LSR of transfer pipelines effect*

One of the regression models to explore the impact of technology transfer pipelines on local innovation capabilities is shown by:

$$\begin{aligned} Y_i = \omega_1 R_f + \omega_2 S_f + \omega_3 R_d + \omega_4 S_d + \beta_1 pergdp + \beta_2 terti + \beta_3 rd \\ + \beta_4 talent + \beta_5 enterp + \beta_6 comp + \beta_7 coll + \beta_8 fdi + \beta_9 RV \\ + \beta_{10} UV + \beta_{11} V + \delta_i + \delta_t + \varepsilon_{it} \end{aligned} \quad (3.50)$$

where δ_i and δ_t are the constant terms, representing individual fixed effects and time-fixed effects, respectively; ε_{it} are the random error terms; i are the total number of patent applications of city i; R_f and S_f represent the relatedness and similarity of foreign technologies imported by city i; and R_d and S_d denote the relatedness and similarity of domestic technologies respectively. *pergdp, terti, rd, talent, comp, enterp, coll, fdi, RV, UV,* and *V* are used as the control variables in the regression analysis, same as the aforementioned equations.

3.5.5.2 *LSR of nodal centrality effect*

The second regression model is used to demonstrate the effect of the network centrality in TTNs on local innovation capabilities and is formulized as:

$$\begin{aligned} Y_i = \omega_1 ind_f + \omega_2 wind_f + \omega_3 ind_d + \omega_4 wind_d + \omega_5 cc_d + \omega_6 bc_d \\ + \beta_1 pergdp + \beta_2 terti + \beta_3 rd + \beta_4 talent + \beta_5 enterp \\ + \beta_6 comp + \beta_7 coll + \beta_8 fdi + \delta_i + \delta_t + \varepsilon_{it} \end{aligned} \quad (3.51)$$

where δ_i, δ_t, ε_{it}, and Y_i are the same as the variables in Equation (3.34); ind_f and ind_d represent the in-degree of city i in the international and national TTNs respectively, and $wind_f$ and $wind_d$ represent the weighted in-degree of city i respectively; cc_d and bc_d are the closeness centrality and betweenness centrality values of domestic city i, respectively. *pergdp, terti, rd, talent, comp, enterp, coll,* and *fdi* are used as the control variables with the same representation as in Equation (3.50) (Table 3.6).

Table 3.6 Main variables and their measurements.

Variable	Definition
InY_i	Log of number of patent applications for city
R_d	Relatedness of domestic imported technology for city
S_d	Similarity of domestic imported technology for city
R_f	Relatedness of transnational imported technology for city
S_f	Similarity of transnational imported technology for city
V	Variety of technology for city
RV	Related variety of technology for city
UV	Unrelated variety of technology for city
Intalent	Log of number of researcher and IT staff for city
Inenterp	Log of number of industrial companies for city
Infdi	Log of FDI for city
Inrd	Log of R&D for city
pergdp	GDP per capita for city
terti	Share of tertiary industry for city
Incoll	Log of number of colleges for city
Incomp	Log of number of listed companies for city

3.5.5.3 *LSR of technology complexity effect*

In order to measure the network effects of transferred technology complexity on local innovation capability, the third equation is written as follows:

$$\begin{aligned} \mathrm{Ln}\, Y_i = {} & \omega_1 Rel_f + \omega_2 Sim_f + \omega_3 Rel_d + \omega_4 Sim_d + \beta_1 pergdp + \beta_2 terti \\ & + \beta_3 rd + \beta_4 talent + \beta_5 enterp + \beta_6 comp + \beta_7 coll + \beta_8 fdi + \beta_9 RV \\ & + \beta_{10} UV + \beta_{11} V + \delta_i + \delta_t + \varepsilon_{it} \end{aligned} \tag{3.52}$$

where ln Y_i is the log transformation of patent application number of city i. The meanings represented by the independent variables are shown in Table 3.6.

3.6 Research Data

Patents not only act as a major indicator of regional innovation capacity but also as an effective vehicle for acquiring external knowledge spillover (Ma & Xu, 2022). A majority of studies on technology transfer rely on secondary sources of patent data, such as patent applications, collaborations, citations, and transfers. However, compared to other sources, patent transfer data is more widely used to investigate directional technology flows and diffusion across regions or cities (Liu *et al.*, 2019; Zhang *et al.*, 2016; Yang *et al.*, 2021). This is because patent transfer data provides a more effective lens for analyzing the geographical relations involved in technology commercialization between the demand and supply sides (Ma & Xu, 2022).

Patent transfer can occur in various ways, but typically fall into one of the three main categories: license (the patent owner permits the use, performance, and sale of intellectual property by another party), ownership (the patent inventor assigns the ownership to other company which contracted the research), and co-ownership (the patent ownership is the share by the company contracted the research and the organization employing the research) (Ortega, 2011). Although licensed patents are the most common form of patent transfer (Ortega, 2011), their significance in analyzing the geographical complexity of technology transfer flows is sometimes overlooked (Yang *et al.*, 2021).

3.6.1 *Data source*

As an open and available source, patent data is the most often used data for capturing technology flow and innovative capability (Park *et al.*, 2013; Ponomariov, 2013; Ma & Xu, 2022; Wang & Lu, 2021; Zhang *et al.*, 2016), because it can be datamined as abundant demand-side and supply-side marketing information (Ma & Xu, 2022).

Consequently, this book employs transaction data of patent license and ownership to characterize technology transfer in order to quantify the spatial flows of technology at global, national, and regional levels. Meanwhile, patent application data is used to measure urban innovation capability from a local perspective. The patent data contains patents for inventions and utility models, as these two categories have consistent IPC classification standards and are of high patent quality.

Using a distributed crawler programme, data on patent transactions between China and other nations and among Chinese cities from 2001 to 2018 were extracted from a global and local perspective. The patent introduction data contains patent application and classification numbers, patent assignor and address information, patent assignee and address information, patent assignment dates, and patent application and grant dates. The patent application data contains the patent application and classification number, patent applicant and address, postal code, and application time. Patent transfer data is obtained from the Patent Information Service Platform of the China National Intellectual Property Administration (PISP of CNIPA) (http://www.cnipa.gov.cn/), a website established by the national statutory patent documentation publisher that updates Chinese patent data weekly with no lag. It includes all publicly accessible patent transfer records in China and China's foreign countries, with a high degree of data integrity. The information on patent applications is obtained from the IncoPat database and *Wanfang* database. Although patent application and transfer data come from different sources, the *Wanfang* database is purchased from the China Intellectual Property Rights Publishing House, resulting in the establishment of a "Patent Information Service Platform". Consequently, the two sorts of data can be regarded as essentially the same source.

The flight data between cities was acquired from OAG database. The data on socioeconomic indicators for each city were obtained from the *China Urban Statistical Yearbook*, the *China Regional Economic Statistical Yearbook*, and the *China Stock Market & Accounting Research* (CSMAR) database, and the statistical yearbooks and statistical bulletins of each prefecture-level city were used to fill in some missing data. Considering the lag, data of cities where patent transfers occurred from 2000 to 2018 were collected (Table 3.7).

Chinese basic geographic data are derived from National Catalogue Service for Geographic Information (www.webmap.cn). The world map data comes from the Resource and Environmental Science and Data Center of the Chinese Academy of Sciences (www.resdc.cn). The domestic basic geographic information data (2021) is in public version and 1:1000000 scale. All the boundaries are not modified.

Table 3.7 Socioeconomic indicators and their data sources.

Variables	Indicators	Data sources
The size of the population	The total population	China Urban Statistics Yearbook
Economic development levels	GDP per capita	China Urban Statistics Yearbook
Industrial structure	Share of tertiary sector	China Urban Statistics Yearbook
R&D investment	Government's R&D investment	China Urban Statistics Yearbook
Technical people	Number of scientific and IT staff	China Urban Statistics Yearbook
Enterprises' intensity	Number of industrial enterprises	China Urban Statistics Yearbook
	Number of listed enterprises	China Stock Market & Accounting Research Database
Universities intensity	Number of universities	China Urban Statistics Yearbook
External investment links	Foreign direct investment	China Urban Statistics Yearbook
External trade links	Imports of goods	China Regional Economic Statistics Yearbook
	Exports of goods	China Regional Economic Statistics Yearbook

3.6.2 *Data proceed*

3.6.2.1 *Temporary division*

Since joining the World Trade Organization in 2001, China's international and technological trade are rapidly increasing. Thus, this year is a more acceptable starting point for research on China's technology transfer dynamics (i.e., Zhang *et al.*, 2016; Duan *et al.*, 2019). To create contrasts along evolutionary paths when comparing a later phase with an earlier one, we evenly divided the period (2001–2018) into three phases, marking an interval every six years, to construct three subnetworks with different dynamics: slow growth (2001–2006), rapid growth (2007–2012), and stable growth (2013–2018).

In addition to conducting a comparative analysis, this book also employs the issuance date of key policies (such as laws and regulations) related to China's technology innovation and transfer as the basis for temporal segmentation (Yang *et al.*, 2021), including the *Measures for the Filing of Patent Exploitation License Contracts* enacted in December 2001, the *National Medium- and Long-term Science and Technology Development Plan* (2006–2020) published in February 2006, the *Implementation Plan on National Technology Transfer Promotion Action* instituted in September 2007, the *Innovation Driven Development Strategy* (IDDS) proposed in November 2012, the *National Medium- and Long-term Plan for Major Science and Technology Infrastructure* issued in February 2013, and the *Specification for Technology Transfer Services* (GB/T34670-2017) implemented in January 2018. These governmental policies are effective in facilitating the transfer (or transformation) of scientific and technological achievement in China, leading to the breakthrough development of TTN. As licensed patents show, China's patent began to increase sharply after 2007 (Zhang *et al.*, 2016), then up to the peak in 2012 (Yang *et al.*, 2021), which further evidence the rationality of our temporal division.

3.6.2.2 *City selection and geocoding*

City is used as the basic analysis unit in the research. Each patent's participating country and city are listed separately by data cleaning, and the city's coordinates are geocoded through Google Map API based on their post codes. The data processed are as follows:

Firstly, when taking city as the unit (including Tianmen, Xiantao, Qianjiang, and autonomous prefectures), two special administrative regions, Hong Kong and Macao, are considered directly as city units, while China's Taiwan Province can be divided into 6 municipalities (Taipei, Kaohsiung, etc.) and 11 county-level cities (patent transfer actors locate within a county) according to its administrative districts.

Secondly, cities in other countries or regions conduct data cleaning by Python and manual filtration. The filtering criteria are as follows: (1) the location of subnational government (e.g., state government in the USA) or top five cities by population size; (2) innovative cities with well-known universities or R&D centers; (3) cities with more than 200,000 population; (4) cities not meeting the above criteria will be grouped into the nearest qualified cities which can be reached within 45 minutes by ground transportation (Matthiessen *et al.*, 2010; Csomós & Tóth, 2016) based on Google Maps; (5) If none of the above criteria is qualified, cities with more than five patent transfer numbers or the only city having transfer record with China in a country are retained. The next step is to establish three spatiotemporal databases at transnational, national, and regional levels by only retaining the data involved intercity transfer.

3.6.2.3 *Patent classification*

For patent categories, this book utilizes data from the IPC classification, a hierarchical coding system assigned directly by national patent offices to classify patents according to their respective technical fields. The IPC code is divided into eight main sections, each of which is divided into three-digit major subcategories and four-digit minor subcategories.

Scholars have reorganized these technology subcategories into meaningful technology sectors, similar to sectoral groupings of products or economic activities (such as the Standard

Table 3.8 Technology sectors and fields.

Sectors	Fields
Electrical Engineering	Motor, instrument, energy; Audiovisual technology; Digital communication; Radio communication; Basic communication processes; Semiconductor category; Computer technology; IT management methods
Instruments	Medical technology; Optics; Mechanical measurement; Mechanical control; Control systems
Mechanical Engineering	Engines, pumps, turbines; Machine parts; Other special machinery; Heat treatment and equipment; Textile and paper machines; Material handling; Traffic engineering; Machine tools
Chemistry	Chemical Engineering; Environmental Technology; Basic Materials Chemistry; Organic Fine Chemistry; Surface Technology, Coatings; Metallurgical Materials; Food Chemistry; Biotechnology; Polymer Chemistry, Polymers; Microstructure and Nanotechnology; Pharmaceuticals
Other sectors	Other consumer goods; Civil engineering; Furniture, toys

Source: Yan (2021).

International Trade Classification (SITC) or the International Standard Industrial Classification (ISIC)). By reclassifying, technical heterogeneity within distinct technology sectors can be reduced, and technological innovation activities can be analyzed more effectively in spatiotemporal comparisons. Using the taxonomy created by Schmoch (2008), this book divides subcategories into 35 technology fields and further groups these technology fields for study into five technology sectors: electrical engineering, instrumentation, mechanical engineering, chemical, and other sectors (Table 3.8).

PART III

Spatial Evolution of Technology Transfer Network

CHAPTER 4

SPATIAL EVOLUTION OF TRANSNATIONAL TECHNOLOGY TRANSFER NETWORK

In the context of globalized knowledge and technology (Audretsch *et al.*, 2014), there is an increasing interest in the flow of technology between China and other countries (Deng & Lu, 2022; Liu & Yan, 2022). Transnational technology flows, viewed as a lens through which globalization can be understood, exhibit complex networking patterns that develop over time. In this vein, a network of international technology transfer is conducted by social network analysis to showcase technology relatedness between China and the rest of the world. Furthermore, this chapter will comprehensively analyze the spatial evolution of China's transnational technology transfer network from two network perspectives: geographical and technological, such as transfer intensities, nodal centralities, pipeline hierarchies, technology categories, diversities and relatedness, etc.

4.1 Evolution of Transfer Intensity

Since China's accession to the World Trade Organization (WTO) in 2001, there has been an increase in the intensity and scope of its transnational technology transfer. In the past 20 years, the intensity of transfer-in and transfer-out has exhibited a similar fluctuating growth that can be divided into three phases: initial stage

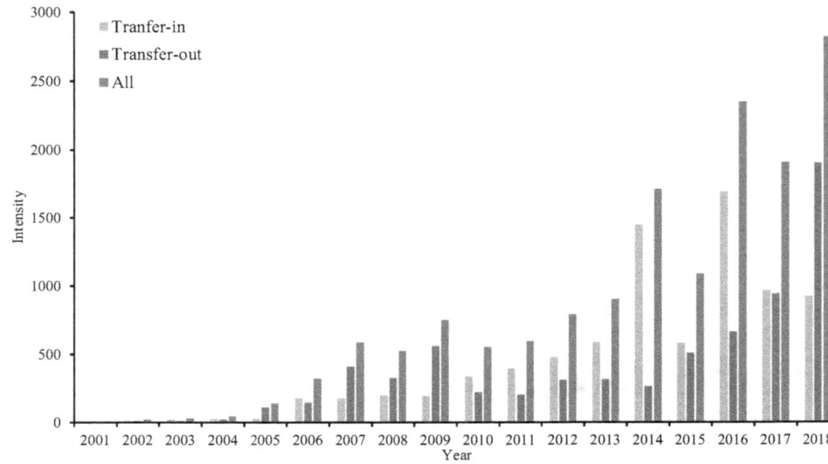

Figure 4.1 Dynamics of China's transnational technology transfer intensity.

(2001–2006), slow growth stage (2007–2012), and rapid growth stage (2013–2018) (Figure 4.1). Moreover, the number of domestic cities involved in transnational technology transfer activities is continuously increasing, but is highly centralized in the southeast coast of China, noticeably located in the three urban agglomerations such as the Beijing–Tianjin–Hebei (BTH) region, the Yangtze River Delta (YRD) region, and the Guangdong–Hong Kong–Macao Greater Bay Area (GBA) region (Figure 4.2).

4.1.1 *The temporal evolution*

4.1.1.1 *Technology transfer-in*

There was a growth of transfer-in technology in fluctuations from 2001 to 2018. Overall, the volume of imported patents in China has increased from less than 10 in 2001 to close to 1,000 in 2018, with an average growth rate far greater than the other regions of the world during the 1994–2017 period (23.91%) (Yang *et al.*, 2021). The dynamics presented a classic periodicity which is classified into the initial stage (2001–2006), slow growth stage (2007–2012), and fluctuating growth stage (2013–2018). During the periods, the

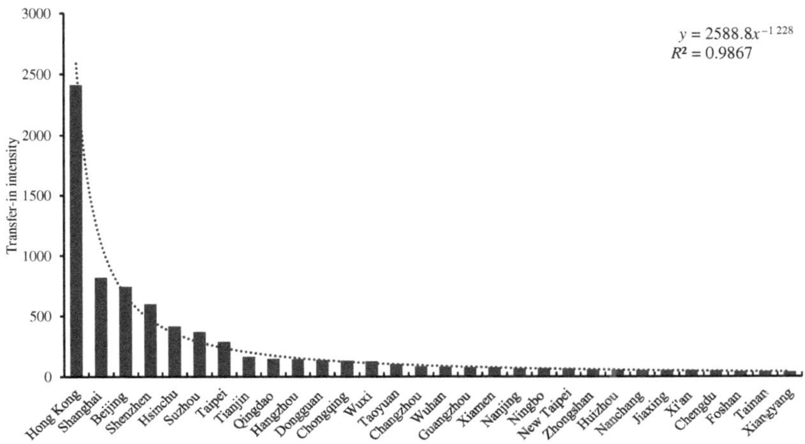

Figure 4.2 Top 30 cities of technology transfer-in intensity.

number of technology transfer-in has increased stably with less than 3% of the decrement in a few years.

However, the intensity was greatly influenced by both domestic and international factors after 2013, which led to larger fluctuations (Figure 4.1). Specifically, its intensity showed a sharp increase in 2014 and 2016 due to the contribution of Hong Kong and the role of intellectual property operating (IPO) firms. As a world-class financial center, Hong Kong is acting as a "gatekeeper" of technology transfer-in between China's mainland and foreign markets (Liu & Yan, 2022), along with technology merger and acquisition (Huang *et al.*, 2021). Over two years, there has been a significant increase in the import of technology to China, with estimates ranging from 6 to 10 times the previous volume, accounting for over 50% of the total volume in China. Furthermore, some IPO companies established by transnational corporations played a brokerage role in "bridging" across-border technology flows in 2016 (Yan, 2021).

4.1.1.2 *Technology transfer-out*

Similarly, the intensity of technology transfer-out also has presented a fluctuating growth, which can be divided into three stages, same with the dynamics of technology transfer-in. During 2001–2006, the Chinese capacity for technology innovation was poor thus presenting low intensity of technology transfer-out. Despite increasing transfer-out during 2007–2012, China still lagged behind global technology giants such as the United States, Japan, and Germany (Duan, 2020). In contrast to the fluctuation of transfer-in intensity, the rate of transfer-out has shown a rapid increase in the third stage owing to China's policy-driven investments in high-tech and emerging industries (Wu *et al.*, 2019).

Compared to the transfer-in, the intensity of technology transfer-out was weaker in maximum years, only exceeding that of transfer-in in 2018 (Figure 4.1). This indicates that China used to have a high dependence on external technology from developed countries or regions, primarily North America, West Europe, and East Asia (Liu & Yan, 2022). Nevertheless, with the recent implementation of the Innovation Driven Development Strategy (IDDS), China's capacity for self-independent innovation has significantly improved. As a result, China is becoming more competitive in the global technology market.

4.1.2 *The spatial evolution*

4.1.2.1 *Technology transfer-in*

In the past 20 years, the size of China's importing technologies had seen an increase in both foreign original cities and domestic terminal cities. In 2001, there was only one terminal city and two original ones. However, by 2018, this number had grown significantly to 67 terminal cities and 118 original cities. Despite this growth, spatial heterogeneity remains an issue due to the present uneven distribution, clustering, and hierarchical properties.

On the one hand, spatial *disparities* and *agglomerations* are observed during the period. The originations of China's imported technologies concentrate on these innovative hubs including the US (i.e., Silicon Valley, New York, Los Angeles, Wilmington, and Boston), the Caribbean (i.e., offshore financial centers such as Roadtown and

Georgetown), West Europe (i.e., London, Paris, and Eindhoven), and East Asia (i.e., Tokyo, Osaka, Seoul). It is worth noting that the pipeline sources have been shifting from the transatlantic axis, namely North America and West Europe, toward East Asia and Southeast Asia such as Japan, Korea, and Singapore playing increasingly important roles (Liu & Yan, 2022). Moreover, domestic cities absorbing foreign technologies are mostly concentrated in Taiwan Province and the southeast coast of China's mainland, particularly in the BTH (Beijing–Tianjin–Hebei), YRD (Yangtze River Delta), GBA (Greater Bay Area), and Northwest Taiwan (such as Taipei, Hsinchu, and Taoyuan). Taipei was the first-tier hub for importing foreign technologies in the initial stage. After that, the terminal cities expanded to Hong Kong and Macao, and then to the coastal cities clustered in the three urban agglomerations (Liu & Yan, 2022).

On the other hand, the transfer-in intensity of domestic cities follows a rank-size rule with a clear *hierarchy*. Its rank-size distribution has presented a power–law distribution as shown in Figure 4.2. Owing to its key position as an international free trade port and financial center, Hong Kong Special Administrative Region has become a primal hub and plays a leading role in China's external technology acquisition. In addition, other transfer centers such as Shanghai, Beijing, and Shenzhen also have rapidly developed as crucial hubs in the transnational technology transfer-in network, which can be contributed to positive innovative milieux, incremental R&D investment, talent pool effect, developed high-tech industries, and enterprising government (Liu & Yan, 2022). However, most of the inland cities are locked in the periphery of the network because of their weaker innovative capacity, absorptive capacity, and high-tech industrial base.

4.1.2.2 *Technology transfer-out*

The volume of transnational technology transfer-out has a similar expansion with the import one in spatial, but its intensity is far lower than the transfer-in. As a latecomer to technology development,

China's exported technologies over time have kept a rapid growth rate in accordance with the import, while both its average size and speed approximately have accounted for one-half of the import. It further indicates that China is rapidly catching up with the world-class leading innovators, but there still remains a gap with them, which is a result of its increasing innovation capacity driven by the IDDS.

Spatially, the original cities exporting technologies have expanded from the coastal area to the inland in the recent 20 years. In the beginning, a few S&T hubs including Taipei, Hong Kong, Shanghai, and Beijing acted as innovative growth poles of national innovation system in exporting technologies internationally, but their export volumes were generally lower with the highest value of less than 10. Then, as subsecondary nodes, some capital cities in Central and Western China such as Wuhan, Xi'an, and Changsha, as well as a few medium-sized subprovincial cities took part in the process over time. The overall exports increased incrementally, while the high-intensity areas were later concentrated in three urban agglomerations and Northwest Taiwan, which had significantly higher intensities than other regions. Taipei, Hsinchu, and Hong Kong emerged as primary technology export centers, while secondary centers emerged in cities on the southeastern coast such as Shenzhen, Changzhou, Shanghai, and Beijing.

In terms of the foreign terminal cities, they presented an expansion from the USA to East Asia and West Europe. In the beginning, the terminal cities which absorbed Chinese technology were highly concentrated in North America, centralizing in some off-shore financial centers and innovation centers, especially in the British Virgin Islands, Chicago and San Jose in the USA. With the continuous extension of the scope, Chinese cities had expanded their terminal cities of export to Europe such as London, Zug, and Leverkusen, as well as Asia such as Tokyo and Singapore. During this period, a significant number of small- and medium-sized cities were involved in the technology transfer process, thereby enhancing their overall capacity to absorb Chinese technology. However, their level of transfer intensity was still far less than the primary cities such as San Jose and Wilmington.

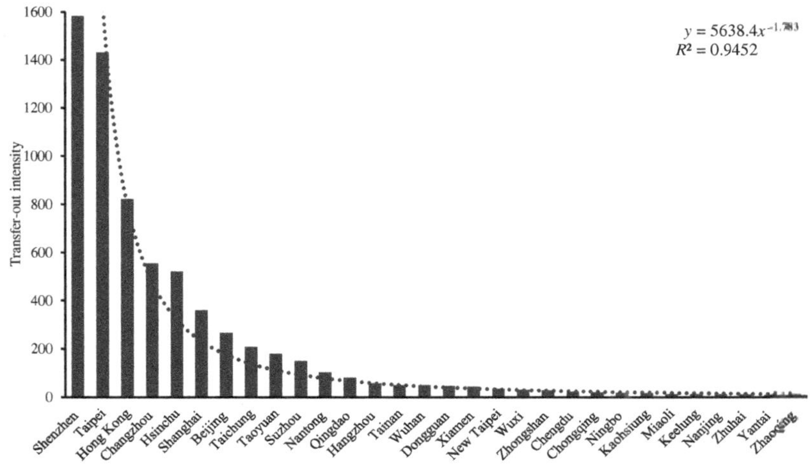

Figure 4.3 Top 30 cities of technology transfer-out intensity.

The intensity of transnational transfer-out in domestic cities from 2001 to 2018 showed a clear hierarchy with a small number of cities dominating the export activities. As shown in Figure 4.3, the rank–size distribution presented a power–law distribution, but compared to the import one, the top city (Shenzhen) was less dominant, and the gap between top 1 (Shenzhen) and 2 (Taipei) was smaller. Shenzhen and Taipei were regarded as the primary export centers in China, followed by Hong Kong, Changzhou, and Hsinchu, with more than 400 exported patents. Most cities were marked inferior as their capacity for technology transfer-out was poor.

4.2 Evolution of Transfer Network

4.2.1 *The whole network*

Transnational technology transfer network in China has witnessed rapid expansion since 2001 with increasing sizes and connectivities in the network. During this research period, more than 500 cities are involved in the transnational technology transfer process which forms over 1,800 relationship pairs. As a result, the network size

Table 4.1 Complexity characters of China's transnational TTN.

Period	Type	Node	Avg. degree	Weight degree	Edge	Network density	Average path length
2001–2006	Overall	98	1.592	5.745	156	0.016	3.432
	Import	75	1.107	3.533	83	0.015	1
	Export	70	1.143	4.257	80	0.017	1
2007–2012	Overall	273	2.469	13.879	674	0.009	3.603
	Import	245	1.963	7.22	481	0.008	1
	Export	143	1.448	14.126	207	0.01	1
2013–2018	Overall	419	3.064	25.644	1284	0.007	3.53
	Import	376	2.471	16.41	929	0.007	1
	Export	219	1.731	20.89	379	0.008	1

has multiplied from 2001 to 2018 with about 4–8 times increment in the number of nodes and edges while the network density has decreased from 0.016 to 0.007. Meanwhile, the network connectivity has also incremental growth, noticeably observed in the average degree and weighted degree which have increased from 1.592 to 3.064 and from 5.745 to 25.644, respectively (Table 4.1).

The network also follows scale-free property that emerges with a power law of degree and cumulative degree distributions. As shown in Figure 4.4, both degree and cumulative degree distributions decay with power-law tails, similar to a large number of technological networks. Owing to the "preferential attachment" mechanism, new nodes transfer technology to those existing nodes with larger numbers of technology flows. Lastly, the well-connected cities present in the network play a dominant role in technology transfer activities and act as technology transfer hubs. However, most of cities are locked in the periphery of the network which has limited participation in these transnational activities. Moreover, the value of the power–law exponent has shown a downward trend since 2001, which means the gap between the nodes weakened under the spatial spread of technology.

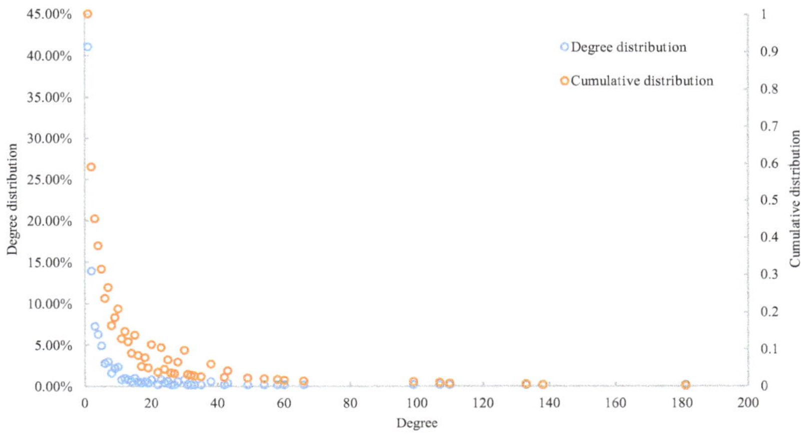

Figure 4.4 Degree and cumulative distributions of transnational TTN.

4.2.2 *The nodes*

Although there is a great increase in the number of cities and their transferred patents involved in the network, the activity is dominated by a few cities acting as a critical pivot. These hub nodes are place-dependent and concentrated in coastal regions of North America, West Europe, and East Asia.

4.2.2.1 *Breadth of technology transfer-in*

The number of both foreign and domestic cities involved in transnational technology transfer has constantly increased from 2001 to 2018. Accordingly, the breadth of their external links, defined by degree centrality, has rapidly expanded. Several cities located on the southeastern coast of the country have emerged as key hubs for importing international technology after establishing connections with numerous foreign cities to maintain their central position in the network among domestic cities (Figure 4.5). Meanwhile, the original cities exporting to China are mainly located in world-class innovative centers, financial centers, and tax-free regions of West Europe, North America, as well as East Asia (Figure 4.6).

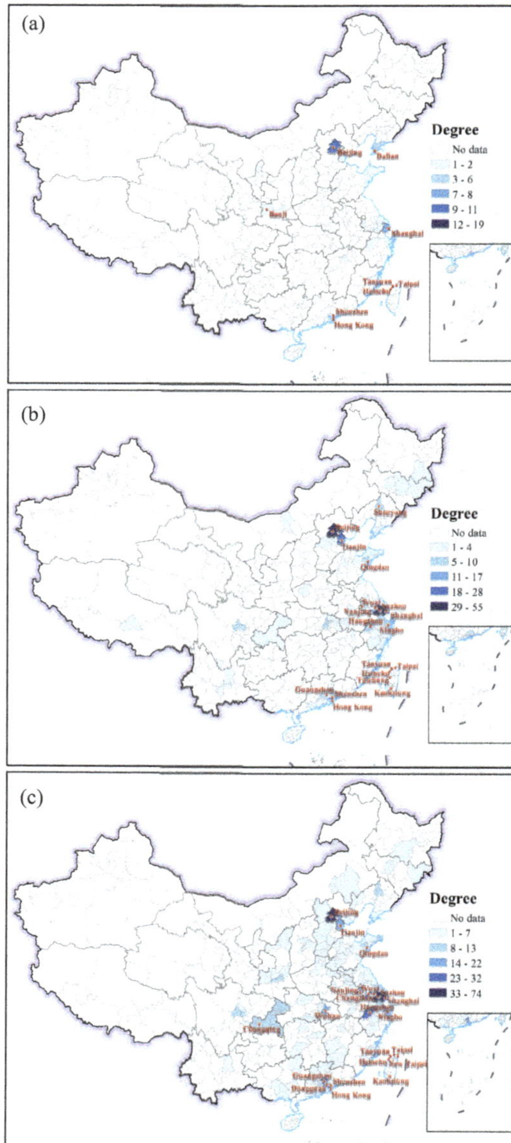

Figure 4.5 Breadth of domestic terminal cities involved in transnational technology transfer-in. (a) 2001–2006; (b) 2007–2012 and (c) 2013–2018.

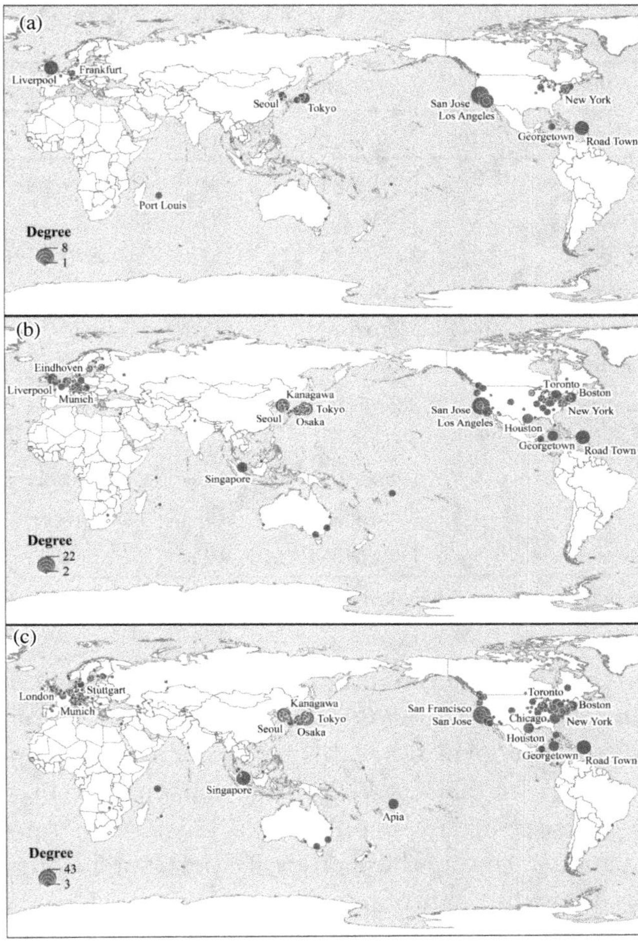

Figure 4.6 Breadth of foreign original cities involved in transnational technology transfer-in. (a) 2001–2006; (b) 2007–2012; (c) 2013–2018.

As illustrated in Figure 4.5 and Table 4.2, domestic hubs of technology transfer-in are mostly concentrated in the southeastern regions, acting as a "sink" of technology flows. Since 2001, Chinese cities have imported an increasing number of foreign patents, but most of them are clustered in the Northeastern Taiwan province

Table 4.2 Degree centrality of domestic terminal cities involved in transnational technology transfer-in.

	2001–2006		2007–2012		2013–2018	
Rank	City	Degree	City	Degree	City	Degree
1	Taipei	19	Shanghai	55	Hong Kong	74
2	Hong Kong	11	Hong Kong	46	Beijing	72
3	Beijing	10	Beijing	43	Shanghai	71
4	Shanghai	8	Suzhou	41	Shenzhen	60
5	Shenzhen	6	Shenzhen	28	Suzhou	57
6	Hsinchu	5	Taipei	22	Hangzhou	32
7	Dalian	4	Tianjin	22	Taipei	29
8	Taoyuan	2	Hsinchu	17	Guangzhou	27
9	—	—	Wuxi	15	Changzhou	24
10	—	—	Guangzhou	10	Tianjin	22
11	—	—	Hangzhou	10	—	—

Note: The degree centrality of cities ranked after 9 during 2001–2006 all presented 1.

and the three urban agglomerations including the BTH, YRD, and GBA. Although an increasing number of western cities have participated in the activity over time, this phenomenon of over-disparity is still evident. From 2001 to 2006, the transfer between domestic cities and overseas occurred only in 26 cities with transnational imported patents. Among these cities, Taipei, Hong Kong, Beijing, Shanghai, Guangzhou, and Shenzhen served as key gateways connecting to a greater number of foreign destinations, while the majority of other cities were only connected to one foreign city. Despite an increasing diffusion of participants, their distribution was still place-dependent during the second stage. Cities with higher transfer size are not well-distributed and locked in the three agglomerations, in which Shanghai, Hong Kong, and Beijing became primary hubs with the most extensive absorption and other cities such as Suzhou and Tianjin rapidly emerged as secondary hubs with a greater breadth of technology import. In the last stage, the number of cities involved in the process increased significantly and their breadth of absorption also expanded many paths

Table 4.3 Degree centrality of foreign original cities involved in transnational technology transfer-in.

	2001–2006		2007–2012		2013–2018	
Rank	City	Degree	City	Degree	City	Degree
1	San Jose	8	San Jose	22	San Jose	43
2	Los Angeles	5	Road Town	16	Seoul	32
3	Road Town	5	Tokyo	14	Singapore	25
4	Liverpool	4	Seoul	13	Road Town	24
5	Tokyo	3	Singapore	11	Tokyo	23
6	New York	3	Munich	10	San Francisco	18
7	—	—	Osaka	10	Georgetown	18
8	—	—	Los Angeles	9	Toronto	17
9	—	—	Boston	9	Osaka	17
10	—	—	New York	9	Boston	15
11	—	—	—	—	New York	15
12	—	—	—	—	Munich	15
13	—	—	—	—	Chicago	15
14	—	—	—	—	Gyeonggi-do	15

Note: The degree centrality of cities ranked 7–18 during 2001–2006 all presented 2.

of technology transfer-in. However, the core cities in the three urban agglomerations such as Hong Kong, Beijing, Shanghai, Shenzhen, and Suzhou still rank at the top in terms of technology transfer-in breadth due to the "preferential attachment" or "Matthew effect".

When it comes to foreign original cities, they have expanded from developed countries or regions in West Europe and North America to emerging economies in East Asia and Southeast Asia (Figure 4.6 and Table 4.3). These technology-exporting cities that are heavily dependent on China are classified into three subcategories: the innovative leaders (centralizing in developed countries of West Europe and North America, such as San Jose, New York, and

Munich), the offshore financial centers (clustered in the Caribbean, such as Road Town in the British Virgin Islands), and the catching-up innovative cities (emerging in East Asia and Southeast Asia, such as Tokyo and Seoul), in which there are a large number of transnational corporations, high-tech leading firms, banking institutions, technology intermediaries, and high-level universities.

In detail, the foreign original cities were highly concentrated in North American regions from 2001 to 2006, such as San Jose, Los Angeles, New York, and Road Town, which generated patent transfer with a large number of Chinese cities. From 2007 to 2012, the original cities expanded into Europe and East Asia, such as Munich, Eindhoven, Seoul, and Tokyo. During this period, cities in North America and East Asia connected closely with China and a few European cities such as Munich also had a strong connection with China. From 2013 to 2018, the number of cities importing international technology increased rapidly, and more Chinese cities established dense connections with San Jose followed by Seoul and Singapore. During this period, Chinese cities tended to import technology from the cities in North America and East Asia, while relatively less from the European cities.

4.2.2.2 *Intensity of technology transfer-in*

The intensity of technology transfer-in, measured by weight degree centrality, had greatly improved since 2001 and a few global cities became the primary origination and destination in the technology transfer-in process.

On the one hand, Chinese cities dominating the process converted from non-mainland cities such as Hsinchu, Taipei, and Hong Kong to the southeastern coastal ones including Shanghai, Beijing, and Shenzhen. The intensity of transnational technology transfer-in has rapidly increased, owing to cities in the eastern coastal area that have participated more actively in transferred-in technology (Table 4.4 and Figure 4.7). From 2001 to 2006, the number of domestic cities involved in the technology transfer-in process was

Table 4.4 Weight degree of domestic terminal cities involved in transnational technology transfer-in.

	2001–2006		2007–2012		2013–2018	
Rank	City	Weight degree	City	Weight degree	City	Weight degree
1	Hsinchu	64	Shanghai	272	Hong Kong	2149
2	Hong Kong	60	Hong Kong	202	Beijing	535
3	Taipei	56	Beijing	184	Shanghai	534
4	Beijing	46	Shenzhen	159	Shenzhen	423
5	Shenzhen	43	Hsinchu	151	Suzhou	259
6	Shanghai	28	Taipei	122	Hsinchu	200
7	Dalian	25	Qingdao	109	Dongguan	128
8	Taoyuan	19	Suzhou	108	Chongqing	128
9	Nanjing	18	Tianjin	43	Hangzhou	122
10	Tainan	18	Wuxi	39	Tianjin	119
11	Baoji	18	—	—	—	—

little and the size of technology transfer-in was also small. In addition, the process was dominated by non-mainland cities such as Hsinchu, Hong Kong, and Taipei, followed by mainland cities like Beijing, Shenzhen, and Shanghai. This stage could be regarded as the initial period when China started to imitate and absorb overseas advanced technology. From 2007 to 2012, eight cities with over 100 transferred-in technology, acting as the convergence centers, were highly concentrated in the eastern coastal area. Among them, Shanghai had replaced Hsinchu as the hub to absorb most of the international technologies, followed by Hong Kong, Beijing, Shenzhen, and Hsinchu. While cities with less than three times transfer-in were mostly located in the Central, Western, and Northeastern China, noticeably in the northwestern regions of Hu's line (an important divide of China's population distribution, firstly founded by Hu Huanyong). From 2013 to 2018, the number of cities with more than 100 times transfer-in increased to 11. Among

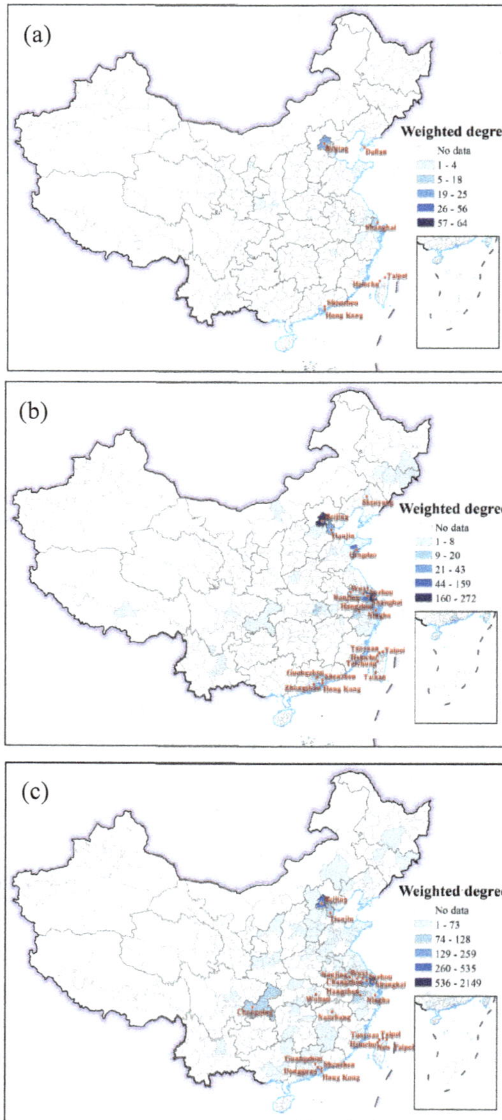

Figure 4.7 Linkage intensity of domestic terminal cities involved in transnational technology transfer-in. (a) 2001–2006; (b) 2007–2012 and (c) 2013–2018.

them, Hong Kong had played a leading role like a "bridge" in the external technology connections, with over 2,000 importing patents accounting for one-third of the total. Key cities in the three urban agglomerations such as Suzhou and Dongguan also had evident promotion. Except for clustering in the coastal area, some secondary centers of technology transfer-in were emerging in the western. In Chongqing, the transfer-in exceeded 100 times (Figure 4.7). Owing to industrial transfers toward the inland, they also presented a stronger absorptive capacity of technology, leading to a slightly western shift of importing technologies.

On the other hand, China's imported technologies are concentrated in the US, West Europe, and East Asia. The foreign original cities in West Europe, North America, and East Asia have close technological communication with the Chinese cities.

The US remains a technology powerhouse in valuable intellectual property and continues to be the global leader in science and technology (Tollefson, 2018). The innovative cities in the USA (i.e., Silicon Valley, Los Angeles, New York, and Boston) loaded the largest intensity of technology exporting to China in the first decade of the 21st century (Table 4.5 and Figure 4.8). However, there is also a recent shift toward the innovative hubs from the North Atlantic Coast (i.e., New York, Boston, Road Town, Liverpool, and Paris) to the North Pacific Coast such as Tokyo, Singapore, Seoul, and San Jose.

Specifically, New York and Paris became the core cities that exported the largest quantity of technology to China during 2001–2006, while most of the original cities kept a weak technologies export to China with one patent. From 2007 to 2012, the number of foreign cities involved in the activities increased and a large number of them were located in the West Europe and North America. Similarly, the originations of China's external technologies were highly locked-in on the coast of the North Atlantic. Except for Osaka, Tokyo, San Jose, and Singapore, most of the top 10 cities with over 40 exporting patents were located on the coast, such as Eindhoven, Espoo, and Derby in West Europe, Road Town in the

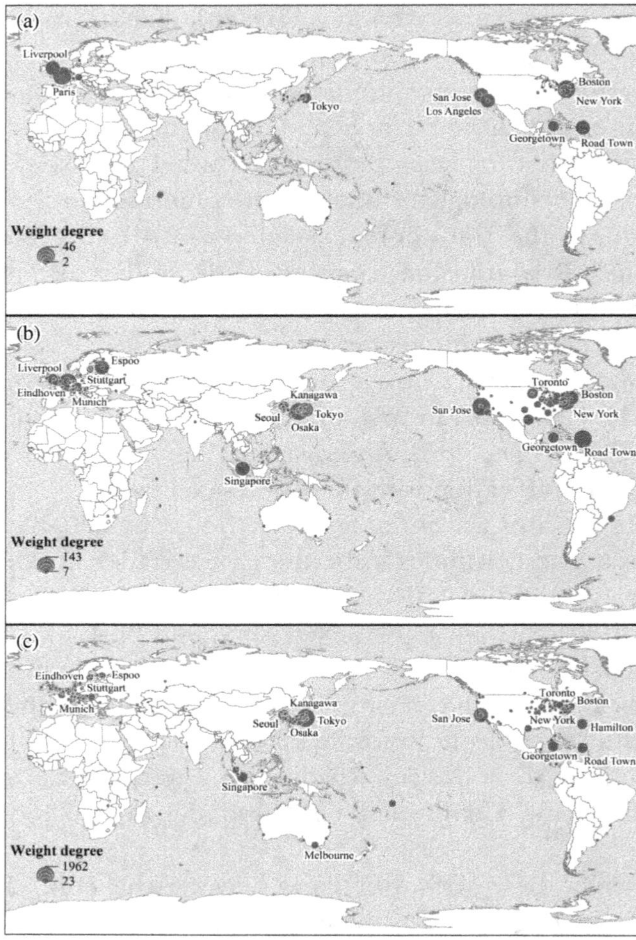

Figure 4.8 The linkage intensity of foreign original cities during transnational technology transfer-in. (a) 2001–2006; (b) 2007–2012; (c) 2013–2018.

Caribbean, as well as New York and Boston in Northeast USA (Table 4.5 and Figure 4.8). Compared to the first two stages, the focus of the third stage continued to move eastward due to the emergence of innovative centers in East Asia and Southeast Asia. In recent times, China has diversified its sources of technology but still

Table 4.5 Weight degree of foreign original cities involved in transnational technology transfer-in.

	2001–2006		2007–2012		2013–2018	
Rank	City	Weight degree	City	Weight degree	City	Weight degree
1	New York	46	Eindhoven	143	Tokyo	1962
2	Paris	43	San Jose	129	San Jose	481
3	Liverpool	28	Road Town	120	Osaka	212
4	Road Town	19	Osaka	120	Road Town	176
5	San Jose	18	New York	107	Singapore	172
6	Los Angeles	18	Singapore	69	New York	172
7	Tokyo	11	Espoo	58	Seoul	155
8	Georgetown	9	Boston	55	Hamilton	119
9	Trenton	8	Tokyo	53	Georgetown	108
10	Boston	6	Derby	44	Munich	90

relies heavily on Japan, South Korea, and Singapore for external technological advancements. Additionally, China's technological dependence is more skewed toward the North Pacific coast as opposed to the North Atlantic coast.

Generally, Chinese external technologies have primarily originated from cities located within the three economies, due to the colocation of advanced S&T infrastructures (e.g., The Silicon Valley, Tokyo, Osaka, Seoul), the clusters of transnational corporations and their branches (e.g., The Silicon Valley, Tokyo, Singapore), and the convenience of offshore finance and low-tax services (e.g., Singapore, Tokyo, New York, Road Town).

4.2.2.3 *Breadth of technology transfer-out*

Since China's entry into the WTO, its growing innovative capabilities have led to an increase in the number of cities involved in technology transfer-out, as well as a broader range of technology destinations. According to nodal degree centrality, the scope of technology flows from Chinese cities to foreign cities has incrementally expanded, while their locations are not well-distributed. One is the origination of transferred-out technology concentrated in Taipei, Hong Kong, Shanghai, and Beijing emerging into international technology hubs for a long time. The other is the destination of technology flows into global cities in North America, West Europe, and East Asia, and further classified into three types: (1) global innovative highlands in the US (i.e., San Jose, Los Angeles, Wilmington, and Chicago) (2) international tax havens and offshore financial centers in some island countries (i.e., Road Town, Georgetown, Apia, Port Louis, and Victoria); (3) global manufacturing bases in East Asia (i.e., Tokyo and Seoul) and West Europe (i.e., Frankfurt, Munich, and Amsterdam) (Figure 4.9).

In terms of domestic cities, China heavily relies on a handful of innovative centers like Hong Kong, Taipei, Beijing, and Shanghai for its technology exports (Figure 4.9). Although there has been an increase in the number of exporting cities since 2001, most of the original cities located in the western region still have rather low export sizes. As such, only a select few cities located on the southeastern coast was able to generate the majority of the technologies being exported. From 2001 to 2006, China's mainland had poor innovative capacity and relied heavily on non-mainland cities such as Taipei, Hsinchu, and Hong Kong to catch up with foreign innovative leaders. From 2007 to 2012, the breadth of export technology expanded to the southeast mainland; Beijing and Shanghai grew into the divergence centers of exporting technology, which was driven by the IDDS since 2006. From 2013 to 2018, with the increasing innovative capacity, Chinese cities had a broader technology transfer toward foreign cities. Due to the greater path dependence, the hubs with stronger transfer-out intensity were centralized in the three urban agglomerations such as the BTH, YRD, and GBA. Among them, Hong Kong, Shanghai, Shenzhen, Beijing, and Suzhou whose outdegree centrality were far more than other cities constantly ranked at the top in the breadth of technology transfer-out (Table 4.6).

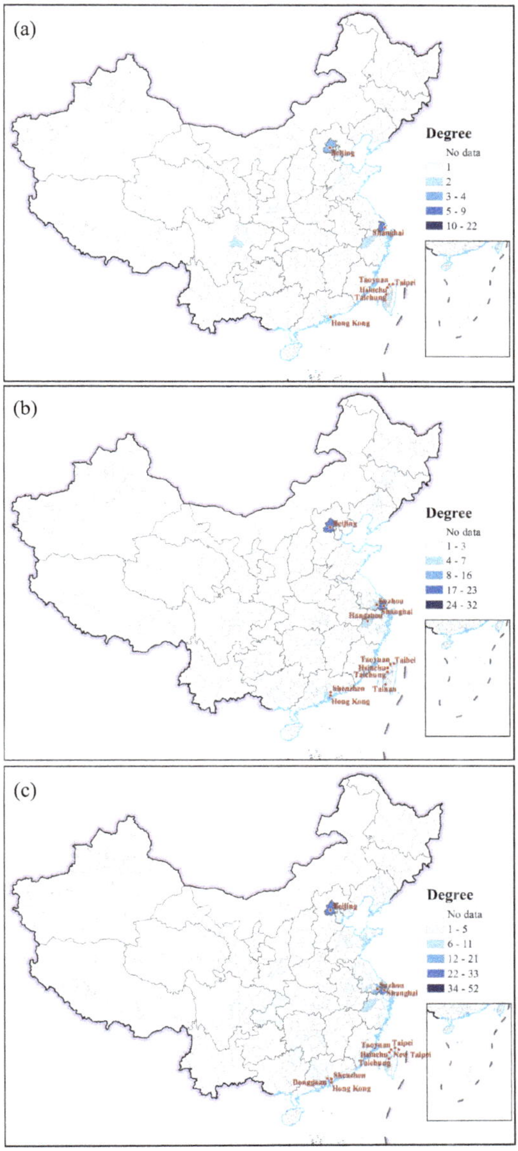

Figure 4.9 Linkage breadth of domestic original cities involved in transnational technology transfer-out. (a) 2001–2006; (b) 2007–2012 and (c) 2013–2018.

Table 4.6 Degree centrality of domestic original cities involved in transnational technology transfer-out.

	2001–2006		2007–2012		2013–2018	
Rank	City	Degree	City	Degree	City	Degree
1	Taipei	22	Taipei	32	Hong Kong	52
2	Hsinchu	9	Hong Kong	31	Shanghai	33
3	Hong Kong	8	Beijing	23	Taipei	33
4	Shanghai	7	Shanghai	19	Shenzhen	30
5	Taoyuan	4	Hsinchu	16	Beijing	24
6	Taichung	4	Taichung	14	Suzhou	21
7	Beijing	3	Hangzhou	7	Taoyuan	15
8	—	—	Taoyuan	6	Taichung	14
9	—	—	Suzhou	6	Hsinchu	14
10	—	—	—	—	Dongguan	11

Note: The degree centrality of cities ranked after 8 was less than 3.

Hong Kong and Taiwan have served as the "gatekeeper" in the transnational technology transfer-out process. However, there is an obvious difference in driving international technology transfer-out between them. Having benefited from the international financial center, Hong Kong plays a crucial role in attracting transnational enterprises from China's mainland and acts as a bridge to export their technologies to other countries. However, Taiwan province is far more innovative than Hong Kong and has become one of the most important innovation highlands owing to a large amount of high-tech enterprises.

Focusing on the terminal cities, it is found that China's technology predominantly flows into the manufacturing bases and financial centers in North America and East Asia (Figure 4.10). During the first stage, most of China's exporting technology flew into North America and clustered in the three innovative hubs including Road Town in the British Virgin Islands, Wilmington, and San Jose in the US, which formed a polycentric structure. Although there was a

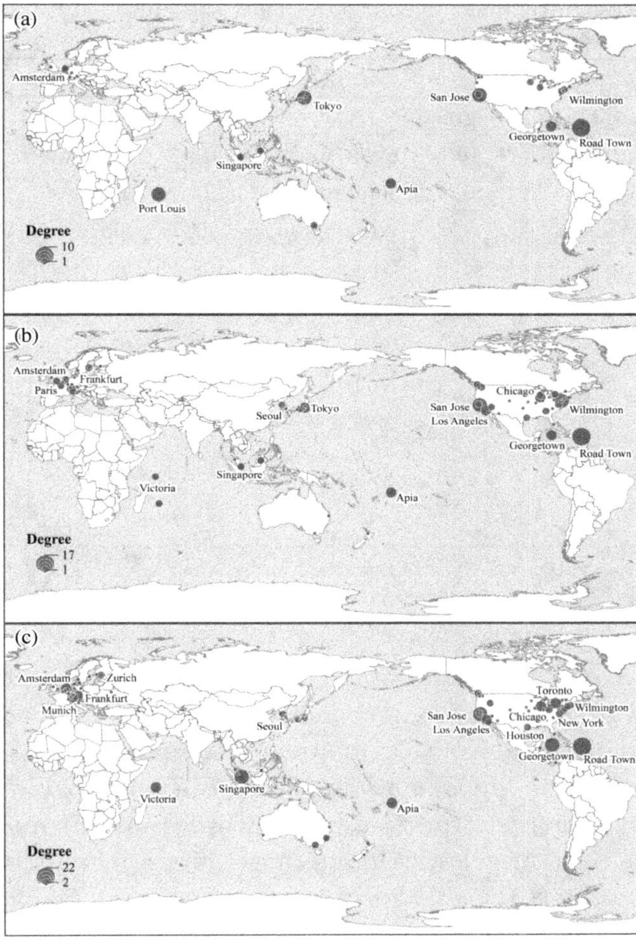

Figure 4.10 Linkage breadth of foreign terminal cities involved in transnational technology transfer-out. (a) 2001–2006; (b) 2007–2012; (c) 2013–2018.

broader connection between Chinese cities and foreign cities from 2007 to 2012, the terminal cities of China's technologies still located in the three hubs resulted from the place-dependent effect. From 2013 to 2018, the role of East Asian cities in Chinese technology transfer-out had been promoted dramatically, leading to new

Table 4.7 Degree centrality of foreign terminal cities involved in transnational technology transfer-out.

	2001–2006		2007–2012		2013–2018	
Rank	City	Degree	City	Degree	City	Degree
1	Road Town	10	Road Town	17	Road Town	22
2	San Jose	5	San Jose	11	Tokyo	16
3	Port Louis	4	Wilmington	9	Singapore	16
4	Tokyo	4	Georgetown	8	San Jose	15
5	Apia	3	Chicago	7	Georgetown	11
6	Georgetown	3	Apia	7	Apia	8
7	—	—	Tokyo	7	Victoria	7
8	—	—	Los Angeles	6	Los Angeles	7
9	—	—	—	—	Wilmington	6
10	—	—	—	—	Chicago	6
11	—	—	—	—	Frankfurt	6

Note: The degree centrality of cities ranked after 7 was less than 3.

emerging connections through path creations. With the rise of Tokyo and Singapore as the terminal cities, China's technology marketplaces had shifted from North America to East Asia and Southeast Asia, noticeably in global cities including offshore tax havens (i.e., Road Town, Georgetown, Apia, and Victoria) and manufacturing bases (i.e., Los Angeles, Wilmington, Chicago, Frankfurt), and innovative centers (i.e., Tokyo, Singapore, San Jose) (Table 4.7 and Figure 4.10).

4.2.2.4 *Intensity of technology transfer-out*

The intensity of technology export had enhanced evidently with the increasing innovative capacity of Chinese cities. According to their weight degree centrality, there are two significant shifts in spatial. One is the shift of domestic technology sources from Taiwan to the GBA (Table 4.8). Another is the shift of foreign technology sinks from North America to Southeast Asia and West Europe (Table 4.9).

Table 4.8 Weight degree of domestic original cities involved in transnational technology transfer-out.

	2001–2006		2007–2012		2013–2018	
Rank	City	Weight degree	City	Weight degree	City	Weight degree
1	Taipei	139	Taipei	920	Shenzhen	1556
2	Hsinchu	48	Hsinchu	273	Hong Kong	586
3	Shanghai	25	Hong Kong	219	Changzhou	554
4	Hong Kong	17	Taichung	150	Taipei	373
5	Taichung	11	Beijing	112	Shanghai	247
6	Kaohsiung	10	Shanghai	88	Hsinchu	199
7	Taoyuan	8	Taoyuan	58	Beijing	146
8	Beijing	7	Hangzhou	37	Suzhou	123
9	Shenzhen	5	Suzhou	26	Taoyuan	113
10	Xiamen	5	Shenzhen	23	Nantong	102

Table 4.9 Weight degree of foreign terminal cities involved in transnational technology transfer-out.

	2001–2006		2007–2012		2013–2018	
Rank	City	Weight degree	City	Weight degree	City	Weight degree
1	Road Town	87	Wilmington	965	San Jose	929
2	Wilmington	51	Road Town	285	Singapore	624
3	San Jose	35	Georgetown	126	Tokyo	570
4	Georgetown	18	San Jose	78	London	380
5	Seri Begawan	16	Apia	77	Georgetown	375
6	Port Louis	8	St. Peter's Port	63	Road Town	206
7	Apia	7	Berne	29	Zug	154
8	Amsterdam	6	Singapore	27	Apia	105
9	Tokyo	5	Chicago	21	Leverkusen	76
10	Chicago	5	Hamilton	20	Sheffield	71
11	Frankfurt	5	—	—	—	—

Since 2001, China's technology transfer-out activity was dominated by original cities in Taiwan province such as Taipei and Hsinchu in the early stage, and later was overtaken by Shenzhen and Hong Kong (Figure 4.11). As shown in Table 4.8, the primary origination had changed from Taipei to Hong Kong. Specifically, China's exporting technology was centralized in Taiwan province from 2001 to 2006, particularly in Taipei and Hsinchu. Benefiting from their developed ICT industries, their quantity of technology export was far more than other cities, accounting for 60% of the total. Although China's technology export volume grew from 2007 to 2012, Taipei and Hsinchu still acted as the hubs of technology diffusion to keep the path dependence of technology generations. Due to the "rich club" or "Matthew effect", the core position occupied by them had been further strengthened, and most of the mainland cities had also been locked in the peripheral network. From 2013 to 2018, new paths of technology outflows were created continuously because of the high-speed increment of innovative capacity in China's mainland. Shenzhen had replaced Taipei to be the primary highland of technology generations. Its innovative capacity was significantly improved with the advantage of innovative milieu, talents pool, R&D investment, and high-tech leading firms. In addition, manufacturing bases in the YRD region such as Changzhou, Suzhou, and Nantong also promoted their technical capacity thus becoming the rising subcenters of technology transfer-out through spatial spillovers of high-tech industries and technologies from Shanghai.

Meanwhile, the concentration of foreign terminal cities had transformed from a monocentric structure led by the North American cities to a polycentric structure with a situation of tripartite confrontation led by North America, Southeast Asia, and West Europe (Figure 4.12). From 2001 to 2006, two cities in North America, Road Town in the Caribbean and Wilmington in the Delaware State of the US, accounted for nearly 50% of the technology import from China, while cities in other regions mostly absorbed less than 10 items. Notable as the world capital of chemical and credit card industries, Wilmington had imported nearly 1,000 Chinese patents from 2007 to 2012 through the frequent links between its headquarters and

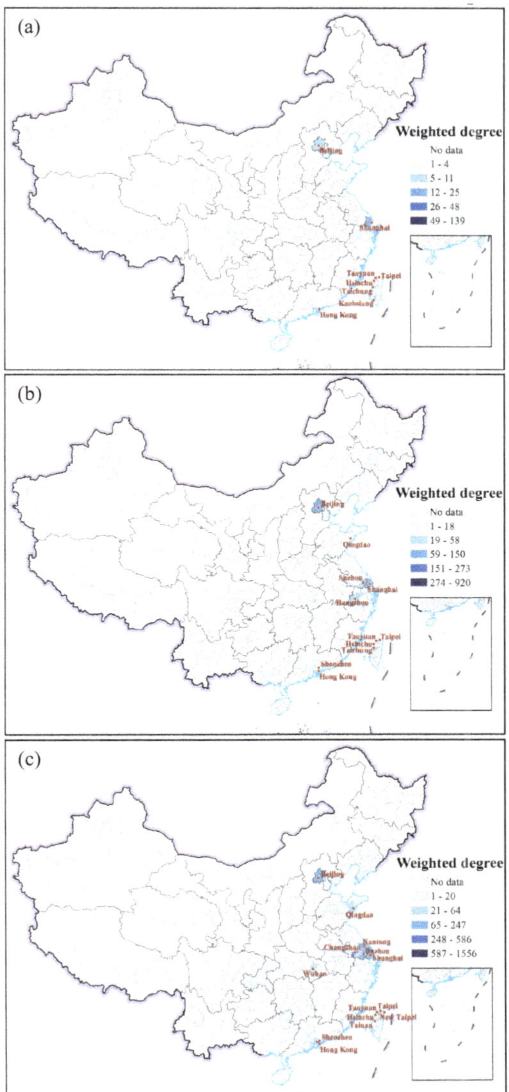

Figure 4.11 Linkage intensity of domestic original cities involved in transnational technology transfer-out. (a) 2001–2006; (b) 2007–2012 and (c) 2013–2018.

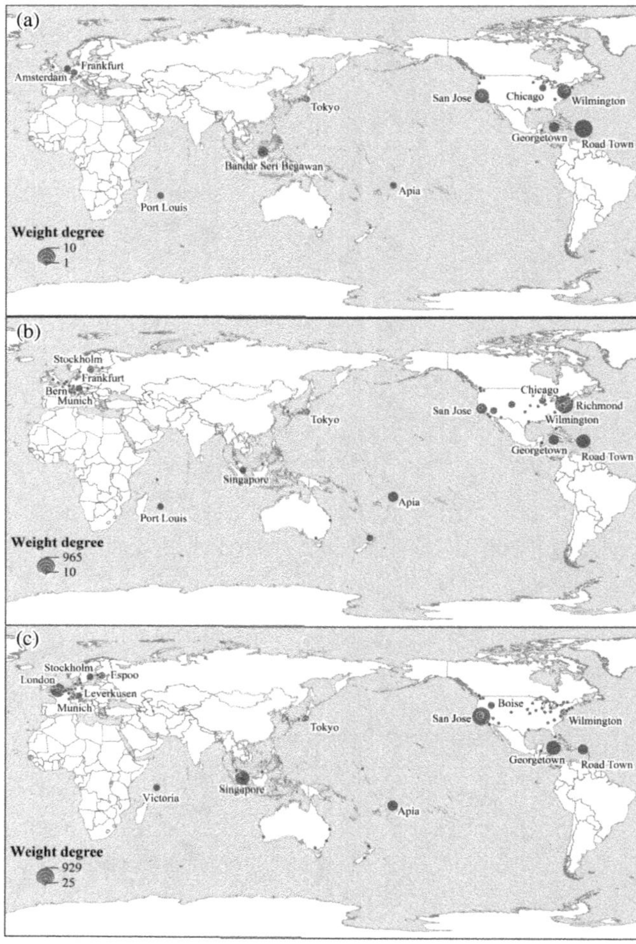

Figure 4.12 Linkage intensity of foreign terminal cities involved in transnational technology transfer-out. (a) 2001–2006; (b) 2007–2012; (c) 2013–2018.

their branches of transnational corporations, accounting for nearly 50% of the total. It became the top destination of China's technology flows, while the second city, Road Town, had fewer than 300. During 2013–2018, Wilmington's status dropped sharply and was replaced by San Jose (Table 4.9). Owing to the rapid growth of high-tech industries, the status of the Western Pacific cities such as Singapore and Tokyo as well as London, Zug, Leverkusen, and Sheffield in Western Europe was promoted swiftly to absorb large numbers of Chinese patents. In this way, the market sites of Chinese technology eventually shifted eastward and formed the triple centric distribution: North America, Western Pacific, and Western Europe, like the three legs of a tripod Figure 4.12.

4.2.3 *The pipelines*

4.2.3.1 *Importing pipelines*

The number of imported patents is not only increasing year by year but the connectivity of transfer-in network is also constantly enhancing. In spatial, the transnational pipelines of China's technology transfer-in have undergone diversified changes in their strength and direction from the Sino-American dominance to the trilateral pattern including links between China and the US, between China and the West Europe, and between China and the West Asia-Pacific. Accordingly, the location of its originations has shifted from monocentricity of the US to polycentricity including Japan and South Korea in East Asia, the US and Road Town in North America, German and Finland in West Europe, and Singapore in Southeast Asia.

As shown in Figure 4.13 and Table 4.10, during 2001–2006, China was greatly dependent on a few innovative centers in North America and West Europe, particularly in the US and the UK. The main importing pipelines were highly concentrated in links between China and the US. The intensity of these pipelines was different with hierarchies. Among them, the Paris–Hong Kong city pair emerged as the most crucial pipeline for China's technology acquisition, operating through the headquarters–branch network of transnational corporations. The New York–Hsinchu pipeline followed closely behind but ranked lower in significance. These two pipelines supported 30% of the total at this stage, while the remaining shared less than 5 times.

From 2007 to 2012, the pipelines of China's technology import were continuously emerging and mainly oriented to North America.

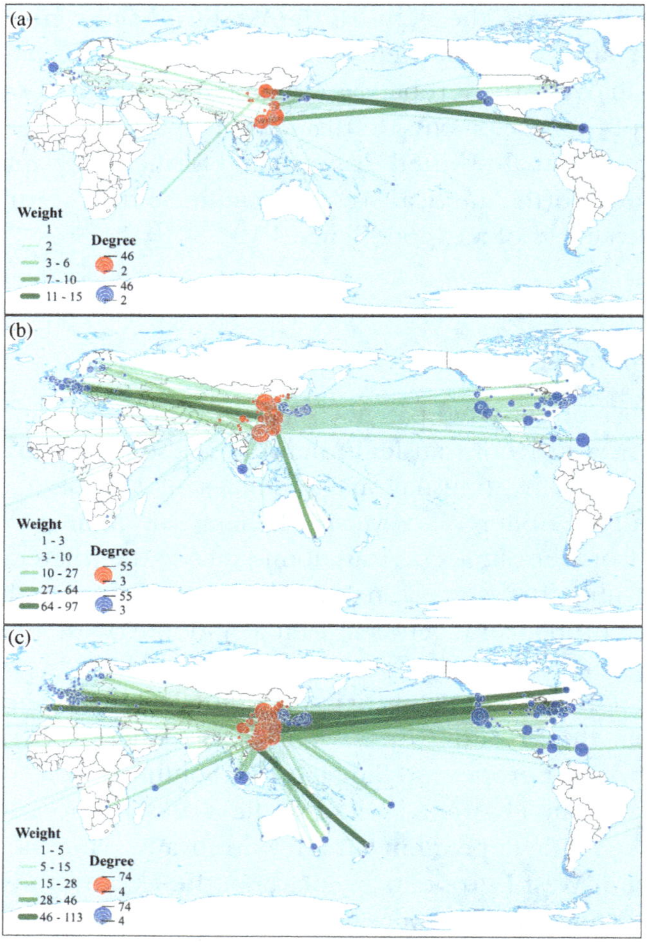

Figure 4.13 Spatial dynamics of transnational technology import pipelines. (a) 2001–2006; (b) 2007–2012; (c) 2013–2018.

However, the leading pipelines had shifted to East Asia and West Europe to avoid larger risks from greater dependence on the minority. Among them, the Osaka–Qingdao pipeline had become the primary linkage with 97 patents, followed by the Eindhoven–Hong Kong pipeline. Although they occupied the top two spots, China's technology still relied highly on the innovative centers in the US such as New York, Boston, and San Jose.

Table 4.10 Main transnational technology import pipelines.

Rank	2001–2006			2007–2012			2013–2018		
	Original cities	Terminal cities	Number	Original cities	Terminal cities	Number	Original cities	Terminal cities	Number
1	Paris	Hong Kong	43	Osaka	Qingdao	97	Tokyo	Hong Kong	1782
2	New York	Hsinchu	34	Eindhoven	Hong Kong	80	Hamilton	Chongqing	113
3	Liverpool	Hsinchu	22	New York	Shanghai	64	New York	Hong Kong	81
4	Road Town	Beijing	15	Espoo	Beijing	55	San Jose	Shanghai	69
5	Los Angeles	Shenzhen	10	Derby	Shenzhen	44	Tokyo	Beijing	64
6	New York	Taipei	10	Sao Paulo	Shanghai	38	Osaka	Shenzhen	63
7	Georgetown	Taipei	8	Boston	Hsinchu	35	Singapore	Hsinchu	62
8	Trenton	Taipei	7	Road Town	Taipei	30	New York	Shanghai	61
9	San Jose	Hsinchu	6	Eindhoven	Hsinchu	27	Eindhoven	Beijing	57
10	Tokyo	Shanghai	6	San Jose	Shenzhen	27	Heidelberg	Tianjin	46
11	—	—	—	—	—	—	Rochester	Beijing	46
12	—	—	—	—	—	—	Espoo	Beijing	46

From 2013 to 2018, the number of China's external pipelines increased significantly, but China's technology was still dependent on a few countries and cities. Because of its developed high-tech industries and global headquarter–branch network, Japan had surpassed the leading position of the United States and had become the chief source of China's patents. As the primary direction, the pipeline between Tokyo and Hong Kong had loaded nearly 30% of the total patent flows at this stage. The high dependence and great concentration were not only depicted by the technology originations but also were evidently shown in the absorptive centers in China. During the period, a few innovative megacities such as Beijing, Shanghai, and Shenzhen in China's mainland were growing into global innovative centers, and replacing the gateway role of Taiwan and Hong Kong, due to their finer innovative milieu and developed high-tech industries.

Moreover, headquarters–branch linkages and technology alliances of transnational cooperation are regarded as the key drivers to enhance the diversification of China's technology import pipelines. For instance, the pipeline of Singapore–Changzhou is led by the internal transfer from the headquarters of *Resound Technology* to its R&D Center; and the Tokyo–Hong Kong pipeline is generated by the technology alliance between National Electrical Code (Nec) Corporation in Japan and Lenovo Innovation Hong Kong Ltd.

4.2.3.2 *Exporting pipelines*

Similar to the import pipelines, the breadth and intensity of China's exporting pipelines have continued to expand, resulting from the enhancement of independent innovation capabilities. Meanwhile, their similarities are also evidently shown in its spatial disparity and dynamics. In the past 18 years, the exporting pipelines have been highly concentrated toward a few urbanized areas in which it has undergone a shift from North America to North America, East Asia, Southeast Asia, and West Europe, forming an obvious polycentricity (Figure 4.14).

From 2001 to 2006, China's technology export was relatively less in size, dominated by cities in Taiwan. Nine of the top 10 pipelines

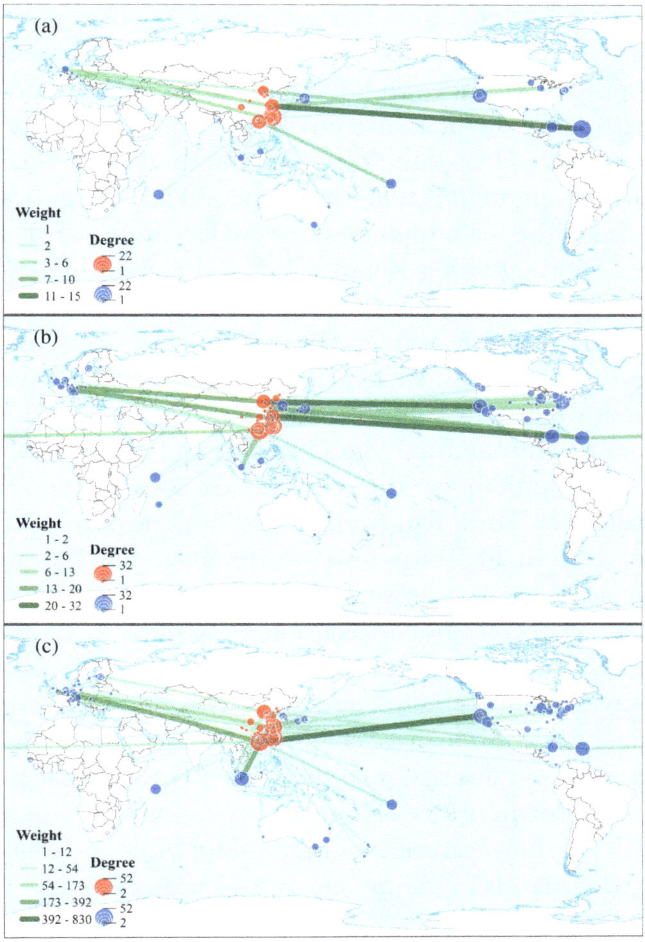

Figure 4.14 Spatial dynamics of transnational technology export pipelines. (a) 2001–2006; (b) 2007–2012; (c) 2013–2018.

originated from Taiwan Province, which is closely related to its innovative capacity. Wilmington and San Jose (the core area of Silicon Valley) in the United States and Road Town in the British Virgin Islands became the major technology pools of Taiwan.

From 2007 to 2012, the intensity and breadth of China's exporting pipelines increased significantly, especially in the East Asian cities. It is evident that China's export pipelines showed a tendency in shifting from North America to East Asia, but the pipelines were dominated by the connection between China's Taiwan and the US.

From 2013 to 2018, in accordance with China's industrial transfer, the center gravity of its technology destinations also accelerated moving eastward, owing to the economic rising of East Asia and Southeast Asia. The size of the main pipelines oriented to them was greater than 300 times, including Hong Kong–Tokyo and Changzhou–Singapore (Table 4.11). Furthermore, a few mainland cities had improved the connectivity significantly and acted as the gateway access to external technology marketplaces, such as Changzhou and Nantong in the YRD, along with Shenzhen in the PRD.

In general, with the increasing breadth and intensity of technology exporting pipelines, their terminals were expanding and shifting from New York, San Jose, and Wilmington innovative centers in the US, Road Town and Georgetown tax havens in the Caribbean, as well as Eindhoven and Frankfurt manufacturing centers in West Europe to global technological hubs in East Asia and Southeast Asia (i.e., Tokyo, Singapore, Seoul, etc.). China's innovation capacity had also been strengthened with several emerging S&T centers (including Shenzhen, Beijing, and Shanghai) and three coastal innovation clusters (including the GBA, the BTH, and the YRD region). Furthermore, Hong Kong's status as a global gatekeeper of technology transfer was increasing, while the role of major technology highlands in Taiwan continued to decline.

4.3 Evolution of Technology Categories

4.3.1 *The temporal evolution*

4.3.1.1 *Importing technology categories*

Based on the IPC classification, China's imported patents are classified into 35 technological categories across five sectors. By examining changes in these subcategories, we can gain insight into China's reliance on important foreign technology fields. Our analysis indicates that China's technology imports have been diversifying and expanding

Table 4.11 Main transnational technology export pipelines.

	2001–2006			2007–2012			2013–2018		
Rank	Original cities	Terminal cities	Number	Original cities	Terminal cities	Number	Original cities	Terminal cities	Number
1	Taipei	Road Town	51	Taipei	Wilmington	587	Shenzhen	San Jose	830
2	Taipei	Wilmington	39	Hsinchu	Wilmington	225	Hong Kong	Tokyo	411
3	Hsinchu	San Jose	16	Taipei	Road Town	147	Changzhou	Singapore	392
4	Taipei	San Jose	13	Taichung	Wilmington	123	Shenzhen	London	379
5	Hsinchu	Seri Begawan	12	Hong Kong	St. Peter's Port	63	Shenzhen	Singapore	173
6	Shanghai	Road Town	12	Hong Kong	Road Town	62	Taipei	Georgetown	157
7	Hsinchu	Wilmington	12	Taipei	Georgetown	58	Changzhou	Zug	153
8	Kaohsiung	Road Town	9	Taipei	Apia	53	Nantong	Tokyo	96
9	Taipei	Georgetown	9	Shanghai	Georgetown	32	Hsinchu	Georgetown	93
10	Taichung	Georgetown	7	Beijing	San Jose	28	Shanghai	Leverkusen	76

over time. The preference for different foreign technology subcategories has varied across different periods, while the import of electrical engineering sector is always in the leading position.

As shown in Figure 4.15, the subcategories of imported technologies are constantly enriched with their increasing number during the three periods. From 2001 to 2006, there were only 30 kinds of imported

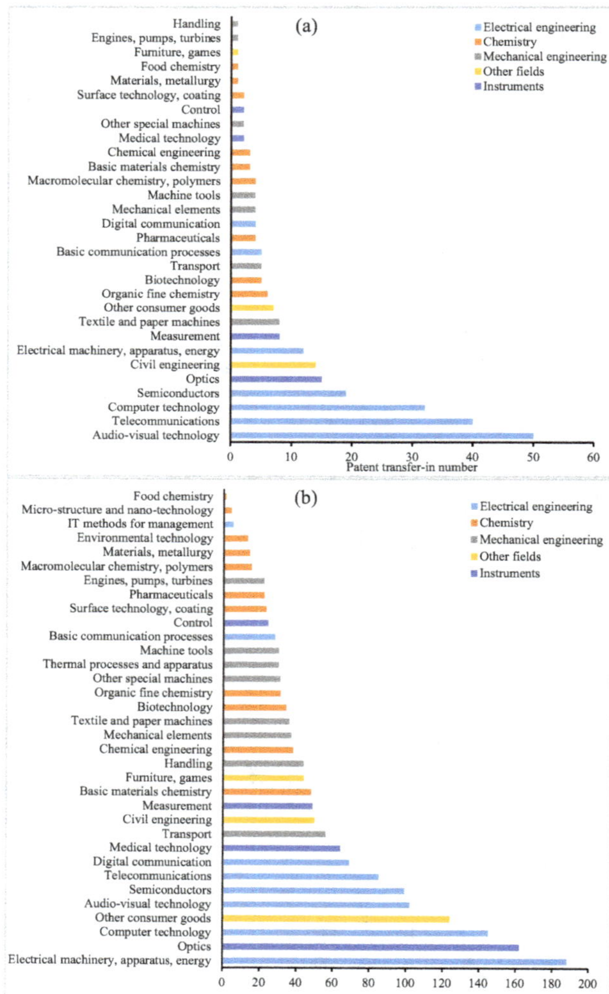

Figure 4.15 Subcategories of China's imported technologies. (a) 2001–2006; (b) 2007–2012 and (c) 2013–2018.

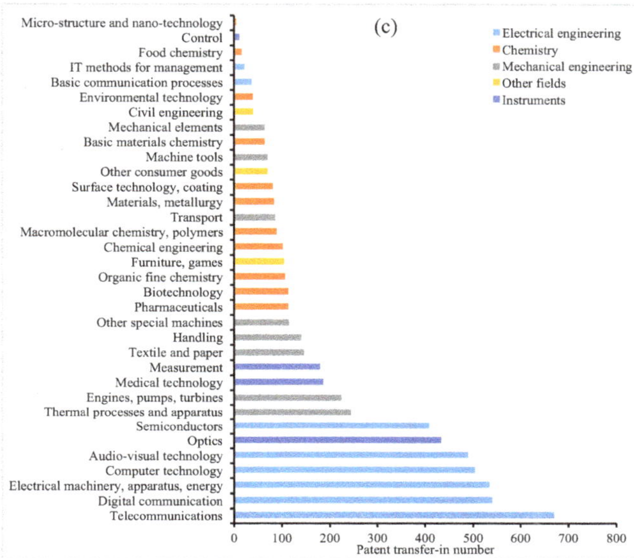

Figure 4.15 (*Continued*)

technologies involved in the transnational TTN, except for IT management, environmental technology, control systems, heat treatment and equipment, microstructure and nanotechnology. Moreover, the number of imported patents in each category was at a low level, in which most of the categories had less than 10 patents. From 2007 to 2012, the scope of the imported patents increased significantly to cover all 35 technical fields in five sectors, while the focus of imported technology was still locked in electrical engineering. Compared to the previous stage, the number of imported patents in 35 subcategories significantly increased from 2013 to 2018, and the difference in patent imports of the top subcategory was relatively reduced.

The patent transfer-in number of electrical engineering and mechanical engineering sectors grew greatly in 18 years, due to China's large-scale industrialization and rapid urbanization. With the increasing technology import, such emerging ICT categories as telecommunications, digital communications, computer technology, audio–visual technology and semiconductors, continue to remain in a leading position (Table 4.12). From 2001 to 2006, China imported the largest number of patents in the electrical engineering

Table 4.12 The number and percentage of imported patents in different sectors.

	2001–2006			2007–2012			2013–2018		
Rank	Sectors	Num.	Per. (%)	Sectors	Num.	Per. (%)	Sectors	Num.	Per. (%)
1	Electrical Engineering	162	61.13	Electrical Engineering	721	40.76	Electrical Engineering	3204	51.93
2	Chemistry	29	10.94	Instruments	301	17.02	Mechanical Engineering	1089	17.65
3	Instruments	27	10.19	Mechanical Engineering	286	16.17	Instruments	854	13.84
4	Mechanical Engineering	25	9.43	Chemistry	243	13.74	Chemistry	809	13.11
5	Other Fields	22	8.30	Other Fields	218	12.32	Other Fields	214	3.47

sector, while presenting a fairly smaller dependence on the rest of the sectors. From 2007 to 2012, the imported patents in the mechanical engineering sector began to improve with transportation as the "bellwether", but the electrical engineering subcategory still dominated the transnational technology import, accounting for more than 40%. Among them, the transfer-in number of the electrical machinery, apparatus, and energy subcategory was listed at the top, mainly resulting from the rapid growth of national electricity consumption. From 2013 to 2018, the electrical engineering sector still remained at the head of the list, implying that its leading role was prominently strengthened by the transformation of power and energy industries. At the same time, the import of mechanical engineering patents had come significantly out in front, such as thermal processes and apparatus, as well as engines, pumps, and turbines, which was driven greatly by the upgrading of equipment and machinery manufacturing industries in current China.

4.3.1.2 *Exporting technology categories*

Same as the variation above, the exporting technology categories are constantly enriched and their sizes have increased from 2001 to 2018 (Figure 4.16). Limited by relatively lower innovative capacity, there are only 28 technical subcategories involved in the transfer-out, and most of the exported patents had less than 10 from 2001 to 2006. From 2007 to 2012, such technology subcategories as basic communication processes and digital communication emerged significantly, while the export size of subcategory that ranked first was much higher than others. From 2013 to 2018, all 35 technology subcategories had evolved in the patent transfer-out, and the

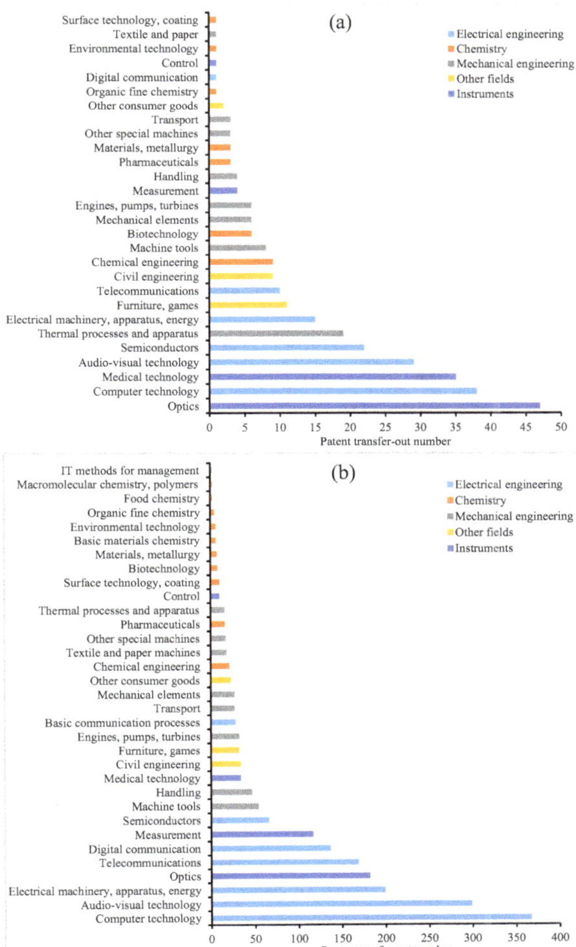

Figure 4.16 Categories of China's exporting technology. (a) 2001–2006; (b) 2007–2012 and (c) 2013–2018.

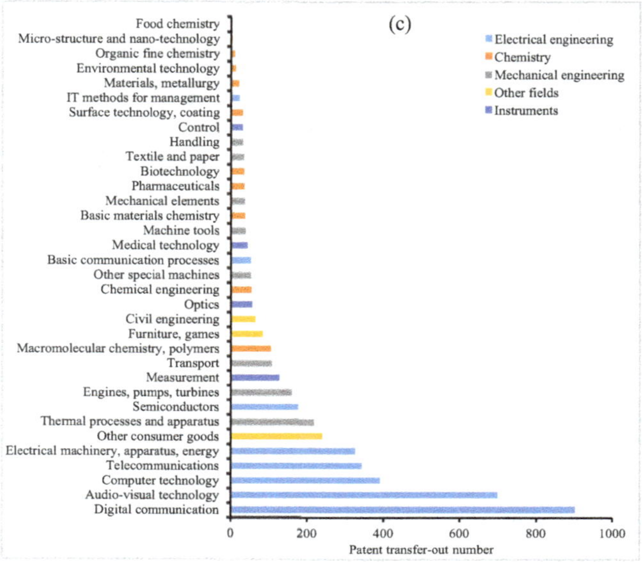

Figure 4.16 (*Continued*)

export size of each subcategory also significantly increased compared to the previous stage, while their gap widened further. As a result, the rank–size distribution of technology subcategories exhibited a primal law distribution and emerged as a scale-free property.

Similar to import, the export number of electrical engineering technology had experienced a significant increase, surpassing instruments as the new leader (Table 4.13). From 2001 to 2006, the main export activities were led by two sectors: electrical engineering and instruments. The size of the top five subcategories belonging to the two sectors was far greater than others, with over 80% of the total. From 2007 to 2012, although the export in other sectors improved increasingly, the electrical engineering sector was still dominant, owing to the great advance in ICT such as computer technology, audio–visual technology, telecommunications, and digital communication. From 2013 to 2018, the electrical engineering sector remained in the first place under the effect of "path dependence", while the other sectors had greatly improved.

Table 4.13 The number and percentage of exported patents in different sectors.

	2001–2006			2007–2012			2013–2018		
Rank	Sectors	Num.	Per. (%)	Sectors	Num.	Per. (%)	Sectors	Num.	Per. (%)
1	Electrical Engineering	115	38.59	Electrical Engineering	1267	62.72	Electrical Engineering	2911	63.63
2	Instruments	87	29.19	Instruments	344	17.03	Mechanical Engineering	675	14.75
3	Mechanical Engineering	50	16.78	Mechanical Engineering	238	11.78	Other Fields	387	8.46
4	Chemistry	24	8.05	Other Fields	89	4.41	Chemistry	338	7.39
5	Other Fields	22	7.38	Chemistry	82	4.06	Instruments	264	5.77

4.3.2 *The spatial evolution*

By portraying the spatial evolution of China's transferred technology categories, this section explores the supply–demand relationship between China and foreign countries. The spatial distribution of transferred patent sectors is notably uneven and agglomerative. Typically, cities with greater diversity and intensity in technology transfers are highly centralized in coastal areas of the developed countries, owing to their superior innovative milieux.

4.3.2.1 *Importing technology sectors*

Because of differences in urban innovative and absorptive capacities, the sectors of China's imported technology varied significantly among cities with different hierarchies and locations.

As illustrated in Figure 4.17, megacities importing more diverse and frequent technologies are mainly centralized in the developed coastal regions, significantly shifting from outside the mainland to the southeast mainland. From 2001 to 2006, most of the transnational patents flew continuously into Taiwan (e.g., Hsinchu, Taipei, and Taoyuan) and Hong Kong. This influx significantly enriched their technology bases. Among them, electrical engineering was the dominant sector of their absorptive patents, far exceeding the size of other sectors.

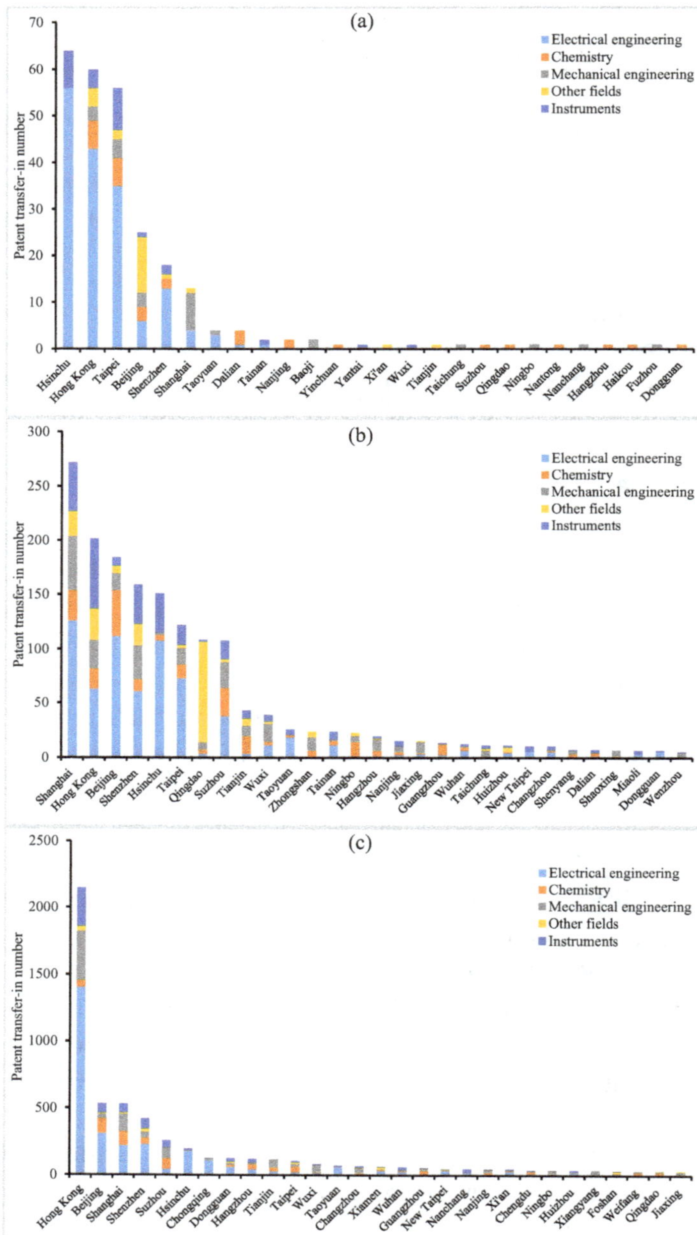

Figure 4.17 Terminal distribution of China's importing technology sectors. (a) 2001–2006; (b) 2007–2012 and (c) 2013–2018.

From 2007 to 2012, a few megacities including Shanghai, Beijing, and Shenzhen rapidly grew into new absorptive centers with diverse and numerous technologies, exceeding Taiwan's innovative centers. In addition, the electrical engineering category saw a large influx of imported technologies during this time, spurred by the prosperity of the electrical industries of these cities.

From 2013 to 2018, some megacities in the middle-west had absorbed more foreign technology from electrical engineering sectors, including Chongqing and Wuhan. Compared to the previous stage, most of the cities had expanded their technological fields and geographical scope of imported patents. During this period, the megacities in the Eastern region continued to primarily absorb patents in the electrical engineering sector. Meanwhile, smaller cities and provincial capital cities with fewer sectors (such as Shijiazhuang) preferred importing technology in chemical and mechanical engineering sectors, same with their leading industries.

There were also significant spatial differences in technology sector and size among the original cities. The primary originations had tended to move from the US and Europe to East Asia. From 2001 to 2006, although the number of original cities was increasing, the electrical engineering and chemistry were still the most extensive source. As shown in Figure 4.18, the electrical engineering originated from New York in the United States and Paris in France, with more than 40 imported patents. Besides that, there were also considerable imports from Liverpool, Los Angeles, and San Jose. Comparatively, the chemical and mechanical engineering patents were mainly imported from Tokyo in Japan.

From 2007 to 2012, the geographical scope of each sector's origins expanded greatly, among which original cities with electrical engineering and mechanical engineering reached nearly 100. However, the main originations of each sector were not well-distributed, as observed in Figure 4.18. Most of them were concentrated in two types of locations: international offshore financial centers located in the Caribbean (i.e., Road Town) and global innovative centers situated in the United States (i.e., San Jose, New York, and Houston), Western Europe (i.e., Eindhoven, Espoo, Liverpool,

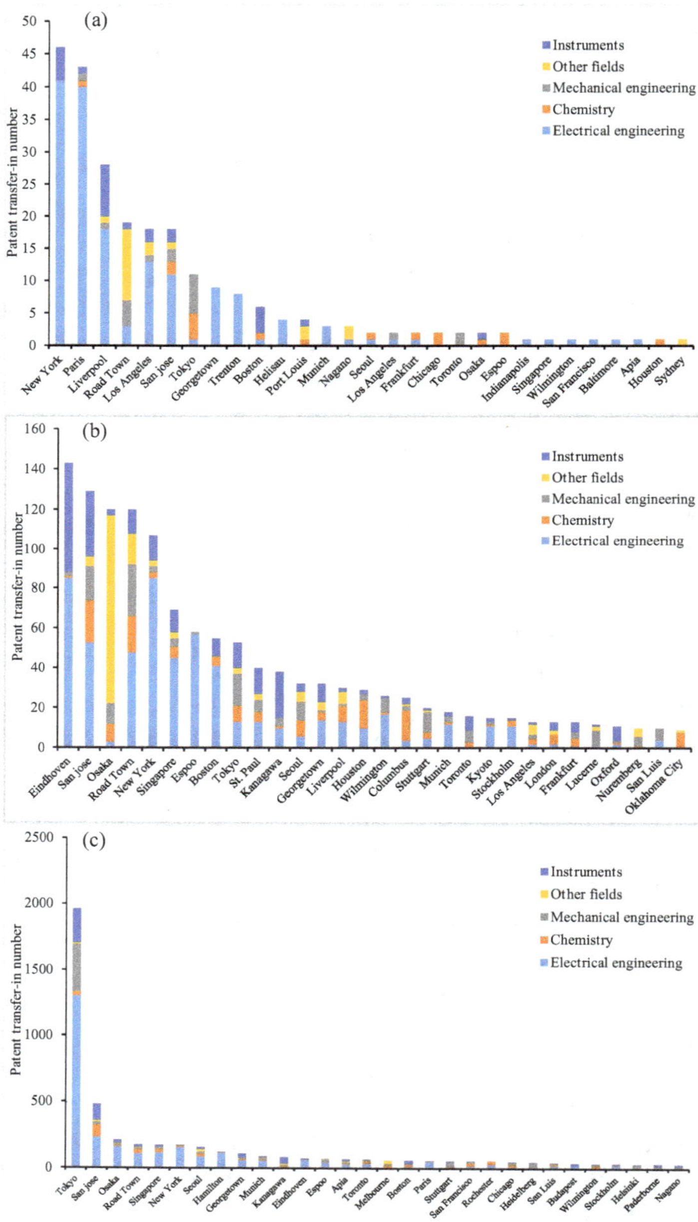

Figure 4.18 Original distribution of China's importing technology sectors. (a) 2001–2006; (b) 2007–2012 and (c) 2013–2018.

and others), and Japan (i.e., Tokyo, Kanagawa, and Osaka). In terms of breadth, 17 international cities have covered all technology sectors but are mainly concentrated in North America and East Asia.

From 2013 to 2018, the geographical scope of the sources of each sector further expanded significantly to more than 100 cities, except for other sectors with only 50 sources. Among them, the chemical sector had the most extensive source, but San Jose was still its primary origination, with nearly 100 patents imported from this city. Although the original cities of electrical engineering were still distributed extensively, the top had given way to Tokyo importing more than 1,300 patents. Similarly, the primary sources of mechanical engineering and instruments had also shifted to Tokyo, implying that Japan had played a dominant role in spilling engineering technologies over to China.

4.3.2.2 *Exporting technology sectors*

There were significant spatial differences in the field of China's exported technology, noticeably in electrical engineering and mechanical engineering.

On the one hand, the originations of these sectors were highly concentrated on only one or two innovative centers shifting from Taipei and Hsinchu in Taiwan province to Shenzhen in Guangdong province (Figure 4.19). From 2001 to 2006, the size of China's transfer-out technology sectors was not only relatively small, but their distribution was also greatly uneven and centralized in Taiwan and Hong Kong. In this stage, such a few innovative cities as Taipei, Hsinchu, Taichung, and Hong Kong not only had larger export size but also had a more diversified technology field. In terms of sector-dominated cities, electrical engineering sector and instruments were dominated by Taipei and Hsinchu, while chemistry was led by Taipei and Xiamen. Mechanical engineering and other sectors were generated mainly from Taipei. Overall, Taipei held a dominant position in both number and size.

From 2007 to 2012, the size and fields of transnational technologies from Chinese cities increased, but Taipei's leading role in

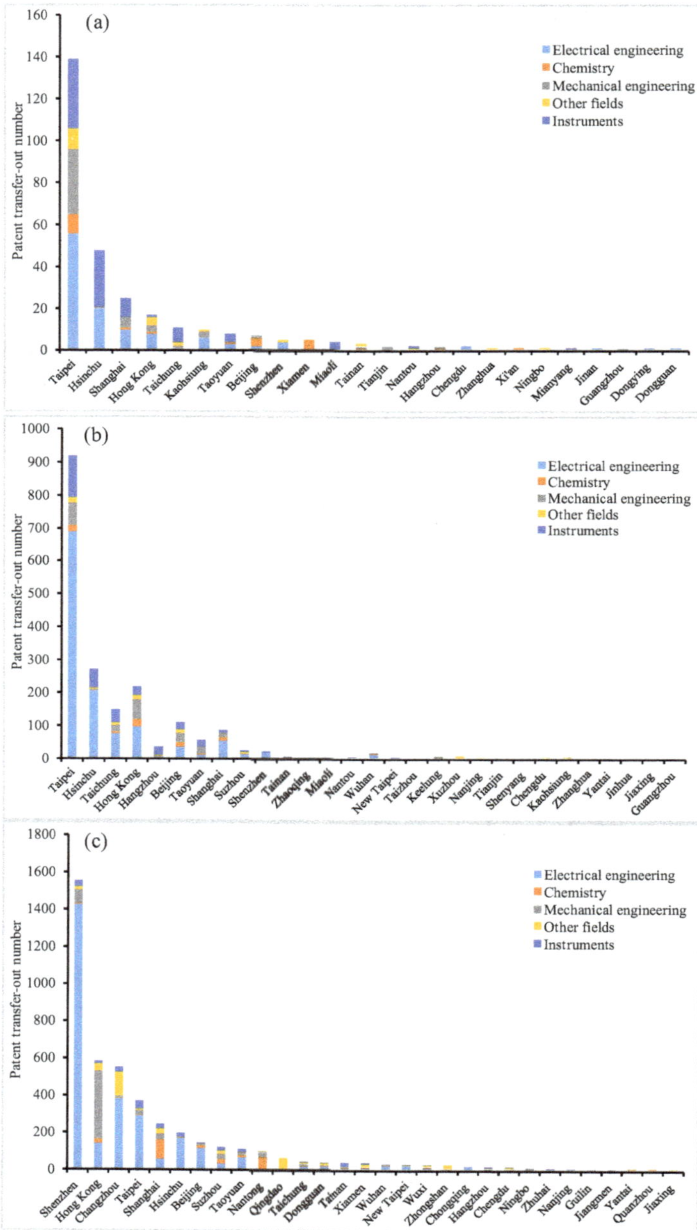

Figure 4.19 Original distribution of China's exporting technology sectors. (a) 2001–2006; (b) 2007–2012 and (c) 2013–2018.

various sectors remained stable due to the effect of path dependence. In this stage, nearly half of the cities exported patents covering only one sector, while the number of cities covering all sectors increased. As a result, some compound technology centers covering most sectors were emerging. In terms of the leading sectors in major cities, electrical engineering sector became a dominant function in Taipei, Hsinchu, Hong Kong, Taichung, Beijing, and Shanghai, while instruments was the primary sector in Taoyuan and Hangzhou. In terms of the leading cities of certain sectors, electrical engineering and instruments were dominated by Taipei and Hsinchu, while chemistry and mechanical engineering came primarily from Hong Kong and Taipei.

From 2013 to 2018, most cities saw an increase in the number of export sectors, and the number of cities covering the five sectors continued to rise. Several coastal cities were rapidly getting involved in technology export activities owing to their increasing innovation capacity and innovative milieux. The dominant position of Taipei and Hsinchu in Taiwan province had been replaced by Shenzhen and Changzhou in the mainland owing to their prominent role in the electrical engineering sector. The electrical engineering sector still played a leading role in patent export in most cities, centralized in Taiwan (i.e., Taipei, Hsinchu, and Taoyuan), Shenzhen, and Changzhou. While the leading sector in Hong Kong had changed from electrical engineering to mechanical engineering because of the mechanical industry transfers from the PRD. Moreover, there were obvious changes in the leading cities of a certain sector. The export of electrical engineering, instruments, and mechanical engineering sectors was mainly dominated by Shenzhen and Hong Kong in the GBA, while the chemical and other sectors were dominated by Shanghai, Nantong, and Changzhou in the YRD.

On the other hand, the terminal cities of exported sectors were not well-distributed with a tendency to shift from North America to East Asia (Figure 4.20). From 2001 to 2006, there were differences in the number of cities to which Chinese patents in certain sectors were exported. The electrical engineering sector had the widest geographical scope of terminal cities, but was dominated by Road

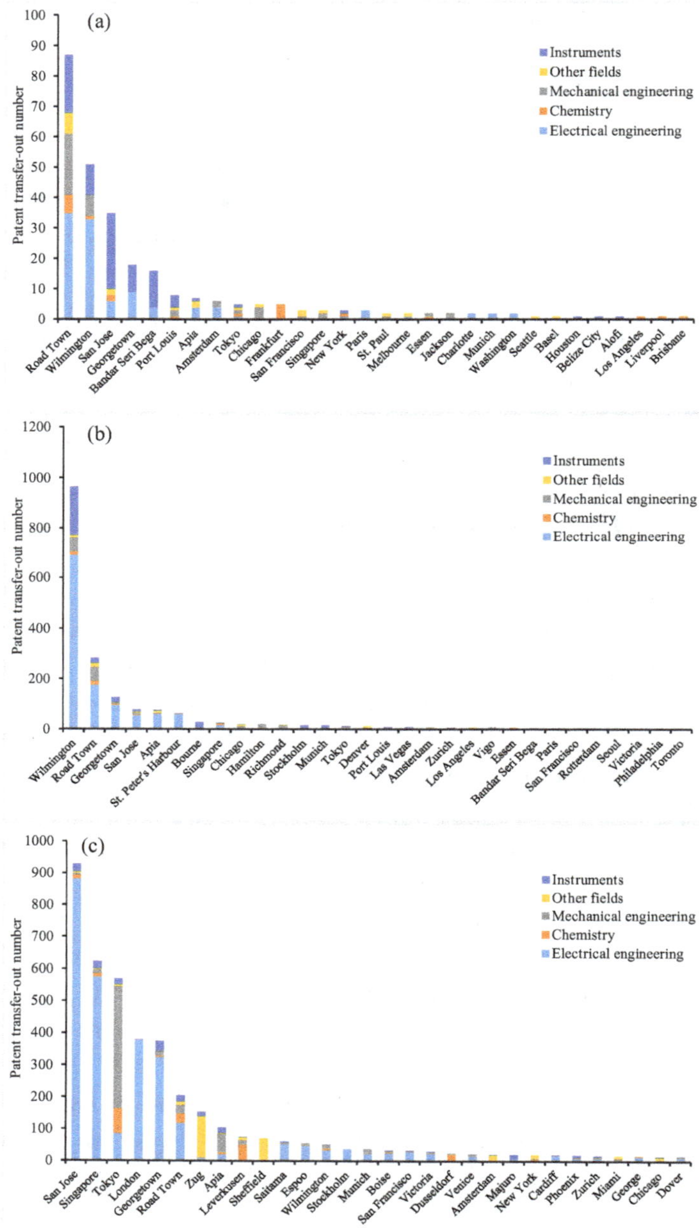

Figure 4.20 Terminal distribution of China's exporting technology sectors. (a) 2001–2006; (b) 2007–2012 and (c) 2013–2018.

Town and Wilmington in North America; followed by the rest of the sectors with a high concentration of transnational exports in Road Town.

From 2007 to 2012, the geographical scope of exported patents in each sector was greatly expanded, with the number of covered cities reaching more than 30 in each sector, while the concentration of their size was also obviously strengthened. Among them, the electrical engineering sector became the most widespread but highly directed to Wilmington, followed by offshore financial centers such as Road Town and Georgetown; the rest of the sectors also flew mainly into Road Town and Wilmington. Overall, the main destinations for all fields in this period were locked in Wilmington and Road Town, which resulted from the "Matthew effect" or "path dependence" of the transnational technology flows.

From 2013 to 2018, a further significant expansion in the terminal scope for each sector, noticeably in some sectors shifting their main destinations to West Asia-Pacific and West Europe. In the electrical engineering sector, the most significant export destination shifted to San Jose, while cities such as Singapore and London also became the major export destinations. In terms of the rest, the principal destinations of mechanical engineering, instruments, and chemistry had shifted from Road Town, Georgetown, San Jose, and Wilmington in North America to Tokyo and Singapore in the West Asia-Pacific; while recent patents of other sectors intensively flew into the West European cities such as Zug in Switzerland, Sheffield in the UK, and Amsterdam in Holland.

4.4 Evolution of Technology Complexity

4.4.1 *Technology diversity*

Technology diversity is a lens capturing the complexity of transferred technology for each city and the capacity of its technology generation and absorption. In general, cities with a more diverse technology usually have a greater ability to learn, apply, and produce multiple sectors of technology.

4.4.1.1 *Importing technology diversity*

The results show that China is experiencing a rise in the diversity of imported technology. Moreover, there is a growing disparity between cities that possess diverse technologies, which has resulted in a shift from three innovative growth poles to multiple urban innovation agglomerations. This shift can be attributed to the integration of regional innovation (Figure 4.21).

From 2001 to 2006, the diversity of imported technology ranged from 0 to 3.24, with more than half of the cities locked in a certain field. Cities with a high diversity of transnational technology were highly concentrated on the eastern coast.

From 2007 to 2012, the diversity of imported technology rapidly increased in all cities. Among them, megacities in the YRD, such as Suzhou, Wuxi, Hangzhou, and Nanjing, had improved significantly, from 0 to more than 3. Meanwhile, a few provincial capitals in the central and western regions, such as Wuhan, Chengdu, Kunming, and Chongqing, had also increased evidently, mostly to more than 2.

From 2013 to 2018, the diversity of imported technology in most cities had shown an ascending trend while only a minority of cities experienced a decline. During this period, the diversity of the central and western provincial capitals such as Xi'an, Changsha, and Chengdu grew up strongly, but the coastal cities such as Dongguan, Weihai, Zhuhai, and Qingdao still dominated the growth trend.

The cities with a more diverse technology landscape shifted from global cities outside of mainland China, such as Taipei and Hong Kong, to key cities within the three urban agglomerations like Shanghai, Guangzhou, and Beijing. Additionally, capital cities in the central and western regions served as supplements, resulting in the emergence of a polycentric structure instead of multiple urban agglomerations with hierarchical innovation.

From 2001 to 2006, cities with a larger diversity of imported technology were highly clustered along the southeastern coast. Among them, Taipei, Hong Kong, Hsinchu, and Taoyuan acted as the hubs importing more diverse technology outside the mainland, while Beijing, Shenzhen, Shanghai, and Dalian with higher diversity grew into the highlands of technology absorption in China's mainland.

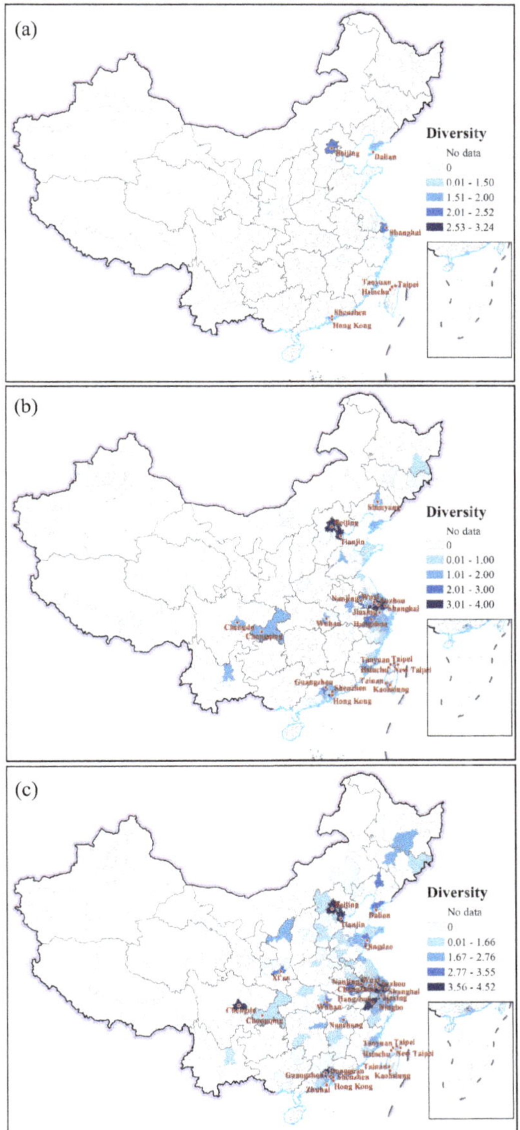

Figure 4.21 Spatial distribution of importing technology diversity. (a) 2001–2006; (b) 2007–2012 and (c) 2013–2018.

From 2007 to 2012, there was an increase in the diversity of imported technology in each city and the status of mainland cities had started to become prominent. Shanghai, Suzhou, and Hong Kong were more diversified, with a diversity index exceeding 4, followed by Shenzhen, Beijing, Taipei, Wuxi, Tianjin, Hangzhou, Nanjing, Guangzhou, and Hsinchu with a diversity index exceeding 3. Overall, the diversity of transnational technology import in Taiwan grew more slowly compared to mainland cities, with some subcore cities of the YRD, the GBA, and the BTH urban agglomerations coming to the fore. The capital cities in central and western provinces also started to import abundant categories of foreign technologies.

From 2013 to 2018, the dominant position of cities in the three major urban agglomerations was further highlighted, and the diversity of transnational technology imports in central and western cities increased significantly. Cities with a high degree of diversity were mostly centralized in the three major urban agglomerations, among which Shanghai, Suzhou, Guangzhou, Beijing, and Shenzhen had a high degree of diversity with the index exceeding 4, followed by cities such as Taipei, Dongguan, Nanjing, and Wuxi. Overall, the diversity of transnational technology imports in China showed a pattern with higher index in the south and east than in the north and west.

4.4.1.2 *Exporting technology diversity*

China's transnational technology exports were increasing in size and categories, and the diversity of foreign technology exported by Chinese cities was also growing (Figure 4.22). From 2001 to 2006, the diversity of transnational export technologies of Chinese cities ranged from 0 to 3.99, which was higher relative to the upper limit of technology import. Nearly half the number of cities with zero diversity only exported technologies in a certain field. Cities with a higher diversity of exported technologies were located on the eastern coast, noticeably in Taiwan province.

From 2007 to 2012, the number of cities exporting technology increased slightly, while there was a slight decrease in the diversity

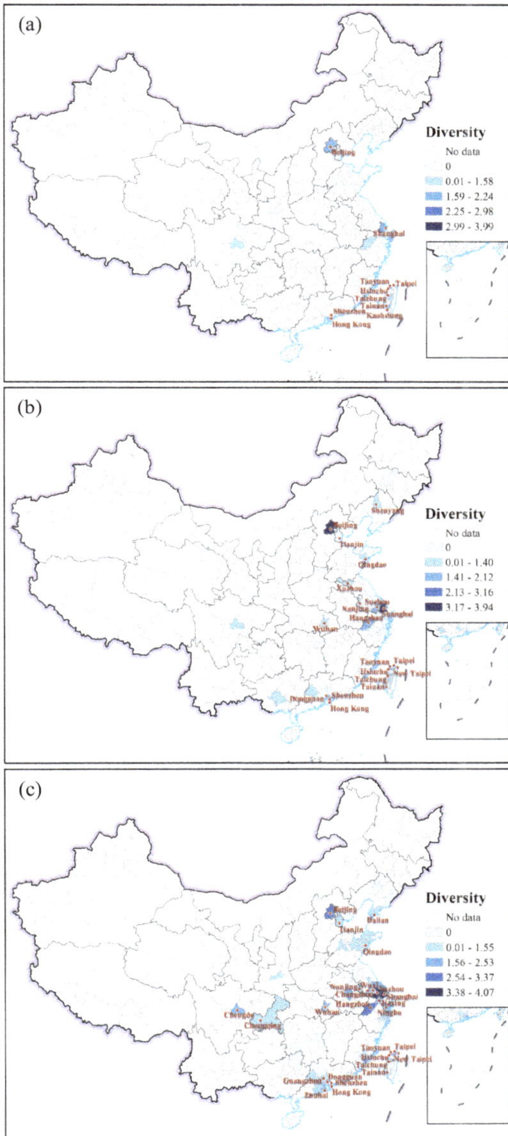

Figure 4.22 Spatial distribution of exporting technology diversity. (a) 2001–2006; (b) 2007–2012 and (c) 2013–2018.

compared to the previous period. More cities exported patents in only one field, though accounting for a smaller proportion overall. However, most cities saw an increase in diversity, except for some of the cities in Taiwan. Among them, cities with a significant increase of export technology diversity were concentrated on several innovative clusters including Taiwan province (i.e., Keelung, New Taipei, Tainan), the YRD (i.e., Shanghai and Suzhou), the BTH (i.e., Beijing), and the GBA (i.e., Dongguan and Shenzhen).

From 2013 to 2018, most cities showed an increment in the diversity of exported technologies with a range from 0 to 4.07. Most of the cities such as Xiamen, Wuxi, Hefei, Changzhou, and Zhuhai showed strong growth in export diversity, indicating that urban innovative capacity and technological strength were growing in trend.

The cities with a larger-level diversity of exported technology in China were highly concentrated on the eastern region, same as the distribution of imported technology. From 2001 to 2006, the number of cities exporting diverse patents to foreign countries was relatively small and highly occupied by Taiwan, followed by Hong Kong, Beijing, Shanghai, and Shenzhen.

From 2007 to 2012, the number of cities exporting diverse patents increased slightly, and their diversity was incrementally dominated by the mainland cities. During this period, cities such as Hong Kong, Beijing, and Shanghai improved evidently to become the most diverse hubs, followed by the five major cities in Taiwan. Overall, cities with a higher diversity of external technology exhibited a high concentration in the innovative cities of Taiwan and the three major urban agglomerations, and relatively poor diversity in the midwest and the northeast.

From 2013 to 2018, higher technology diversity was still dominated by the core cities in the three urban agglomerations and the innovative cities in Taiwan. In this period, the status of major cities in the YRD had further improved, especially in Shanghai, Suzhou, and Hangzhou. Hong Kong and Taiwan (including Taichung, Taipei, Taoyuan, etc.) also had a diverse technology base. On a word, the diversity of the southeastern cities was much higher than that of cities in the central and western regions, with many small- and medium-sized cities ranking at the top of the list. However, several provincial cities in the central and western provinces, such as Wuhan and Xi'an, were emerging as new technology centers with abundant technology base.

4.4.2 *Technology relatedness*

Through the relatedness of technologies, one can understand the relationship between external and internal technologies in each city. Technologies that are interrelated but not exactly similar to each other are more likely to be imported and absorbed by buyers, thus enhancing their technological innovation capability. As can be seen from Figure 4.23, the number of cities importing related technologies was increasing, but their distribution varied considerably between cities. There is an obvious concentration on several innovative hubs including Beijing, Shanghai, and Shenzhen, forming the triple growth poles of national innovation system where transnational imported technologies were strongly linked to local technology bases.

From 2001 to 2006, the range of technological relatedness was 0–0.0509, with only Shenzhen, Beijing, Shanghai, and Nanjing importing related technologies. Limited by local innovation capacity, most of the cities had imported transnational technologies which were more different from or similar to local technologies rather than interrelated to them.

From 2007 to 2012, the threshold of relatedness range decreased, because international innovative cities such as Beijing, Shanghai, and Shenzhen had imported large amounts of more different or similar technologies. However, majority of the cities showed an increasing trend of technological relatedness, owing to their growing absorptive ability.

From 2013 to 2018, even after benefiting from these innovative hubs like Beijing, Shanghai, Shenzhen, and Suzhou, the whole relatedness range was still declining. Similar to the previous stage, the total number of Chinese cities importing related technologies was increasing, implying an extensive increment in the absorptive ability of most cities.

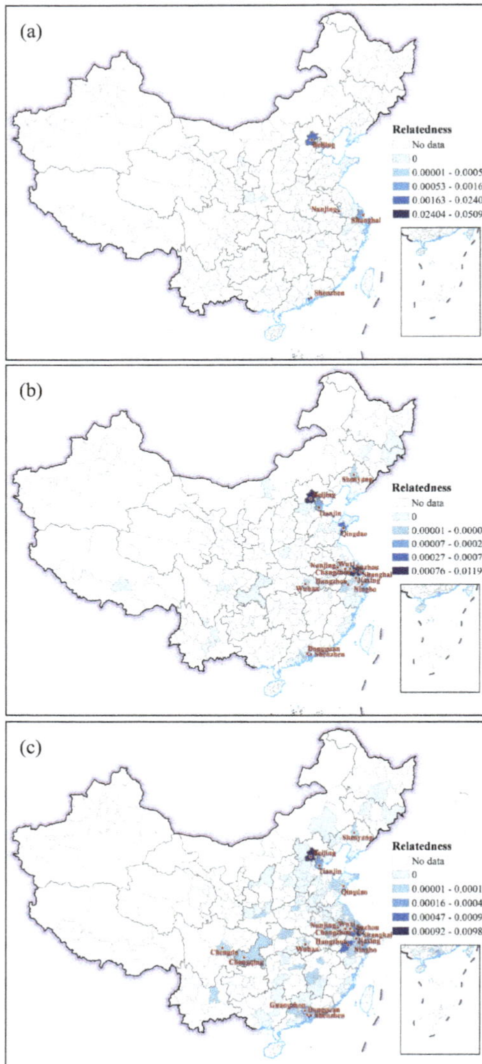

Figure 4.23 Spatial disparity of importing technology relatedness. (a) 2001–2006; (b) 2007–2012 and (c) 2013–2018.

4.4.3 *Technology similarity*

The evaluation of technologies similarity can be used to understand the demand-side level for similar technologies in different cities.

If a city has a higher similarity in imported technologies, it means that the city has imported more homogeneous technologies. However, the learning and application of similar technologies cannot significantly enhance the technological innovation capacity of these cities.

As can be seen from Figure 4.24, the similarity of transnational imported technologies was increasing, and the cities that imported more homogeneous technologies were expanding from the eastern coastal region to the central and western regions. From 2001 to 2006, cities importing more similar technology were clustered in a few innovative megacities, particularly in Beijing, Shenzhen, and Shanghai with the most homogeneous technologies (the similarity values were 10.53, 9.92, and 8.82, respectively). In addition, there were fewer than 10% of the total cities with positive similarity, indicating that these cities imported more homogeneous technologies rather than interrelated ones. However, the average degree of technology similarity sharply increased from 2007 to 2012, resulting in about 1/3 of the cities getting involved in importing similar technologies due to their stronger technology lock-in. From 2013 to 2018, the threshold range of technology similarity continued to increase, which meant the similar degree of imported technology was further aggravated. At this stage, a total of 100 cities had imported technologies similar to their knowledge bases, indicating that Chinese cities' absorptive capacity of foreign homogeneous technologies was stronger than related technologies.

The cities with higher similarity of imported technologies had expanded from the eastern coastal areas dominated by Shenzhen, Beijing, and Shanghai to the central and western regions due to the western shift of manufacturing industries in the last two decades. From 2001 to 2006, a few core cities were located in the hubs of the southeastern coast including Beijing, Shenzhen, Shanghai, Dalian, Hangzhou, Nanjing, and Qingdao, with their technological similarities above 5. From 2007 to 2012, the similarity between urban external technology and internal technology bases had been increasing, mainly owing to the role of MAR externalities. Those cities importing similar technologies were clustering around several urban agglomerations such as the YRD (i.e., Shanghai, Suzhou,

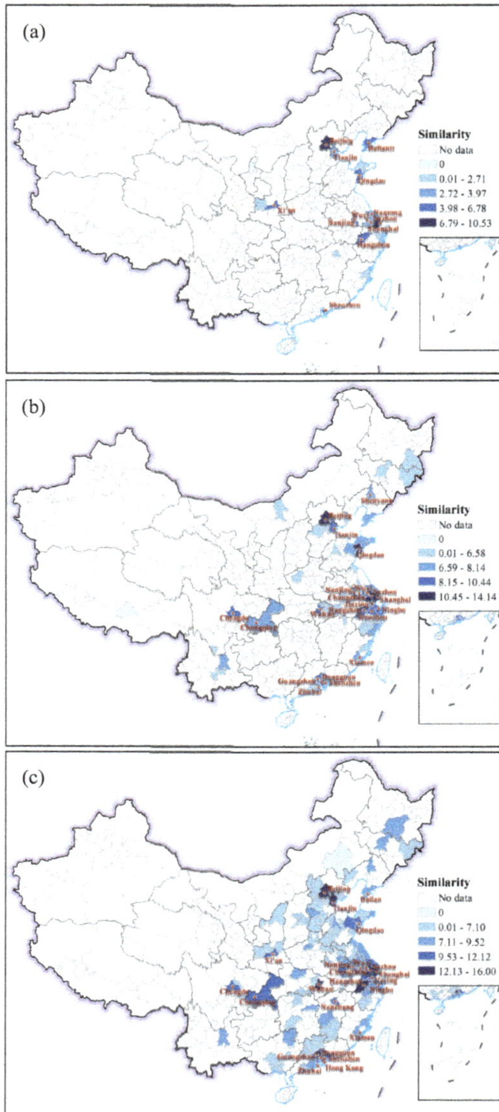

Figure 4.24 Spatial disparity of importing technology similarity. (a) 2001–2006; (b) 2007–2012 and (c) 2013–2018.

Nanjing, Ningbo, Wuxi, and Hangzhou), the BTH (including Beijing and Tianjin), and the GBA (including Guangzhou and Shenzhen). Until 2013, in addition to the concentration on the southeastern coast, some innovative cities from the midwestern regions also showed a higher similarity of technologies, noticeably in these provincial capitals such as Wuhan, Chengdu, Hefei, Xi'an, and Nanchang. Obviously, innovative hubs with greater technology similarity are observed evidently as a spatial shift from the eastern coast to the central and western regions, which is highly dependent on the high speed transportation and large-scale industrial transfer in China since the 21st century.

4.5 Summary

This chapter examines the spatiotemporal dynamics of the intensity and breadth of transnational technology transfer, technology transfer network, and transferred technology structure from both import and export perspectives. The main findings are as follows.

Since China's accession to the WTO in 2001, its foreign technology exchanges have continued to intensify, and transnational technology transfer has achieved a leap in development, while its geographical scope has shown a significant expansion. Overall, the intensity of China's transnational technology transfer has shown fluctuating growth in the past two decades, and its technology import can be roughly divided into the initial period (2001–2006), the booming period (2007–2012), and the fluctuating period (2013–2018). Similarly, technology exports can also be classified into three phases including the initial period (2001–2006), the fluctuating period (2007–2012), and the rising period (2013–2018).

The technology transfer network continues to expand as numerous network indicators, due to the globalization of China's enterprise and their technology market. Spatially, the geographical scope of both China's technology import and export has been continuously enlarging. The scope of domestic cities in the network has also experienced an evident expansion from Hong Kong and Taiwan to its southeastern coast, then to some provincial capitals in inland regions. And the sourcing cities have gradually expanded from the United States to West Europe, then to East Asia and Southeast Asia. Japan, South Korea, and Singapore have gradually become the primary sources of China's imported technologies, owing to their emerging innovative megacities such as Tokyo, Osaka, and Seoul. Meanwhile, the original cities in China have likewise undergone a spatial shift from the coast to the inland, while its terminal cities have changed from the United States to East Asia and West Europe.

Moreover, the intensity of technology transfer in each city follows a clear rank-size law emerging hierarchy. Along with the year-on-year growth of China's foreign technology trade, the spatial complexity of the network had been increasing. In the last decade, the disparity of China's imported originations had weakened and evolved from the bicentricity with double cores of West Europe and North America to the polycentricity including East Asia, North America, West Europe, and Southeast Asia. Similarly, China's exported pipelines have oriented from the monocentric hub — North America — to the polycentric regions including North America, East Asia, Southeast Asia, and West Europe.

The fields and sectors to which China's transnational technology transfer belongs were constantly enriched. There were differences in China's preference for foreign technology import and export at different times, with the electrical engineering sector growing faster in terms of import and always in a dominant position, while the mechanical engineering sector had improved significantly. On the other hand, in terms of transnational patent export, the electrical engineering sector had replaced instruments as the dominant sector, while other sectors had also improved significantly.

Spatially, the coastal cities in Eastern China had a more diverse and balanced structure of technology transfer fields, with cities such as San Jose and Tokyo playing a leading role in transferring a wide range of fields. In terms of domestic demand in China, patents in the electrical engineering sector were the main sector of patents absorbed by major cities in Eastern China, while small- and medium-sized cities imported fewer fields and preferred to obtaining patents in the electrical engineering and mechanical engineering sectors. The chemistry sector had the most widely original places, but San Jose was always the main source city for this sector; the sourcing cities for electrical engineering had shifted from New York and Paris to Tokyo, San Jose, Osaka, New York, and Singapore, and the main source cities for mechanical engineering and instrument had also shifted to Tokyo; the number of patents imported in other sectors was relatively small and geographically concentrated on Melbourne and Seoul. From the perspective of China's transnational patent export, the leading role of cities in Taiwan, mainly in Taipei, was gradually replaced by mainland cities. For example, the dominant cities of engineering, chemistry, mechanical engineering, and instruments have changed from Taipei to Shenzhen, Shanghai, Hong Kong and Shenzhen, respectively. It demonstrates the position of mainland cities in the export of patents in various sectors had significantly enhanced.

When it comes to the evolution of technology complexity, the diversity of external technologies had increased, and so did its relatedness and similarity to local technologies, with Beijing, Shenzhen, and Shanghai performing most prominently in all three categories of indicators. This indicated that they had a greater absorption capacity for all types of technologies.

Firstly, the diversity of transnational imported technologies had been increasing, and the capacity of cities to apply a variety of external technologies had enhanced. The cities with a high degree of diversity in the import of transnational technologies had developed from the "triple-core" pattern of Shenzhen, Beijing, and Shanghai to a polycentric pattern with Shanghai, Suzhou, Beijing, Shenzhen, Qingdao, and Chengdu as the main involvers.

Secondly, the number of cities importing related technologies was also increasing, but the trend of the relatedness of external technology in different cities showed an evident difference: the spatial distribution of cities with related transnational technologies has formed a polycentric pattern with Beijing, Shanghai, and Shenzhen as the core, but the relatedness of core cities such as Beijing, Shanghai, and Shenzhen were decreasing.

Thirdly, the similarity between transnational and local technologies showed an increasing trend, and the cities' capacity to absorb external homogenized technologies had increased. Cities with a high degree of homogenization of transnational technologies had expanded from the eastern coastal regions, mainly Shenzhen, Beijing, and Shanghai, to the central and western regions in China.

CHAPTER 5

SPATIAL EVOLUTION OF NATIONAL TECHNOLOGY TRANSFER NETWORK

In the era of a knowledge-based economy, cities are increasingly recognizing the importance of acquiring external knowledge as a key driver of economic growth. In China, the government has implemented measures to stimulate technology transfer between cities and promote the exchange of ideas. By analyzing patent transaction data from China's State Intellectual Property Office (SIPO) spanning from 2008 to 2015, researchers have been able to map out the landscape of technology transfer. The findings reveal a trend toward increasing intercity patent transfer, with a growing number of cities and city pairs involved over time. Two notable features of this network are the dominant role played by certain key cities, resulting in a "diamond-shaped" pattern that extends geographically, and the hub-and-spoke structure of the network with Beijing, Shanghai, and Shenzhen acting as first-tier hubs for technology transfer (Gui *et al.*, 2019a).

Compared with the last chapter, Chapter 5 focuses on the technology transfer network at a domestic scale. It analyzes the intensity, categories, and relatedness of intercity technology transfer in China, focusing on the centrality of nodes and the connectivity of pipelines in the technology transfer networks, along with the community of subnetworks consisting of the hubs and their hinterworlds.

5.1 Evolution of Transfer Intensity

5.1.1 *The temporal dynamics*

Since the 21st century, China's intercity technology exchange has continuously enhanced, and the size of technology transfer has steadily improved. The intercity patent transfer has experienced three stages: initial growth (2001–2006), steady growth (2007–2012), and fluctuating growth (2013–2018). Figure 5.1 shows the annual variation and growth rate of intercity patent transfer. There had been over 300,000 patents transferred between cities in China in 18 years. From 2001 to 2006, the number of patents transferred was far smaller than the patent licensing (Yang *et al.*, 2021), with only 3,000 total times and less than 30% of the transfer rate during the six years. From 2007 to 2012, more than 32,000 patents were transferred, and the growth rate remained stable above 0.5, benefiting from the implementation of the NSTDP since 2006. From 2013 to 2018, the size of patent transfer showed a trend of fluctuating growth, with a total of more than 267,000 transferred patents during the six years. The significant increase in domestic patent transfers observed in 2013 can be largely attributed to the emphasis on Science and Technology (S&T) innovation as a strategic means to enhance social productivity and

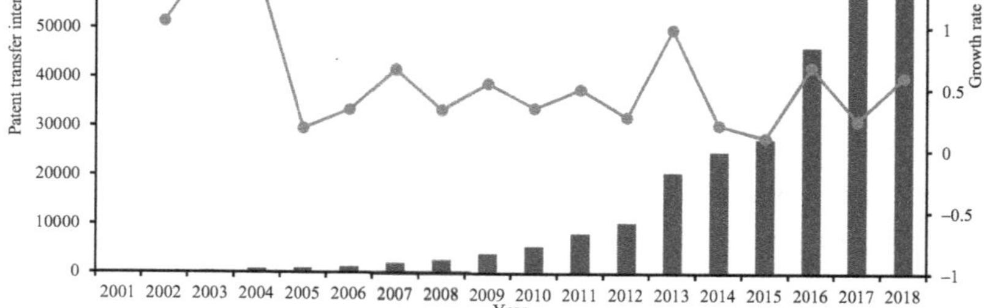

Figure 5.1 Dynamics of national technology transfer intensity (2001–2018).

overall national strength, which was highlighted at the 18th National Congress of the Communist Party of China toward the end of 2012. Furthermore, the substantial increase in China's funding in scientific research during this period also played a crucial role. The low growth rate in 2014 and 2015 was mainly because China's economy had stepped into the "new normal", with a sluggish economy, insufficient innovation activity, and declining patent demand.

5.1.2 *The spatial evolution*

5.1.2.1 *Technology transfer-in*

The rank–size distribution of intercity technology transfer-in activities and weight degree centrality from 2001 to 2018 presents a significant power–law phenomenon and follows the scale-free property with hierarchical organization. Minority of the cities had become the absorption pool of domestic technology, but most of them were at the edge of the network and had little technology linkage with other cities. Figure 5.2 illustrated that the scope and

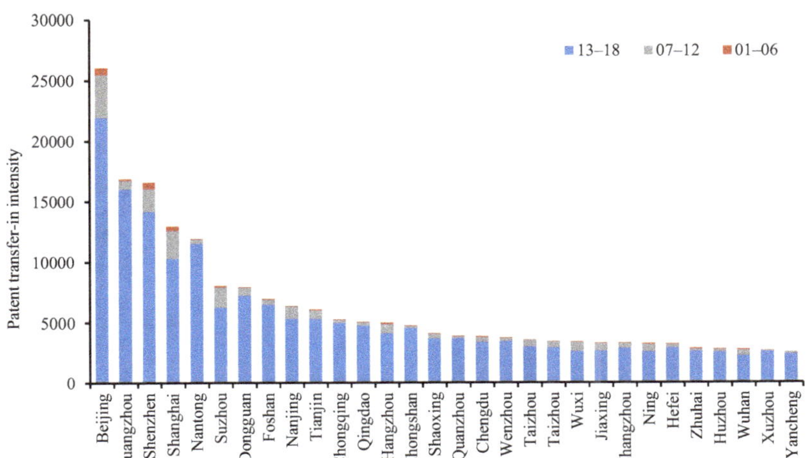

Figure 5.2 Rank–size distribution of national technology transfer-in (2001–2018).

intensity of technology transfer-in activities were continuously expanding, while they were concentrated on the top core megacities including Beijing, Guangzhou, Shenzhen, and Nantong. The potential driving mechanism is the opening of a regional innovation system that innovation networks evolved toward hyperlocalization, facilitating the integration of local and external knowledge (Asheim *et al.*, 2002). In this term, cities with a high degree of openness tend to be the source of knowledge flows.

As shown in Figure 5.3, the degree distribution and accumulative degree distribution of domestic technology transfer activities presented the obvious phenomenon of power–law distribution similar to transnational ones (Figure 5.3). It demonstrated that the domestic technology transfer-in network exhibits a scale-free structure and emerges as a high heterogeneity. Minority of the innovative megacities had acted as "the reservoir" of technology inflows, but most of them were locked in the periphery of the network due to a lack of technology linkage with other cities. The network was dominated by a few absorptive hubs with more frequent external linkage and more

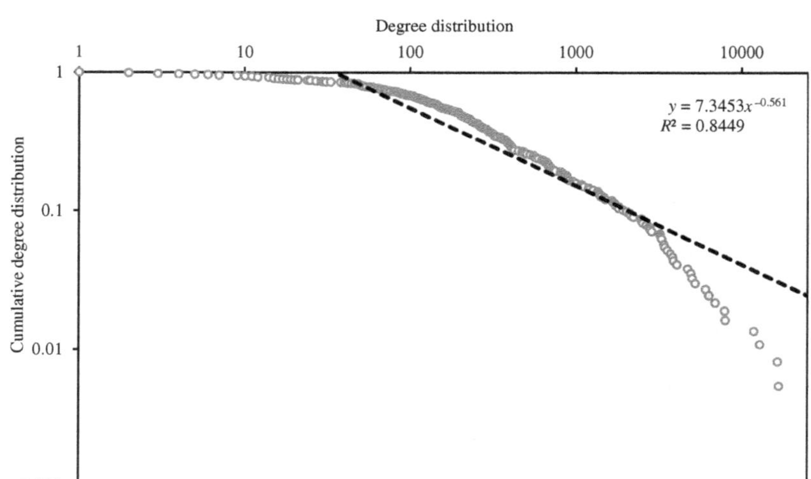

Figure 5.3 Degree and cumulative degree distributions of national technology transfer-in (2001–2018).

strong absorptive capacity, including Beijing, Guangzhou, Shenzhen, Shanghai, and Nantong accounting for above 1/4 of the total.

Moreover, according to the power–law index, its scale-free nature was found to be gradually weakened (the absolute value of power–law index is 0.87, 0.71, and 0.57 in the three periods, respectively) and weaker than the transnational technology transfer-in network in the later stage. It further showed that the domestic technology transfer-in activities were better distributed than the transnational ones. It is not surprising that within a country, due to the similar culture and language (Lawson & Lorenz, 1999), it is easier to build trust among the actors thus acquiring and absorbing new knowledge from other cities more effectively (Morrison *et al.*, 2013). In other words, a variety of knowledge often originates from outside the region and country, and cities with higher intensity of intercity technology introduction have access to more diverse and innovative knowledge.

5.1.2.2 *Technology transfer-out*

Same as the transfer-in, the rank–size distribution of technology transfer-out intensity from 2001 to 2018 also presented a significant power–law phenomenon (Figure 5.4). Few coastal megacities not

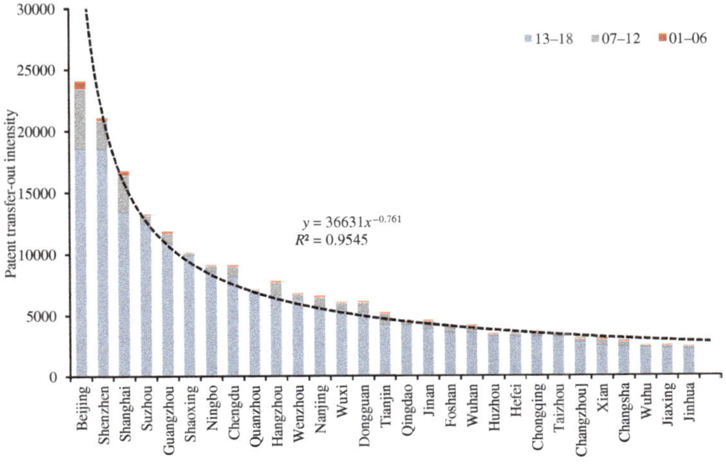

Figure 5.4 Rank–size distribution of national technology transfer-out (2001–2018).

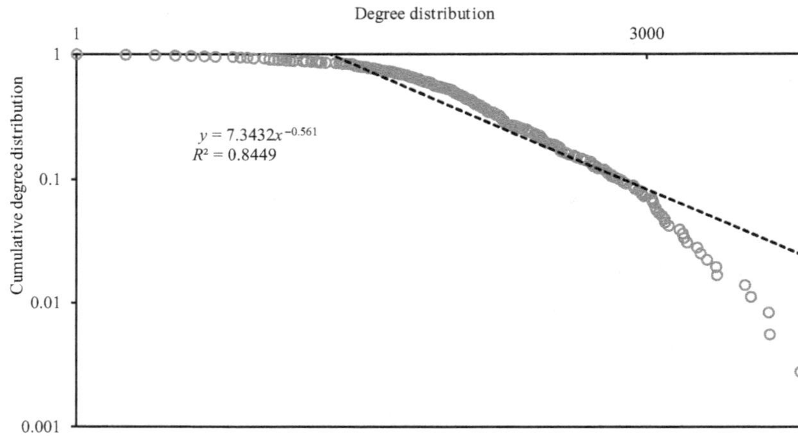

Figure 5.5 Degree and cumulative degree distributions of national technology transfer-out (2001–2018).

only acted as absorptive technology growth poles but also had become the technology pools spilling spatially over the midwest along with industrial transfer between them. Specifically, about 28% of the total patents were contributed by the top five cities such as Beijing, Shenzhen, Shanghai, Suzhou, and Guangzhou with more than 10,000 transferred-out patents. This layout fully demonstrates Bathelt *et al.* (2020) that complex technologies are spatially sticky under global economic background. Theory holds that ubiquitous technologies are readily available while complex technologies always agglomerate in innovative cities and are difficult to spread (Sorenson *et al.*, 2010). Technology flows from regions with similar capacities are beneficial for innovative regions to absorb external novel technologies to effectively pursue regional diversification (Feldman *et al.*, 2015).

Meanwhile, the scale-free property of technology transfer-out activities was depicted by the degree and accumulative degree distribution. During the three periods, these curves followed power–law distribution with scale-free property, as shown in Figure 5.5. Differing from the transfer-in, the scale-free index of the transfer-out network gradually increased from 1.90 in 2001–2006 to 6.11 in

2013–2018. It indicated that the concentration of domestic transfer-out technology was not only greater than that of technology transfer-in in China but also higher than transnational transfer activities.

5.2 Evolution of Transfer Network

5.2.1 *The nodes*

5.2.1.1 *Breadth of technology transfer*

The breadth of cities in the domestic TTN is measured by nodal degree centrality. As Tables 5.1 and 5.2 revealed, the scope of urban technology linkages had similarly shown an increasing trend at city scale. And cities with greater breadth were concentrated on the three urban agglomerations of southeastern coast, particularly in Beijing, Shenzhen, Shanghai, and Guangzhou.

Temporally, the overall trend in the breadth of urban technology transfer continued to grow (Tables 5.1 and 5.2). During the period 2001–2006, 30–40% of total cities had generated intercity technology flows, while the size of technology exchange among cities was rather small. During 2007–2012, about one-half of the cities had participated in intercity technology trade, but less than 3% of the cities had degree centrality above 50. During 2013–2018, the number of cities involved in the process had exceeded 350, including more than 150 cities with a degree centrality exceeding 50. It showed an increasingly extensive exchange among cities in China.

Spatially, the connectivity of intercity technology transfer had been highly concentrated on the four innovative hubs including Beijing, Shanghai, Guangzhou, and Shenzhen. Among cities with patent rights transactions existed in 2001–2006 and only Beijing, Shenzhen, and Shanghai had a breadth of both transfer-in and transfer-out above 40. In addition, some provincial capitals such as Hangzhou, Chengdu, Guangzhou, Nanjing, and Shenyang also had a relatively high breadth. From 2007 to 2012, in addition to the four megacities, other coastal cities including Hangzhou, Dongguan, Tianjin, Suzhou, Wenzhou, Nanjing, and Ningbo showed a high

Table 5.1 Top 50 cities of in-degree centrality in national technology transfer-in network.

	2001–2006		2007–2012		2013–2018	
Rank	City	Degree	City	Degree	City	Degree
1	Beijing	111	Beijing	189	Beijing	305
2	Shenzhen	58	Shenzhen	140	Shenzhen	265
3	Shanghai	58	Shanghai	127	Guangzhou	240
4	Guangzhou	35	Guangzhou	93	Shanghai	233
5	Hangzhou	26	Hangzhou	72	Chongqing	216
6	Chengdu	24	Dongguan	71	Dongguan	214
7	Nanjing	22	Tianjin	65	Nantong	210
8	Ningbo	18	Suzhou	65	Foshan	199
9	Tianjin	17	Wenzhou	65	Qingdao	195
10	Suzhou	17	Nanjing	64	Zhongshan	188
11	Dalian	16	Ningbo	61	Quanzhou	186
12	Jinan	15	Chengdu	59	Shaoxing	184
13	Xi'an	15	Foshan	58	Tianjin	181
14	Foshan	15	Wuhan	57	Chengdu	174
15	Harbin	15	Chongqing	51	Hangzhou	173
16	Shenyang	14	Wuxi	50	Xuzhou	168
17	Wenzhou	14	Taizhou	50	Suzhou	166
18	Wuxi	14	Changsha	49	Taizhou	165
19	Changsha	13	Qingdao	47	Taizhou	162
20	Chongqing	13	Quanzhou	46	Ningbo	159
21	Wuhan	12	Xiamen	45	Nanjing	159
22	Fuzhou	12	Jinhua	43	Wenzhou	158
23	Changchun	11	Hefei	42	Changsha	151
24	Xiamen	11	Zhengzhou	40	Hefei	149
25	Kunming	11	Xi'an	39	Xi'an	149
26	Zhuhai	11	Fuzhou	39	Nanchang	149
27	Qingdao	10	Jinan	38	Jinhua	148
28	Shantou	10	Changzhou	38	Yancheng	145
29	Jining	10	Wuhu	37	Zhengzhou	142

Table 5.1 (*Continued*)

	2001–2006		2007–2012		2013–2018	
Rank	City	Degree	City	Degree	City	Degree
30	Dongguan	9	Shenyang	36	Wuhan	142
31	Zhongshan	9	Dalian	36	Wuxi	139
32	Tonghua	9	Zhongshan	34	Jiaxing	137
33	Yantai	8	Yangzhou	34	Jinan	132
34	Guiyang	8	Taizhou	34	Jiangmen	132
35	Taizhou	8	Harbin	33	Zhuhai	131
36	Huzhou	8	Kunming	33	Tangshan	131
37	Jinhua	8	Shaoxing	33	Yangzhou	131
38	Zhengzhou	7	Yantai	32	Fuzhou	130
39	Taizhou	7	Weifang	32	Huizhou	125
40	Jiaxing	7	Changchun	30	Changzhou	124
41	Siping	7	Jiaxing	30	Xiamen	119
42	Shijiazhuang	6	Guiyang	29	Yantai	116
43	Shaoxing	6	Nantong	29	Weifang	111
44	Daqing	6	Shijiazhuang	28	Harbin	110
45	Jiangmen	6	Taiyuan	28	Shijiazhuang	110
46	Zibo	6	Nanchang	28	Jining	110
47	Yangzhou	6	Urumqi	28	Urumqi	109
48	Haikou	6	Zhuhai	27	Kunming	107
49	Taian	6	Haikou	26	Baoding	107
50	Guilin	6	Jiangmen	25	Nanning	106

breadth of transfer-in, while a few provincial capitals in the West such as Chengdu also presented a higher breadth of transfer-out. Between 2013 and 2018, 18 cities had transfer-out breadth of over 200, with widespread intercity technology linkages due to the enhancement of its innovative capacity.

Urban agglomerations had shown a similar strong capacity of technology absorption. In the early years, cities were not very active in technology trade and were led by provincial-level municipality and capitals such as Beijing, Shanghai, Guangzhou, and Shenzhen. As a whole, most of the cities ranking in the top 50 list in terms of degree centrality were located in the three major urban agglomerations in China, especially in the YRD and GBA. This trend was becoming more pronounced, with the Chengdu–Chongqing (abbreviated as *Chengyu*) urban agglomeration continuing to rise in the later stages, which indicated the importance of urban agglomeration in domestic technology linkage.

Table 5.2 Top 50 cities of out-degree centrality in national technology transfer-out network (2001–2018).

	2001–2006		2007–2012		2013–2018	
Rank	City	Degree	City	Degree	City	Degree
1	Beijing	92	Beijing	217	Beijing	304
2	Shenzhen	44	Shanghai	149	Shenzhen	288
3	Shanghai	43	Shenzhen	133	Shaoxing	282
4	Guangzhou	37	Guangzhou	99	Chengdu	280
5	Chengdu	33	Chengdu	84	Guangzhou	276
6	Shenyang	26	Hangzhou	84	Ningbo	270
7	Jinan	23	Nanjing	75	Suzhou	260
8	Hangzhou	21	Tianjin	75	Shanghai	255
9	Tianjin	19	Wuhan	75	Wenzhou	252
10	Wuhan	19	Changsha	71	Quanzhou	247
11	Changsha	19	Xi'an	69	Tianjin	226
12	Changchun	17	Wuxi	66	Hangzhou	224
13	Xi'an	16	Suzhou	64	Qingdao	217
14	Dalian	13	Jinan	63	Dongguan	214
15	Dongguan	13	Shenyang	62	Wuxi	211
16	Qingdao	13	Dongguan	61	Nanjing	209
17	Yantai	13	Chongqing	60	Chongqing	203
18	Chongqing	13	Foshan	57	Huzhou	200
19	Foshan	12	Ningbo	54	Hefei	198

Table 5.2 (*Continued*)

	2001–2006		2007–2012		2013–2018	
Rank	City	Degree	City	Degree	City	Degree
20	Harbin	12	Dalian	53	Foshan	192
21	Fuzhou	11	Wenzhou	52	Taizhou	189
22	Nanjing	11	Changchun	52	Zhengzhou	175
23	Suzhou	11	Harbin	51	Changzhou	172
24	Guiyang	10	Zhengzhou	46	Wuhan	172
25	Xiamen	10	Hefei	45	Jinhua	171
26	Taiyuan	10	Shijiazhuang	43	Wuhu	171
27	Wenzhou	10	Yantai	43	Jinan	169
28	Hefei	9	Qingdao	42	Taizhou	169
29	Lanzhou	9	Kunming	39	Xi'an	168
30	Nanning	9	Taizhou	39	Jiaxing	167
31	Ningbo	9	Zhuhai	38	Changsha	155
32	Anshan	8	Jinhua	36	Huizhou	151
33	Baoji	8	Xiamen	36	Anqing	148
34	Kunming	8	Lanzhou	35	Lishui	147
35	Nanchang	8	Taiyuan	34	Nantong	146
36	Shijiazhuang	8	Zhongshan	34	Fuzhou	145
37	Taizhou	8	Changzhou	32	Harbin	145
38	Weihai	8	Fuzhou	32	Nanning	145
39	Wuxi	8	Nanchang	30	Zhongshan	143
40	Zhengzhou	8	Guiyang	27	Bengbu	138
41	Changzhou	7	Guilin	27	Yantai	138
42	Urumqi	6	Shaoxing	27	Quzhou	137
43	Zhongshan	6	Huizhou	26	Nanchang	135
44	Baoding	5	Jiaxing	26	Longyan	134
45	Chifeng	5	Nantong	26	Dalian	131
46	Daqing	5	Yangzhou	25	Zibo	129
47	Jilin	5	Zibo	25	Shenyang	128
48	Jiangmen	5	Baoding	23	Xuzhou	128
49	Loudi	5	Huzhou	23	Chuzhou	126
50	Luoyang	5	Xuzhou	23	Zhenjiang	124

5.2.1.2 *Intensity of technology transfer*

The intensity of domestic technology transfer can be measured by weight degree centrality. According to Tables 5.3 and 5.4, it can be found that the intensity of both transfer-in and transfer-out technologies had been expanding over time and showed a trend of exponential growth in a few cities under the effect of path dependence. These cities including Beijing, Shenzhen, and Shanghai had been the "anchor regions" of Chinese patent transfer activities (Seo, 2019).

Table 5.3 Top 50 cities of weight in-degree centrality in national technology transfer-in network (2001–2018).

	2001–2006		2007–2012		2013–2018	
Rank	City	Weight indegree	City	Weight indegree	City	Weight indegree
1	Beijing	548	Beijing	3543	Beijing	21953
2	Shenzhen	505	Shanghai	2296	Guangzhou	16027
3	Shanghai	333	Shenzhen	1913	Shenzhen	14152
4	Guangzhou	114	Suzhou	1658	Nantong	11563
5	Suzhou	86	Nanjing	984	Shanghai	10315
6	Hangzhou	78	Hangzhou	756	Dongguan	7257
7	Shenyang	66	Wuxi	742	Foshan	6512
8	Chengdu	58	Guangzhou	722	Suzhou	6288
9	Tianjin	54	Tianjin	714	Nanjing	5366
10	Ningbo	51	Yantai	712	Tianjin	5321
11	Dongguan	51	Dongguan	645	Chongqing	5008
12	Foshan	49	Ningbo	587	Qingdao	4761
13	Zhuhai	48	Jiaxing	581	Zhongshan	4562
14	Wuhan	47	Taizhou	543	Hangzhou	4139
15	Xi'an	37	Taizhou	476	Quanzhou	3713
16	Nanjing	36	Wuhan	469	Shaoxing	3702
17	Harbin	30	Jinan	440	Wenzhou	3453
18	Changsha	29	Foshan	424	Chengdu	3376
19	Changchun	27	Chengdu	415	Taizhou	3017

Table 5.3 (*Continued*)

	2001–2006		2007–2012		2013–2018	
Rank	City	Weight indegree	City	Weight indegree	City	Weight indegree
20	Huzhou	27	Changzhou	379	Taizhou	2950
21	Qingdao	26	Shaoxing	375	Hefei	2920
22	Zhongshan	26	Nantong	350	Changzhou	2899
23	Jinan	25	Changsha	304	Jiaxing	2698
24	Fuzhou	24	Shenyang	295	Wuxi	2632
25	Dalian	23	Hefei	294	Zhuhai	2630
26	Wenzhou	23	Xiamen	283	Ningbo	2613
27	Wuxi	23	Qingdao	274	Xuzhou	2604
28	Chongqing	22	Dalian	267	Huzhou	2566
29	Xiamen	22	Wuhu	259	Yancheng	2384
30	Jilin	22	Wenzhou	258	Wuhan	2247
31	Shaoxing	19	Chongqing	251	Xiamen	2172
32	Zibo	18	Shijiazhuang	242	Changsha	2160
33	Shantou	17	Xi'an	240	Jiangmen	2047
34	Daqing	17	Huzhou	204	Jinhua	2027
35	Jining	16	Mianyang	204	Huizhou	1954
36	Changzhou	16	Liaocheng	200	Zhengzhou	1751
37	Xianyang	16	Zhuhai	186	Yangzhou	1651
38	Yantai	15	Jinhua	181	Nanchang	1643
39	Jinhua	15	Xiangyang	179	Fuzhou	1571
40	Jiaxing	15	Zhenjiang	178	Jinan	1561
41	Yangzhou	15	Quanzhou	174	Weifang	1515
42	Guilin	15	Zhongshan	171	Xi'an	1441
43	Kunming	14	Zhengzhou	166	Bengbu	1352
44	Guiyang	14	Changchun	166	Huaian	1337
45	Xinyu	13	Haikou	165	Suqian	1298
46	Taiyuan	12	Nanchang	161	Wuhu	1246
47	Tonghua	10	Yancheng	152	Langfang	1240
48	Taizhou	10	Zhangzhou	152	Guiyang	1189
49	Taian	10	Yangzhou	148	Tangshan	1127
50	Taizhou	9	Huizhou	146	Shenyang	1123

Table 5.4 Top 50 cities of weight out-degree centrality in national technology transfer-out network (2001–2018).

	2001–2006		2007–2012		2013–2018	
Rank	City	Weight outdegree	City	Weight outdegree	City	Weight outdegree
1	Beijing	628	Beijing	4836	Beijing	18533
2	Shanghai	305	Shanghai	2961	Shenzhen	18525
3	Shenzhen	183	Shenzhen	2285	Shanghai	13398
4	Shenyang	133	Hangzhou	1002	Suzhou	12493
5	Guangzhou	120	Tianjin	876	Guangzhou	10849
6	Chengdu	106	Dongguan	824	Shaoxing	9851
7	Nanjing	78	Guangzhou	797	Ningbo	8544
8	Xi'an	72	Chengdu	797	Chengdu	8127
9	Changsha	57	Nanjing	753	Quanzhou	6894
10	Tianjin	56	Shenyang	678	Hangzhou	6642
11	Wuhan	53	Suzhou	634	Wenzhou	6304
12	Hangzhou	51	Jinan	604	Nanjing	5651
13	Suzhou	44	Wuhan	518	Wuxi	5460
14	Foshan	44	Wuxi	500	Dongguan	5169
15	Changchun	39	Foshan	490	Tianjin	4161
16	Chongqing	35	Ningbo	484	Qingdao	4136
17	Jinan	34	Xi'an	459	Jinan	3824
18	Qingdao	30	Changsha	396	Foshan	3523
19	Yantai	29	Wenzhou	383	Wuhan	3475
20	Zhongshan	28	Dalian	367	Huzhou	3255
21	Dalian	27	Xiamen	300	Hefei	3199
22	Shijiazhuang	26	Taizhou	291	Chongqing	3190
23	Zhengzhou	24	Chongqing	281	Taizhou	3149
24	Taiyuan	24	Nantong	275	Changzhou	2738
25	Guiyang	22	Harbin	248	Xi'an	2426
26	Baoding	22	Fuzhou	243	Changsha	2309
27	Weihai	21	Xianyang	238	Wuhu	2270
28	Lanzhou	21	Shijiazhuang	234	Jiaxing	2223
29	Cangzhou	21	Shaoxing	220	Jinhua	2111

Table 5.4 (*Continued*)

	2001–2006		2007–2012		2013–2018	
Rank	City	Weight outdegree	City	Weight outdegree	City	Weight outdegree
30	Wenzhou	20	Changzhou	220	Zhengzhou	2110
31	Kunming	20	Laiwu	212	Huizhou	2014
32	Harbin	20	Zhuhai	202	Zhongshan	1742
33	Huzhou	19	Qingdao	200	Anqing	1722
34	Dongguan	19	Jinhua	199	Taizhou	1628
35	Yueyang	17	Zibo	196	Fuzhou	1616
36	Fuzhou	17	Hefei	186	Nantong	1545
37	Daqing	17	Zhongshan	180	Zhenjiang	1537
38	Wuxi	16	Changchun	175	Shenyang	1486
39	Nanchang	16	Taiyuan	172	Nanchang	1436
40	Baoji	15	Yantai	169	Bengbu	1419
41	Nanning	14	Zhengzhou	161	Harbin	1375
42	Taizhou	13	Xiangyang	152	Yantai	1314
43	Zibo	12	Jiaxing	152	Dalian	1311
44	Zhuhai	12	Huaian	147	Xiamen	1222
45	Ningbo	12	Zhenjiang	137	Changchun	1209
46	Jiaxing	12	Langfang	128	Lishui	1166
47	Daqing	12	Tangshan	125	Shijiazhuang	1129
48	Binzhou	12	Lanzhou	123	Longyan	1073
49	Changzhou	12	Kunming	121	Huaian	1038
50	Xiamen	11	Guiyang	119	Zibo	1035

Specifically, the general increase in the intensity of urban transferred patent was large. From 2001 to 2006, most of the top 50 listed cities had transferred patents of less than 500, except for Beijing and Shenzhen. Among them, Beijing had become the first-tier national center of technology inflows and outflows, and Shenzhen had acted as an absorptive highland with more than 500 imports. From 2007 to 2012, technology import and export had been led by four megacities including Beijing, Shanghai, Shenzhen, and Suzhou with more than 1,000 patents. From 2013 to 2018, five cities had emerged into the first-tier group, such as Beijing, Guangzhou, Shenzhen, Nantong, and Shanghai, importing or exporting more than 10,000 patents.

The core region of patent transfer-in shifted from the eastern to the southeastern coast. It could be found that most cities in its early stage were located in the southeastern regions of the Hu Line (the Hehei–Tengchong Line), including Shenyang, Harbin, and Changchun in the northeast, Beijing, Tianjin, and Qingdao in the north, Shanghai, Suzhou, Hangzhou, and Ningbo in the east, as well as Shenzhen, Guangzhou, Dongguan, and Foshan in the south. However, at a later stage, the transferred intensity of cities in the northeast continued to decline, and only Beijing, Tianjin, and Qingdao were still active in northern China. In contrast, cities in Eastern and Southern China continued to rise, including Nantong, Nanjing, Zhongshan, Quanzhou, and Shaoxing, reflecting their improved capacity for domestic technological absorption.

The leading roles of the three major urban agglomerations of the YRD, the GBA, and the BTH were evident, and the *Chengyu* urban agglomeration started to develop. Most cities in the three major urban agglomerations demonstrated notable success in domestic patent transfer. Some cities in the YRD region also exhibited an active interaction. Notably, the city located west of Chongqing has made significant strides, leading to the emergence of *Chengyu* as a new center for technological convergence.

5.2.1.3 *Closeness centrality of cities*

The closeness centrality is used to evaluate the degree to which different cities are technologically connected to other cities (Table 5.5). The higher the closeness centrality of a city, the shorter the distance it is connected to other cities and it tends to behave more centrally in the network. Overall, technology transfer between cities is becoming more frequent and their linkages are becoming closer, and the flow of technology within the country continues to increase.

Beijing, Shanghai, Guangzhou, and Shenzhen had been at the heart of the network, but Shanghai's central position had declined. Beijing was the most connected city in the network, and Shenzhen

Table 5.5 Top 50 cities of closeness centrality in national technology transfer network (2001–2018).

	2001–2006		2007–2012		2013–2018	
Rank	City	Closeness	City	Closeness	City	Closeness
1	Beijing	0.67	Beijing	0.80	Beijing	0.90
2	Shanghai	0.56	Shanghai	0.68	Shenzhen	0.86
3	Shenzhen	0.55	Shenzhen	0.68	Guangzhou	0.83
4	Guangzhou	0.53	Guangzhou	0.62	Shaoxing	0.83
5	Chengdu	0.51	Hangzhou	0.59	Chengdu	0.83
6	Hangzhou	0.50	Tianjin	0.58	Shanghai	0.81
7	Shenyang	0.49	Chengdu	0.58	Ningbo	0.80
8	Nanjing	0.49	Dongguan	0.58	Suzhou	0.79
9	Tianjin	0.49	Changsha	0.58	Quanzhou	0.79
10	Changsha	0.48	Wuhan	0.57	Wenzhou	0.78
11	Xi'an	0.48	Wenzhou	0.57	Chongqing	0.77
12	Wuhan	0.48	Nanjing	0.57	Qingdao	0.76
13	Jinan	0.48	Suzhou	0.57	Dongguan	0.76
14	Changchun	0.47	Foshan	0.57	Tianjin	0.75
15	Dalian	0.47	Wuxi	0.56	Hangzhou	0.75
16	Suzhou	0.47	Xi'an	0.56	Foshan	0.73
17	Ningbo	0.47	Chongqing	0.56	Nantong	0.73
18	Wuxi	0.47	Ningbo	0.56	Nanjing	0.73
19	Foshan	0.47	Jinan	0.55	Wuxi	0.72
20	Chongqing	0.47	Shenyang	0.55	Taizhou	0.71
21	Harbin	0.47	Dalian	0.54	Taizhou	0.71
22	Wenzhou	0.46	Qingdao	0.54	Hefei	0.71
23	Fuzhou	0.46	Taizhou	0.54	Zhongshan	0.70
24	Zhongshan	0.46	Hefei	0.54	Huzhou	0.70
25	Guiyang	0.46	Harbin	0.54	Zhengzhou	0.70
26	Zhuhai	0.46	Yantai	0.54	Xi'an	0.70
27	Xiamen	0.46	Xiamen	0.54	Jinhua	0.69
28	Dongguan	0.46	Zhengzhou	0.54	Changzhou	0.69
29	Qingdao	0.45	Jinhua	0.54	Changsha	0.69

(*Continued*)

Table 5.5 (*Continued*)

	2001–2006		2007–2012		2013–2018	
Rank	City	Closeness	City	Closeness	City	Closeness
30	Kunming	0.45	Fuzhou	0.54	Jinan	0.68
31	Yantai	0.45	Shijiazhuang	0.53	Wuhan	0.68
32	Changzhou	0.45	Kunming	0.53	Jiaxing	0.68
33	Shantou	0.45	Zhuhai	0.53	Huizhou	0.68
34	Lanzhou	0.44	Zhongshan	0.53	Nanchang	0.67
35	Shijiazhuang	0.44	Changchun	0.53	Fuzhou	0.66
36	Taizhou	0.44	Changzhou	0.53	Xuzhou	0.66
37	Hefei	0.44	Quanzhou	0.53	Wuhu	0.65
38	Nanning	0.44	Lanzhou	0.53	Yancheng	0.65
39	Nanchang	0.44	Yangzhou	0.53	Harbin	0.65
40	Taian	0.44	Shaoxing	0.52	Zhuhai	0.64
41	Tonghua	0.44	Nanchang	0.52	Nanning	0.64
42	Yangzhou	0.44	Taiyuan	0.52	Yantai	0.64
43	Taizhou	0.44	Jiaxing	0.52	Quzhou	0.64
44	Nantong	0.44	Guiyang	0.52	Weifang	0.63
45	Haikou	0.44	Haikou	0.52	Bengbu	0.63
46	Huzhou	0.43	Taizhou	0.52	Lishui	0.63
47	Baoji	0.43	Nantong	0.51	Dalian	0.63
48	Maoming	0.43	Wuhu	0.51	Shenyang	0.63
49	Jining	0.43	Urumqi	0.51	Tangshan	0.63
50	Siping	0.43	Huizhou	0.51	Shijiazhuang	0.63

and Guangzhou remained in the top four. During the period from 2013 to 2018, Shanghai experienced a decline in its ranking as it was surpassed by Shaoxing and Chengdu. Despite an increasing trend in Shanghai's closeness centrality, which rose from 0.56 to 0.81 during this time frame, Shaoxing demonstrated remarkable growth with a centrality score of 0.43 in 2001–2006, which increased to 0.52 in 2007–2012, finally reaching an impressive score of 0.83 in 2013–2018. This dynamic growth highlights Shaoxing's active engagement in technology communication.

The closeness centrality of cities has experienced continuous growth, indicating an improvement in technological connectivity and proximity between them. During 2001–2006, approximately 90% of the cities had a closeness centrality score below 0.45, with a range of 0.23–0.67. During 2007–2012, the range increased to 0.32–0.80, and about 27% of cities scored below 0.45. From 2013 to 2018, the range extended to 0.26–0.9, and only 4.4% of cities scored below 0.45. Notably, cities such as Shaoxing, Suzhou, Chongqing, Wenzhou, Qingdao, and Hefei demonstrated significant growth trends across all phases and exhibited robust technological development dynamics.

5.2.1.4 *Betweenness centrality of cities*

The betweenness centrality can represent a city's ability to control the network. The higher the betweenness centrality of a city, the greater the city acts as a bridge to connect to more cities that are not directly connected with each other, so that plays a more important role in the diffusion of technology. As can be seen from Table 5.6, the betweenness centrality of core cities demonstrates a decreasing trend, whereas subcore and peripheral cities exhibit an increasing trend. This phenomenon signifies the development of a more extensive network with more intensified linkages for technology transfer, which has spread to previously underdeveloped regions.

The ability of the core cities to control the network was generally weaker compared to the subcore and peripheral cities. Among the key cities, including Beijing, Shenzhen, Chengdu, and Nanjing, there was a consistent decline in their betweenness centrality, which is indicative of a weakening ability to control over time. Similarly, Shanghai, Guangzhou, Hangzhou, and Chongqing showed a declining trend in betweenness centrality during 2013–2018, despite a slight increase from 2007 to 2012. Overall, this trend indicates a weakening ability to control these core cities. However, subcore cities such as Suzhou, Tianjin, Qingdao, Ningbo, Wenzhou, and Haikou exhibited continuous growth in betweenness centrality, suggesting active involvement in domestic technology transfer networks,

Table 5.6 Top 50 cities of betweenness centrality in national technology transfer network (2001–2018).

	2001–2006		2007–2012		2013–2018	
Rank	City	Betweenness	City	Betweenness	City	Betweenness
1	Beijing	16081.50	Beijing	15882.79	Beijing	4750.33
2	Shenzhen	4314.54	Shenzhen	5470.18	Shenzhen	2292.38
3	Shanghai	4184.30	Shanghai	4608.79	Guangzhou	2220.56
4	Chengdu	2996.85	Guangzhou	2723.39	Shaoxing	2215.48
5	Guangzhou	2521.45	Chengdu	2328.24	Haikou	1942.86
6	Hangzhou	1133.12	Kunming	1697.96	Chengdu	1852.64
7	Nanjing	1076.71	Hangzhou	1696.27	Shanghai	1700.11
8	Jinan	1037.82	Dongguan	1525.12	Suzhou	1484.59
9	Fuzhou	1036.69	Wuhan	1428.29	Tianjin	1480.60
10	Xi'an	1032.07	Tianjin	1326.99	Qingdao	1402.68
11	Foshan	968.60	Changsha	1214.71	Quanzhou	1333.81
12	Guiyang	953.21	Lanzhou	1208.01	Ningbo	1280.76
13	Wuhan	925.07	Chongqing	1082.65	Chongqing	1057.79
14	Chongqing	915.22	Xi'an	1031.04	Wenzhou	1050.68
15	Kunming	850.51	Foshan	994.93	Hangzhou	1000.71
16	Yantai	827.95	Urumqi	929.71	Dongguan	898.35
17	Tonghua	805.25	Guiyang	923.64	Changzhou	888.17
18	Tianjin	791.99	Wenzhou	849.71	Foshan	842.90
19	Changsha	763.71	Shenyang	822.09	Jinan	780.99
20	Harbin	686.88	Changchun	809.11	Nantong	773.51
21	Ningbo	610.73	Suzhou	765.86	Xi'an	772.93
22	Dalian	603.36	Ningbo	757.20	Lhasa	759.98
23	Nanning	597.06	Nanjing	744.70	Taizhou	739.01
24	Xiamen	517.65	Wuxi	703.54	Nanjing	725.01
25	Shenyang	438.19	Zhengzhou	641.40	Urumqi	698.40
26	Taiyuan	375.25	Taizhou	639.59	Kunming	688.19
27	Lanzhou	364.79	Haikou	544.34	Fuzhou	674.21
28	Qingdao	356.78	Guilin	519.84	Lanzhou	670.03
29	Suzhou	297.42	Jinan	493.07	Changsha	602.68
30	Shaoxing	294.38	Harbin	485.06	Taizhou	578.50

Table 5.6 (*Continued*)

	2001–2006		2007–2012		2013–2018	
Rank	City	Betweenness	City	Betweenness	City	Betweenness
31	Taizhou	294.04	Qingdao	445.21	Zhongshan	578.13
32	Anshan	285.64	Shijiazhuang	441.12	Wuxi	573.02
33	Wenzhou	281.16	Loudi	400.10	Hefei	558.45
34	Jinzhou	278.66	Dalian	378.98	Zhengzhou	523.89
35	Hefei	272.32	Zhongshan	358.22	Jinhua	522.91
36	Hohhot	268.20	Fuzhou	330.96	Wuhan	483.69
37	Changchun	267.94	Xiamen	303.26	Huzhou	480.82
38	Handan	265.19	Lishui	297.35	Jiaxing	429.60
39	Luzhou	260.67	Taiyuan	294.21	Shihezi	426.26
40	Mudanjiang	260.61	Jinhua	274.86	Huizhou	421.06
41	Yiyang	260.00	Qianjiang	274.44	Xuzhou	412.85
42	Xining	260.00	Yangzhou	270.91	Harbin	404.14
43	Fuxin	260.00	Nanning	265.74	Changjiang	363.00
44	Anqing	260.00	Xianyang	262.52	Nagchu	363.00
45	Xiangxi	260.00	Hefei	261.46	Nanchang	351.84
46	Hengyang	199.10	Yantai	245.38	Dalian	345.67
47	Nanchang	197.11	Zhuhai	241.08	Nanning	319.43
48	Sanming	179.44	Zhenjiang	223.49	Longyan	291.56
49	Dongguan	166.14	Quanzhou	220.35	Wuhu	254.98
50	Dandong	164.13	Xining	198.44	Yancheng	249.43

and promoting sustainable technology development in the neighboring cities. Additionally, many peripheral cities like Lhasa and Zunyi also showed an increasing trend in betweenness centrality, indicating an improved ability to control the network.

During a period of time, the gap in the ability to control between cities was wide but eventually narrowed, representing a more balanced technological development of cities. In the period 2001–2006, the gap in control ability between cities was the largest, with a range of 0–16081.50 in betweenness centrality. Around 37% of cities were 0 and 79% of cities had less than 100 betweenness centrality. In the period 2007–2012, the range of betweenness centrality was

0–15882.79. The gap between cities was slightly lower than in the previous period, with 11.7% of cities having a betweenness of 0 and 80.7% of cities being less than 100. In the period 2013–2018, the range was 0–4750.33, which showed the gap between cities was still large but significantly lower than in the previous period. About 5.2% of cities had a betweenness centrality of 0, and 78% of cities had less than 100.

5.2.2 *The pipelines*

5.2.2.1 *Spatial evolution of pipelines*

From 2001 to 2018, the domestic technology transfer network underwent significant evolution, resulting in a denser network with an increasing number of nodal cities involved in the process. Despite this growth, its spatial heterogeneity remained relatively apparent. As can be seen from Figure 5.6, the network showed a trend of increasing core nodes and expanding main pipelines. The eastern region, particularly the three major urban agglomerations, exhibited robust technological connections and served as technology growthpoles of the domestic technology transfer network.

Specifically, the eastern region had a significant advantage while the central and western regions were gradually getting involved. From 2001 to 2006, the domestic technology transfer network was relatively sparse, and the overall size of transferred patents was smaller. The national innovation system was mainly led by technology flows among the three innovative growthpoles including Beijing, Shenzhen, and Shanghai. Most of the cities in the central, western, and northeastern regions were locked in the periphery of the network. From 2007 to 2012, though the focus of the network remained in the eastern region, the central and western regions began to gain prominence in the network owing to the gradient transfer of industries from coastal to inland China (Zheng *et al.*, 2013). From 2013 to 2018, the eastern region was still at the core of the network along with its increasing importance and density, while the central and western regions also became more active with Chengdu and Chongqing emerging as the subcore nodes in the network. Urumqi

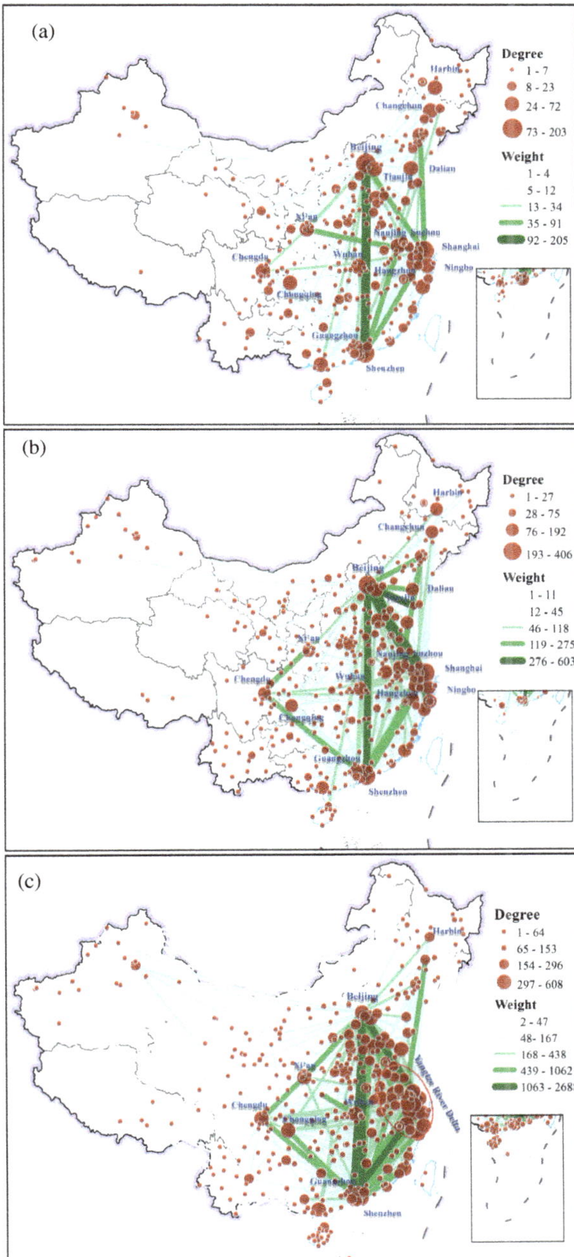

Figure 5.6 Spatial evolution of domestic technology transfer network (2001–2018). (a) 2001–2006; (b) 2007–2012 and (c) 2013–2018.

was also actively participating in the network and becoming an important node in the regional network.

The configuration of the network has transformed over time, shifting away from a polycentric model centered around Beijing, Shanghai, and Shenzhen toward a more clustered structure. Notably, this new structure has emerged around a series of innovative clusters with high-level technology transfer activities such as the YRD, GBA, BTH, and Chengyu urban agglomerations. From 2001 to 2006, Shenzhen–Beijing was the largest city pair in terms of patent transfer size, with Shenzhen introducing 205 patents from Beijing (Table 5.7). This was followed by Shanghai–Shenzhen, Beijing–Shanghai, Shanghai–Beijing, Shenyang–Shanghai, and Xi'an–Shanghai, with more than 40 transferred patents. It could be found that the three innovative growthpoles of Beijing, Shanghai, and Shenzhen dominated the entire technology transfer network during this period,

Table 5.7 Top 20 city pairs toward domestic technology transfer.

	2001–2006		2007–2012		2013–2018	
Rank	City-pair	Number	City-pair	Number	City-pair	Number
1	Beijing–Shenzhen	205	Beijing–Yantai	603	Shenzhen–Dongguan	2688
2	Shanghai–Shenzhen	91	Shanghai–Suzhou	522	Shanghai–Beijing	1868
3	Beijing–Shanghai	78	Beijing–Shanghai	500	Shenzhen–Beijing	1410
4	Shanghai–Beijing	51	Beijing–Shenzhen	452	Shaoxing–Guangzhou	1397
5	Shenyang–Shanghai	42	Tianjin–Taizhou	359	Beijing–Tianjin	1386
6	Xi'an–Shanghai	41	Beijing–Wuxi	341	Shanghai–Suzhou	1268
7	Nanjing–Shenzhen	39	Beijing–Tianjin	275	Shenzhen–Guangzhou	1252
8	Shanghai–Shenyang	39	Shanghai–Beijing	266	Beijing–Guangzhou	1204

Table 5.7 (*Continued*)

Rank	2001–2006 City-pair	2001–2006 Number	2007–2012 City-pair	2007–2012 Number	2013–2018 City-pair	2013–2018 Number
9	Shenzhen–Dongguan	34	Dongguan–Jiaxing	264	Beijing–Shanghai	1062
10	Shenzhen–Wuhan	25	Shenyang–Nanjing	262	Guangzhou–Foshan	959
11	Guangzhou–Foshan	23	Shenzhen–Dongguan	259	Tianjin–Beijing	945
12	Shenzhen–Beijing	23	Shenzhen–Beijing	246	Suzhou–Shanghai	898
13	Beijing–Guangzhou	20	Laiwu–Jinan	210	Shenzhen–Nanjing	896
14	Beijing–Tianjin	20	Hangzhou–Shaoxing	197	Beijing–Shenzhen	880
15	Shenyang–Shenzhen	20	Beijing–Suzhou	177	Shenzhen–Huizhou	854
16	Foshan–Jilin	19	Nanjing–Suzhou	150	Quanzhou–Guangzhou	806
17	Guangzhou–Shenzhen	19	Shenzhen–Suzhou	150	Suzhou–Nantong	789
18	Baoding–Beijing	18	Dalian–Beijing	144	Shanghai–Shenzhen	786
19	Beijing–Zhuhai	18	Jinan–Liaocheng	142	Shanghai–Nantong	784
20	Chengdu–Beijing	18	Shenyang–Beijing	140	Shenzhen–Shanghai	784

while non-coastal cities led by emerging industries, such as electronic information, aeronautics and astronautics, and biomedicine industries, were also actively introducing advanced technologies, particularly in Shenyang and Xi'an. From 2007 to 2012, the Beijing–Yantai pipeline conducted the largest volume of patent transfer, owing to the Yantai high-tech park introducing large amounts of patents into the emerging industries from Beijing. In addition, the intensity of patent transfer pair of Shanghai–Suzhou, Beijing–Shanghai, Beijing–Shenzhen, Tianjin–Taizhou, and Beijing–Wuxi was also relatively large, with 300–600 transferred patents. It demonstrated that in addition to the three technology growthpoles of Beijing, Shanghai, and Shenzhen, emerging innovative cities such as Suzhou, Tianjin, Taizhou, and Wuxi had acted vibrantly in the network, so that the structure of the network gradually expanded from the three growthpoles to the three urban agglomerations. During the 2013–2018 period, Shenzhen–Dongguan was the largest city pair in terms of patent transfer volume. Because of their integration, Dongguan had imported more than 2,500 patents from Shenzhen through local production networks. In addition, Shanghai–Beijing, Shenzhen–Beijing, Shaoxing–Guangzhou, Beijing–Tianjin, Shanghai–Suzhou, Shenzhen–Guangzhou, Beijing–Guangzhou, and Beijing–Shanghai were also city pairs taking more than 1,000 transfers. In this way, the technology flows happened within or between the three major urban agglomerations, indicating their dominant role in the national technology transfer system. When it came to the western region, the *Chengyu* urban agglomeration was gaining prominence in the network and was regarded as a rising secondary technology growthpole of the network, owing to the growth of "production footprints" in emerging industries (Yeung, 2008).

5.2.2.2 *Organization modes of pipelines*

Technology flows between headquarters and its branches, foreign enterprises and local enterprises, along with foreign enterprises and joint ventures were the main forms of domestic technology transfer. From 2001 to 2006, nearly 80% of technology transfer between Beijing and Shenzhen came from the licensing transactions between *Harbour Network Limited* and *Huawei Technologies Limited. Harbour Network Limited* is a subsidiary established in Beijing whose headquarters are located in Hong Kong, and it reflected the technology cooperation between foreign and local enterprises. From 2007 to 2012, the patent transfer between Beijing and Yantai was mainly from *LG Electronics (China) R&D Centre Co., Ltd.* to *Inspur LG Digital Mobile Communication Co., Ltd.* Among them, *LG Electronics* is a wholly foreign-owned high-tech enterprise established in China by *LG Electronics of Korea*, while *Inspur LG Digital Mobile Communication Co., Ltd.* is a Sino-South Korea joint venture between *Inspur Group Ltd.* and *LG Electronics Co.* This kind of technology cooperation between foreign enterprises and joint ventures, was an important catching-up path for China to learn advanced foreign technology. From 2013 to 2018, Shenzhen–Dongguan patent transfers were mainly conducted from *Huawei Terminal (Shenzhen) Co., Ltd.* to *Huawei Terminal (Dongguan) Co., Ltd.*, reflecting the flows of key enabling technologies between headquarters and their branches due to the licensing exclusivity of key and core technologies (Van den Berghe & Guild, 2008).

5.2.3 *The hubs*

5.2.3.1 *Hierarchical evolution of hubs*

The nodes in the network exhibited a distinct pyramid-shaped hierarchy that was well-established. From 2004 to 2018, Chinese cities displayed a similar pyramidal structure in terms of technology imports and exports, with clear hierarchies at the national, regional, and local levels. These technology transfer hubs followed a spatial disparity with higher density and greater concentration, remaining relatively stable over time while also allowing certain flexibility (Figure 5.7).

Statistically, the well-developed hierarchy follows the scale-free property and the rank-size law. In terms of the hierarchy statistics, during 2004–2018, the rank–size distribution of domestic technology transfer hubs toward both technology transfer in and out exhibits the scale-free character. Its power–law fitted better than the exponential one, with a goodness of fit ranging from 0.95 to 0.98. In addition, the technology transfer hubs shifted from a Pareto distribution toward an equilibrium, which is well evident by the *q*-values variation of both transfer-out and transfer-in hubs.

The hierarchy seemed to be relatively solidified, with significant "lock-in effect" and "hierarchical stickiness" owing to the path dependence of its evolution. During the past 15 years, the

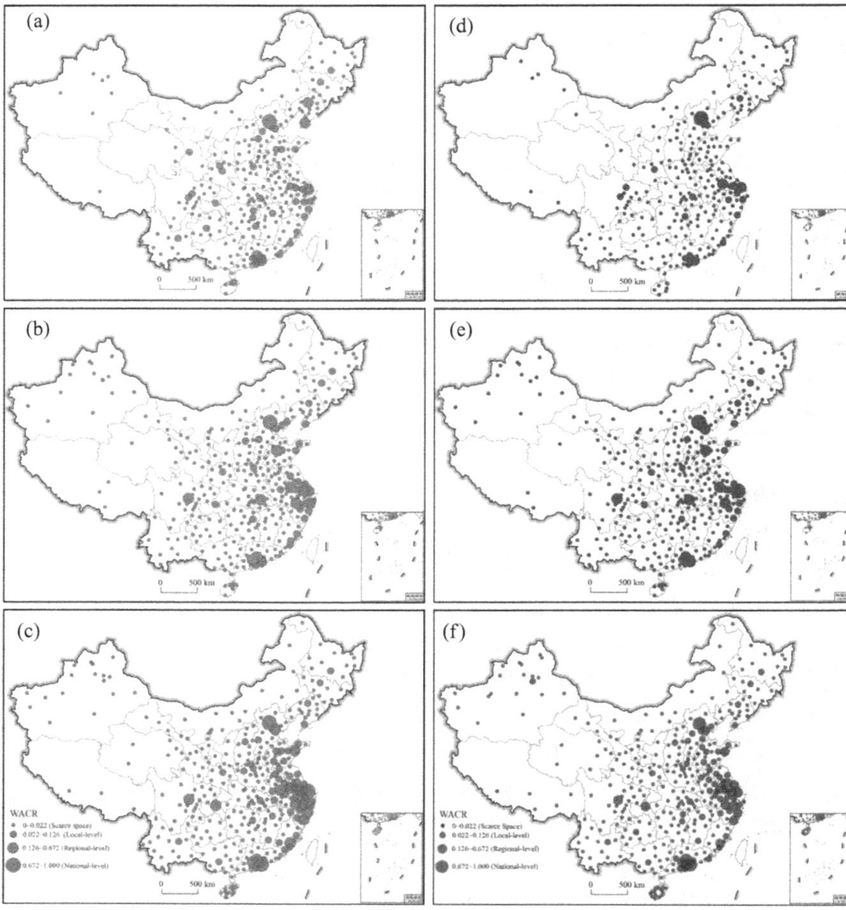

Figure 5.7 Spatial evolution of hubs' hierarchy in national technology transfer network (2014–2018). (a) 2004—2008 (Outflow); (b) 2009—2013 (Outflow); (c) 2014—2018 (Outflow); (d) 2004—2008 (Inflow); (e) 2009—2013 (Inflow) and (f) 2014—2018 (Inflow).

hierarchical structure of the network has throughout remained scale-free property. National and regional-level transfer hubs are heavily concentrated in the YRD and the PRD in spatial, while local-level hubs are scattered throughout central and western provincial capitals as well as eastern subprovincial cities. This distribution pattern resembles a piecemeal concentration at higher levels and a dotted arrangement at lower levels. In this way, the spatial evolution of technology transfer hubs revealed an obvious "place-dependence" or "spatial stickiness". As a result of the "preferential attachment" mechanism, the innovative cities including Beijing, Shanghai, Shenzhen, and Guangzhou, tend to occupy the top positions in the pyramidic hub system, typically exhibiting a Matthew effect. Municipalities, provincial capitals, and subprovincial cities in Eastern China possess robust technology production and absorption capabilities. It is worth noting that many prefecture-level cities in the central, western, and northeastern regions fall into technology transfer hubs, experiencing what may be described as "innovation averse regions" due to the strong filter effect of local social conditions on innovation spillover (Rodríguez-Pose, 1999).

In spatial, technology transfer hubs were emerging with jump-start growth (Wang *et al.*, 2023a). Some eastern technology hubs had continued to emerge, undergoing significant frog leaps and showing some mutability. From 2004 to 2018, the regional-level technology transfer hubs could be seen highly concentrated in the YRD and the GBA regions (e.g., Shaoxing, Wenzhou, Suzhou in the YRD, and Dongguan, and Foshan in the GBA). Over the past five years, cities such as Shenzhen, Shanghai, and Guangzhou have experienced rapid increases in their innovation capacity, transitioning from regional to national levels. Guangzhou, in particular, has seen remarkable progress, rising from a local to a national-level center of technology imports, which is strongly related to its increasing demand of advanced technologies under the upgrading of manufacturing industries and the growth of emerging industries. At the same time, some innovative cities in Central and Western China (such as Chengdu, Hefei, and Xi'an) had significantly increased their R&D capability and upgraded to become regional and local hubs for technology transfer.

Specifically, in terms of technology outflows, the emergence of technology production centers has followed the law of regional agglomeration, leading to the polycentric shifting from three centers to four cores over the past 15 years. This shift followed the locational rules of dominant flows, geographical proximity, and socioeconomic directionality. Beijing had become the primary hub of technology production and diffusion in the national innovation system, and was the largest patent source for major cities in the neighboring or developed provinces such as Hebei, Shandong, Henan, and Jiangsu. Meanwhile, Shanghai, Guangzhou, and Shenzhen were regarded as the secondary nodes and acted mostly as the primary highlands exporting technology to manufacturing centers in their neighboring provinces such as Zhejiang, Jiangsu, Anhui, Fujian, Jiangxi, Hunan, and Yunnan. Therefore, it could be seen that technology transfer had a certain distance constraint. The national technology transfer system diverged into two subgroups: the Guangzhou subnetwork and the Beijing subnetwork, with the superior flow of technology presenting a multilevel hub-and-spoke organizational structure with four cores such as Beijing, Guangzhou, Shanghai, and Shenzhen. Among them, Guangzhou has undergone significant development and serves as the hub of neighboring provinces such as Guangdong, Zhejiang, and Guizhou. Similarly, in the Beijing subnetwork, Beijing spills technology over the eastern, northeastern, and most of the central provinces.

Over the 15-year period, the production highland of advanced technology shifted from north to south, with a high concentration in the YRD and the PRD regions. Northern cities showed a declining trend in technology diffusion, with the number of nodes connected with Beijing falling to 60, overtaken by Guangzhou. The YRD became an "innovation prone region" owing to better innovative milieux and weaker social filter (Rodríguez-Pose, 1999), with about half of the top 20 cities located in the YRD region. Guangzhou, with its advantages of abundant scientific and educational resources and good innovation ecology, had become a newly dominant hub. It maintained close technology linkages with the neighboring developed cities as well as overflowing technology to the central and western cities such as Chongqing, Guizhou, and Haikou.

In terms of technology inflows, there was a tendency for technology absorptive centers to be polycentric and paired with production centers, which spread from the three major urban agglomerations to the coastal area spatially. Over the past 15 years, the urban system constituted of technology absorptive hubs had become polycentric and networked, organized by the hub-and-spoke mode including four hubs of Beijing, Shenzhen, Shanghai, and Guangzhou with four subnetworks (Figure 5.8). The four subnetworks overlapped with the BTH, the YRD, and the PRD urban agglomerations. The technology transfer nodes at the edge of the network were mostly located in the provinces surrounding the three urban agglomerations, mainly driven by geographical proximity.

In general, over the past 15 years, the spatial distribution and network organization of technology production and absorption centers have shown remarkable homogeneity. Firstly, technology absorption and production centers remain highly colocated and coaggregation in spatial, which is strongly related to their economic development, innovative capabilities, and social conditions. Secondly, these hubs have retained stronger technology ties with each other and developed a typical collocational assortativity, followed by the “preferential attachment” mechanism and “rich-club effect”. Thirdly, technology flows among these hubs are mainly organized by three modes: contagious diffusion, jump diffusion, and hierarchical diffusion, which were deeply influenced by geographical, technological, and economic proximities.

5.2.3.2 *Functional evolution of hubs*

During the period 2004–2018, some cities fostered innovative milieu and social conditions of open innovation, resulting in a thriving innovation ecosystem for technology production and absorption, and enhancing collaboration and knowledge-sharing among industry players. Both types of hubs were dominated by complex functions with technological complementarities through technology agglomeration and diffusion. However, technology production (or divergence) centers and technology absorption (or convergence) centers are led by a single function account for a relatively small proportion (Figure 5.9).

During the past 15 years, China’s technology transfer hubs have predominantly adopted an agglomeration-and-diffusion model. This approach has gained traction due to its effective facilitation of

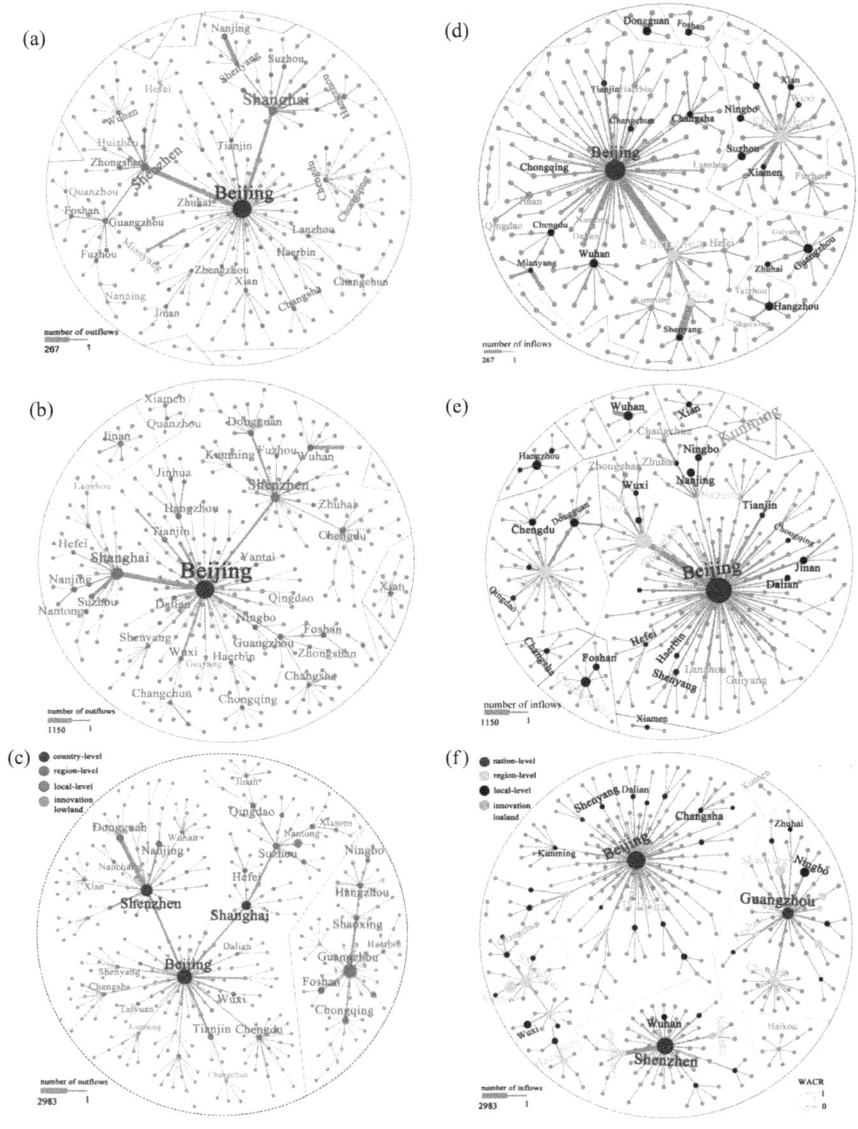

Figure 5.8 Dominant flow dynamics of national technology transfer network (2004–2018). (a) 2004—2008 (Outflow); (b) 2009—2013 (Outflow); (c) 2014—2018 (Outflow); (d) 2004—2008 (Inflow); (e) 2009—2013 (Inflow) and (f) 2014—2018 (Inflow).

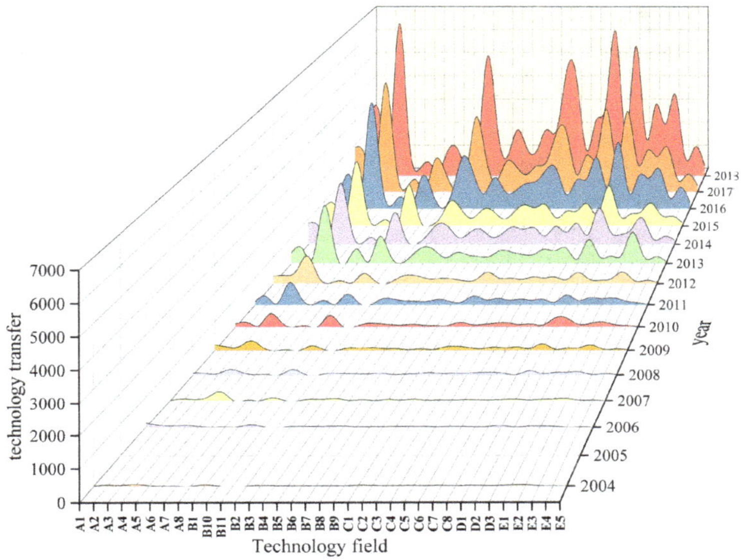

Figure 5.9 Technology diversity dynamics of China's technology transfer hubs (2004–2018).

Notes: (1) *Electrical Engineering*: Digital Communications (A1), Computer Technology (A2), Radio Communications (A3), Electrical Machinery, Instruments, Energy (A4), Audio-visual Technology (A5), IT Management Methods (A6), Semiconductors (A7), Basic Communication Processes (A8); (2) *Chemistry*: Pharmaceuticals (B1), Organic Fine Chemistry (B2), Chemical Engineering (B3), Metallurgical Materials (B4), Biotechnology (B5), Basic Material Chemistry (B6), Food Chemistry (B7), Polymer Chemistry, Polymer (B8); Environmental Technologies (B9), Surface Technologies (B10), Coatings, Microstructure and Nanotechnology (B11); (3) *Mechanical Engineering*: Heat treatment and equipment from (C1), Material Handling (C2), Other Special Machinery (C3), Engines, Pumps, Turbines (C4), Mechanical parts (C5), Traffic Engineering (C6), Machine Tools (C7), Textile and Paper machines (C8); (4) *Other Fields*: Civil Engineering (D1), Other Consumer Goods (D2), Furniture and toys (D3); (5) *Instruments*: Medical Technology (E1), Mechanical measurement (E2), Optical (E3), Mechanical Control (E4), Control System (E5).

matching technology supply with demand, coupled with a strong commitment to openness. The proportion of agglomeration-and-diffusion hubs with both technology clustering and diffusing functions rose from 31.9% in 2004–2008 to 58.3% in 2014–2018, gradually becoming the leading force in the national technology transfer system. The number of technology convergence hubs also increased, from 8.5% at the beginning to 25% at the end, as upgrading and shifting of industries had led to a rising demand for technology in cities, making them a technology "pool". However, the proportion of technology divergence hubs declined significantly, from 31.9% to 16.7%, as the technology spillover effect from cities slowed down.

Technology hubs were mainly concentrated in the eastern coastal region and showed a certain "place dependence" and "path breakthrough" effect. On the one hand, technology hubs were "spatially sticky" and "temporally continuity". In this way, they had maintained their functional roles over a 15-year period and were almost highly concentrated in the eastern coastal cities which mainly included two groups: cities with strong technological innovation capabilities (mainly at the national level such as Beijing and Shenzhen) and cities with a strong open economy (give priority to regional or local level such as Tianjin, Xiamen, and Yantai). On the other hand, several early technological innovation centers and traditional industrial bases have undergone a shift toward agglomeration-and-diffusion hubs, leveraging technological transformation and upgrading. Among them, manufacturing cities including Dongguan, Zhengzhou, Dalian, Shenyang, and Shijiazhuang, have transitioned from divergence-led hubs to complex hubs owing to strong demand for technological advancement. Meanwhile, Jiaxing and Nanjing could convert them from convergence-led hubs to complex hubs, as the integration of YRD urban agglomeration and the implementation of high-tech industry strategies lead to a sustained increase in technological production capacity.

There had been a modest rise in the number of technology convergence-led hubs at the local level, predominantly in well-established traditional manufacturing cities with strong industrial foundations and significant technology demand. These convergence hubs have typically transformed from divergence hubs such as Guangzhou, Guiyang, and Tangshan, which are located near regional technology production centers that serve as "reservoirs" for their technology absorption. The number of technology divergence hubs had declined significantly and was concentrated in the YRD region (including Shanghai, Ningbo, Wuxi, Huzhou, Hefei, etc.), where the local innovative capacities had significantly increased.

The functional fields measured by technology classification were dominated by electrical engineering, chemistry, and mechanical engineering. The overall structure had shifted from two sectors of chemistry and mechanical engineering to a multi-sectoral dominance. Over the 15-year period, China's domestic technology transfer hubs had been highly concentrated in a number of specific technology fields and sectors and maintained a steady growth (Figure 5.9). In terms of technology fields, the equipment manufacturing industry, such as motor-instrument-energy (A4, 10.4%), civil engineering (D1, 7.2%), and machine tools (C7, 5.8%) (Figure 5.10), consistently led the way in terms of patent transfers and became the dominant fields of technology transfer hubs. Manufacturing automation, digitization, and greening were developing rapidly, with the communications equipment industry becoming a strategic pioneer, leading to control systems (E5), digital communications (A1), and environmental technologies (B9) ranking among the fastest growing. In contrast, transferred technologies for pharmaceuticals (B1) had continued to decline due to the long R&D period, high costs, and low success rates, as well as the transformation and upgrading of the national pharmaceutical industry.

At the same time, the functional sectors of China's technology transfer hubs had evolved from single and simple to integrated, and from dual-sector driven to multi-sector led. The proportion of single-sectoral hubs continued to decline, with chemistry, and mechanical engineering, and other sectors being their most dominant combination patterns (Figure 5.11). From the perspective of technology export, the overall structure shifted from dual-sector to multisector dominant. From 2004 to 2008, the leading technology sectors were highly concentrated in electrical engineering, chemistry, and mechanical engineering, and chemistry plus mechanical engineering (Changchun, Harbin, etc.). From 2009 to 2018, the technology diversity in technology transfer hubs increased

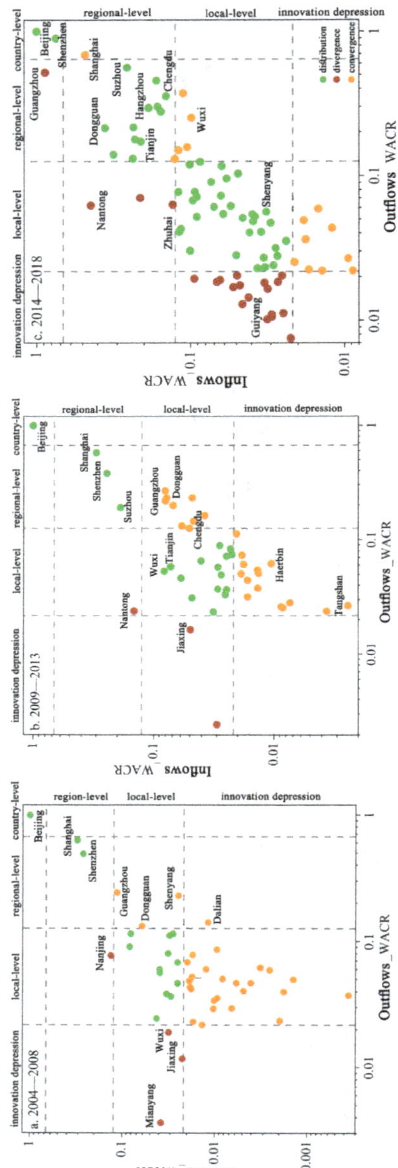

Figure 5.10 Functional evolution of China's technology transfer hubs (2004–2018).

significantly, as the proportion of multi-sectoral patterns increased rapidly, such as "chemistry plus mechanical engineering" and "electrical engineering plus chemistry".

In terms of technology import, it was dominated by dual-sectoral and three-sectoral technology groups (Figure 5.11). From 2004 to 2008, the dual-sector of "electrical engineering+" was the dominant pattern, followed by the multisector dominant pattern. Between 2009

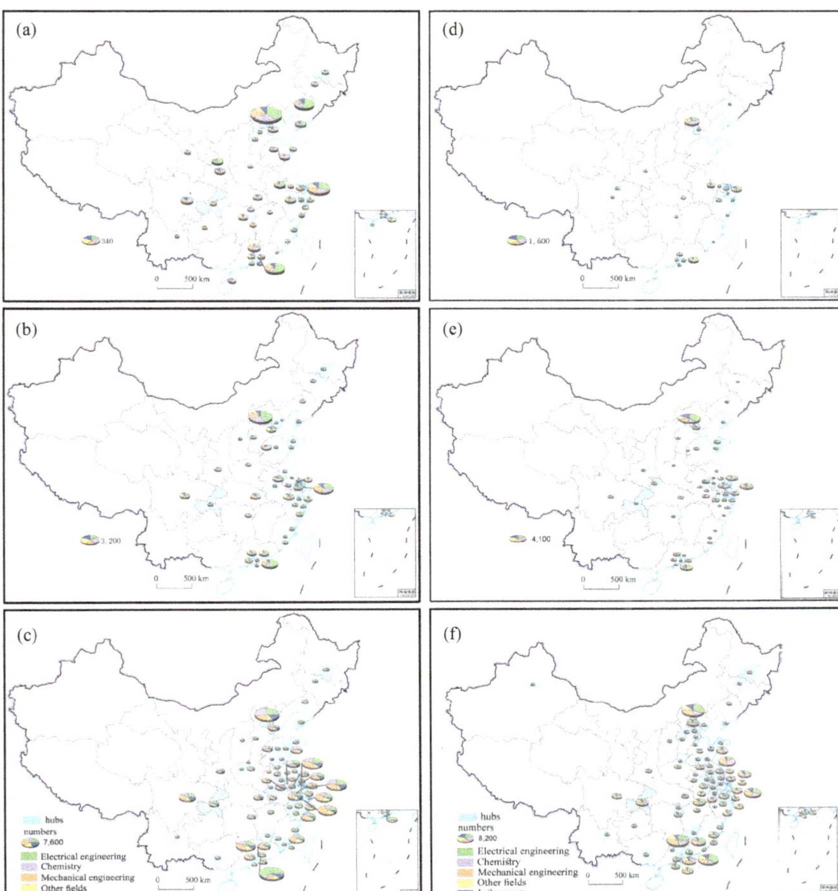

Figure 5.11 Dominant sectoral evolution of China's technology transfer hubs (2004–2018). (a) 2004–2008 (Outflow); (b) 2009–2013 (Outflow); (c) 2014–2018 (Outflow); (d) 2004–2008 (Inflow); (e) 2009–2013 (Inflow) and (f) 2014–2018 (Inflow).

and 2013, there was a relatively even distribution between dual-sector and three-sector dominant modes, including "electrical engineering+", "mechanical engineering+", and "mechanical engineering+ chemical+". Moving into 2014 to 2018, there is a shift in the primary sectors, resulting in the prevalence of "chemical+mechanical engineering+", "mechanical engineering+other sectors+", and "chemical + other sectors+" (Figure 5.11).

Over the 15-year period, enterprises played a more prominent role than universities and research institutions in driving the evolution of functional types within technology transfer hubs. From 2004 to 2018, there was an increase in the proportion of enterprise transferors and transferees at both national- and regional-levels, indicating that high-tech enterprises gradually emerged as the backbone of technology transfer. The national-level saw an increase from 8.14% to 16.71%, while the regional-level presented a growth from 15.48% to 27.17%.

Differing from the two scales, the percentage of transferors decreased from 15.8% to 10.57% at the local level. Meanwhile, universities and research institutions had a lower proportion of technology transfer volume, at no more than 6%. They tended to be stronger in transferring out patented technology rather than absorbing it, mainly acting as producers rather than consumers within the technology transfer ecosystem. At the three stages, the top-ranked companies of transferors and transferees were mainly in the fields of electronics, information and communication technology, and automobile manufacturing, particularly in *Lejin Electronics*, *Huawei*, *Semiconductor Manufacturing International Corporation* (SMIC), *Zhongxing Telecom Equipment* (ZTE), and *Dongfeng Motor*, which were the most active companies in technology transfer. Due to the exclusivity of patented technology, the headquarters–branch organization of enterprises was the most important way of technology transfer, along with technology flows from individuals to enterprises.

5.2.3.3 *Spatial evolution of hubs*

The formation and evolution of technology transfer hubs is a process in which innovation elements are continuously accepted, learned, imitated, and reinvented through spatial diffusion (Jolly, 1997). In this process, along with industrial transfer, some divergent hubs begin to import advanced foreign technologies to maintain innovation capability and competitiveness, while other convergent hubs achieve technological diffusion through the process from imitation to reinvention. As a consequence, both these types gradually transformed into complex hubs with technological agglomeration and diffusion. In a study on the evolution of hubs, Zhao (2020) divided them into the initial stage, the strengthening stage, the rising stage of industrial competition and cooperation, and the maturity stage of cyclical accumulation. Accordingly, technology transfer hubs were also classified into three stages: functional separation in the initial stage, functional integration in the developmental stage, and functional synergy in the mature stage (Figure 5.12) (Wang *et al.*, 2023a).

In the initial stage, the city system of technology transfer was mainly led by divergent-based hubs and complemented by complex

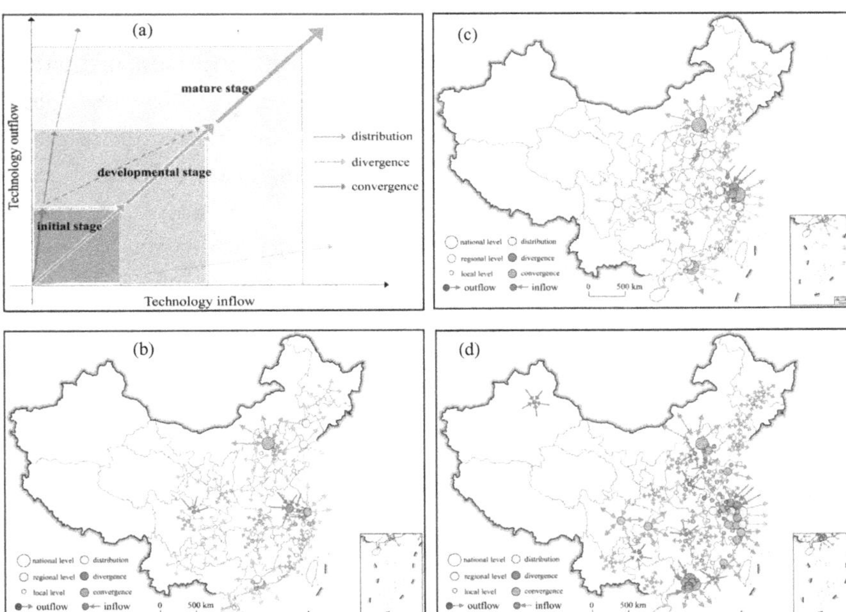

Figure 5.12 Evolutionary stages and modes of technology transfer hubs. (a) Evolutionary stage; (b) initial stage 2004–2008; (c) developmental stage 2009–2013 and (d) mature stage 2014–2018.

ones. The divergent hubs conducted technology diffusion and overflow to the surrounding areas through industrial gradient transfer, and were concentrated on the central and western regions and the northeast (e.g., Zhengzhou, Lanzhou, Shenyang, Changchun, etc.). Complex hubs with strong technology agglomeration and diffusion were centralized in the eastern coastal regions, noticeably in municipalities directly under the Central Government (e.g., Beijing, Shanghai), provincial capitals (e.g., Hangzhou, Wuhan), and sub-provincial cities (e.g., Xiamen, Suzhou), owing to their social filter conditions (innovation-prone societies) (Rodríguez-Pose, 1999). Convergent hubs with developed economic levels, faster industrial transformation and upgrading, and greater technology inflows were agglomeratively distributed in the YRD (e.g., Nanjing, Wuxi, Jiaxing, etc.)

As technology transfer hubs grew into the development stage, their functional evolution showed either consistent progression or abrupt changes. Traditional technology transfer hubs including Beijing, Shanghai, and Shenzhen achieved a balance between technology spillover and absorption by undergoing internal industrial transformation and outward industrial transfer, resulting in typical local dependency. Meanwhile, divergent and convergent hubs evolved from single to composite types, experiencing "path breakthroughs" and functional integration in opposing directions. Divergent-based hubs such as Chongqing and Shenyang renewed their own technologies, upgraded industries, and began relying on external technologies to become regional hubs for integrated technology production and consumption. Convergent-based hubs like Mianyang and Wuxi improved their technological capability and underwent the "introduce, absorption, and reinvention" process, eventually exporting new technologies to the outside.

Mature stage was represented by functional synergy. Many hubs evolved into diffusion and agglomeration functions, achieving synergistic development between technological acquisitions and spillovers. Technology centers and innovation-limited places like Weifang, Xuzhou, Nanning, and Linyi transformed through paths of industrial gradient upgrades and transfer, integrating into regional innovation networks for high-quality open innovation. However, some hubs like Nantong and Zibo maintained their divergent and convergent patterns due to "path dependence" and "technology lock-in", but they shrunk in number and scale over time.

5.2.4 *The hinterworlds*

5.2.4.1 *Scope evolution of hinterworlds*

From 2004 to 2018, the hinterworlds of technology transfer hubs at different levels experienced varying degrees of expansion. According to the scope of their hinterworlds, these hubs roughly followed the order: national level>regional level>local level, along with core hinterworld>peripheral hinterworld. However, only a few local-level (e.g., Zhongshan, Huizhou) and regional-level (e.g., Dalian, Quanzhou, and Tianjin) exporting hubs had larger peripheral hinterworlds than their core ones. About 20% of the importing hubs exhibited this pattern, mainly characterized by significant proximity to the core hinterworld but a wider breadth of peripheral hinterworlds. Some local-level importing hubs in Central and Western China (e.g., Mianyang, Shiyan, and Urumqi) and some complex hubs in Northeast China (e.g., Changchun and Harbin) had larger hinterworlds than their respective regions, due to greater geographical distance or closer technology ties between those hubs and their hinterworlds.

Majority of the cities became the hinterworlds of the four technology growthpoles such as Beijing, Shanghai, Shenzhen, and Guangzhou, owing to their leading innovative and absorptive capabilities. The scope of their hinterworlds expanded rapidly, particularly in their core hinterworlds (Figure 5.13). From 2004 to 2008, the size of their importing hinterworlds was significantly larger than that of those exporting hinterworlds, due to technology backflow of enterprises in manufacturing industries from their R&D branches to their headquarters. However, the scope of their exporting hinterworlds had enlarged rapidly and exceeded that of their technology imports in recent decades, owing to their increasing innovative capability and technology spillover.

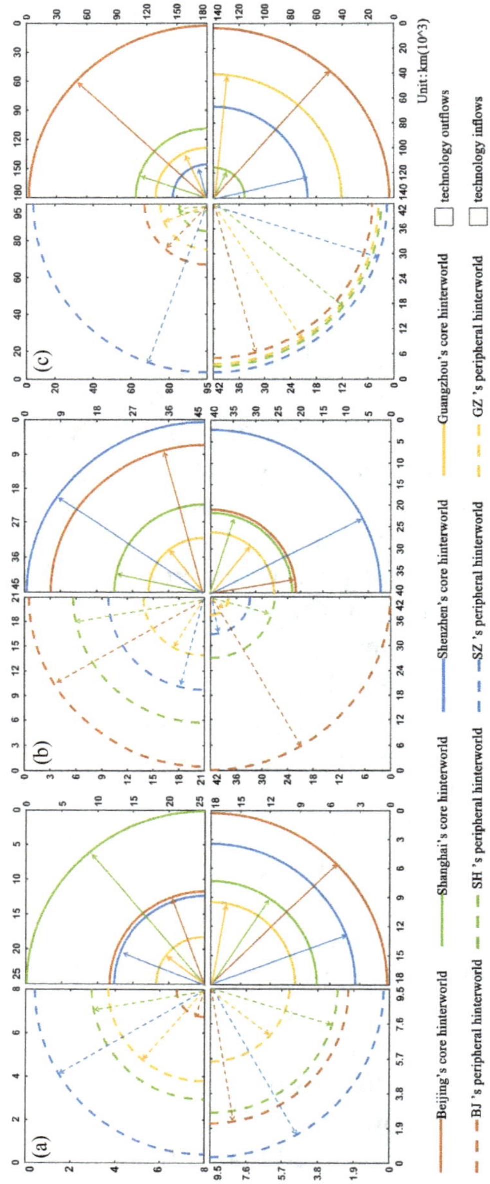

Figure 5.13 Hinterworld dynamics of Beijing, Shanghai, Shenzhen, and Guangzhou (2004–2018). (a) 2004–2008; (b) 2009–2013 and (c) 2014–2018.

As specifically shown in Figure 5.13, the capacity and effect of technology diffusion and spillover in Beijing had increased sharply from 2009 to 2018, leading to the enlarging scope of its hinterworld for technology export exceeding that of import. As a result, it further became the technology growth pole of the national innovation system. During the period, Shanghai had always been a technology export-dominated center, but its core hinterworld had shrunk significantly. Similarly, the size of Shanghai's exporting hinterworlds had also significantly exceeded that of technology import. In recent years, Beijing and Shenzhen had significantly expanded their technological capacity and competitiveness, resulting in a rapid reduction of Shanghai's hinterworlds. Slightly differing from Shanghai, there was a shift of technology flows from an import-led to an agglomeration-and-diffusion balanced pattern in Shenzhen, owing to its improving innovation capacity from 2004 to 2013. However, between 2014 and 2018, Shenzhen had conversely changed from a balanced pattern to an import-led pattern, with a significant improvement in its technology absorption and transformation capacity and a continuous increase in external technology demand (Figure 5.13). In the past five years, Guangzhou's technology market had expanded rapidly, benefiting from the integration of the GBA. As a result, the size of its core hinterworld had approached that of Beijing, and its peripheral hinterworld even surpassed that of Beijing.

5.2.4.2 *Spatial evolution of hinterworlds*

During 2004–2018, there was a colocation of spatial agglomeration and diffusion in their peripheral hinterworlds for majority of the technology transfer hubs. Technology flows between hubs and their hinterworlds developed spatial complexity, which included both gradient and anti-gradient laws, resulting in a complex pattern of contact diffusion, jump diffusion, and hierarchical diffusion (Figure 5.14).

Firstly, the organization of their hinterworlds followed the law of distance decay, due to the effect of geographical proximity. Their hinterworlds exhibited a high degree of concentration in the neighboring cities of the four hubs. Technology flows between them organized by hub-and-spoke network, primarily presented a contact

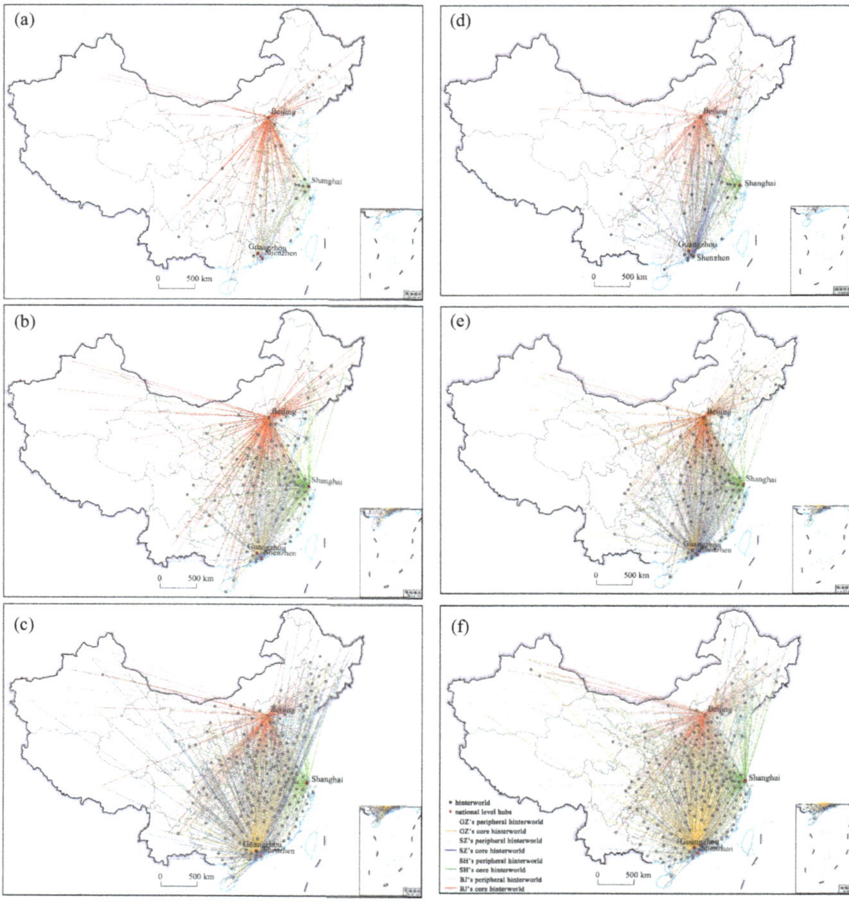

Figure 5.14 Spatial evolution of the hinterworlds of Beijing, Shanghai, Shenzhen, and Guangzhou (2004–2018). (a) 2004–2008 (Outflow); (b) 2009–2013 (Outflow); (c) 2014–2018 (Outflow); (d) 2004–2008 (Inflow); (e) 2009–2013 (Inflow) and (f) 2014–2018 (Inflow).

diffusion driven by geographical proximity. However, the hinterworlds also encompassed cities across provincial boundaries, where agglomeration and diffusion of technology followed the "inverse gradient law" and the "gradient law", resulting in the coexistence of hierarchical diffusion and jump diffusion of intercity technology transfers. Thus, the hinterworlds were characterized by a more extensive and balanced pattern of technology diffusion. Only a few cities (e.g., Zhuhai, Zhongshan, and Shenyang) had their hinterworlds concentrated in the peripheral areas. Furthermore, the hinterworlds of different-level hubs varied significantly in scope, which were influenced largely by both geographical distance and network organization. Among them, the hinterworlds of national-level hubs covered a vast majority of cities across the country through close and long-distance links. In contrast, regional-level hubs' hinterworlds were largely dominated by nearby cities (with patent transfer volume in cities within 300 km accounting for approximately 29%) and medium to long-distance links spanning urban agglomerations (with peaks of 1,000–1,200 km). While local-level hubs' hinterworlds were characterized by a predominance of local links.

Secondly, the hinterworlds of the four hubs were primarily concentrated in the southeast of Hu Line (Heihe–Tengchong Line). Simultaneously, the national innovation system was evolving into a hub-and-spoke network structure with four major hubs — Beijing, Shanghai, Shenzhen, and Guangzhou. From 2004 to 2018, technology production and diffusion centers had been highly concentrated in the three innovative urban agglomerations of the BTH, the YRD, and the PRD along the eastern coast. These agglomerations gave rise to four major technology growthpoles — Beijing, Shanghai, Guangzhou, and Shenzhen, which were reinforced by the "Matthew Effect" mechanism of the network and continued to strengthen the polycentric nature organized by the "hub-and-spoke" network.

The hinterworld surrounding these national-level hubs exhibits clear unevenness and overlap in spatial, with a concentration in the southeastern half. From 2004 to 2018, the hinterworlds of the four hubs were largely overlapped and concentrated in the southeast — mirroring the spatial disparity of population and economy density, which strengthened the emergence of technology and economic complex centers. However, with the shift, transformation, and upgrading of industries, new technology centers have emerged in the central and western regions in recent years. As a result, the hinterworlds of these four hubs have expanded to Northwest and Southwest China.

The core hinterworlds of the four hubs remained relatively stable between 2004 and 2018, with a concentration in their peripheral areas. However, technology diffusion was primarily led by the effect of geographical proximity. Additionally, those core hinterworlds exhibited spatial division during this period. Among them, Beijing's hinterworlds were largely concentrated in the key cities of neighboring provinces including Hebei, Shandong, Henan, and Liaoning. Similarly, Shanghai's exporting and importing hinterworlds were both highly centralized in the Delta region and extended to the middle and upper reaches of the Yangtze River. Meanwhile, Shenzhen's hinterworld was rapidly expanding, while its core hinterworlds were concentrated in southern China, which overlapped with Guangzhou's core hinterworld.

Thirdly, from 2004 to 2018, the hinterworlds of the four national hubs were overlapped and competitive in spatial to some extent. The core hinterworlds remained a regional division to a degree, while the peripheral ones (particularly in the northeast and northwest) became focal points for competition among the four hubs. Notably, Beijing's hinterworlds showed a considerable contraction during this period, particularly in the northeast, north, northwest, and southwest regions of China. Some of Beijing's core hinterworlds (e.g., the Northeast, Henan, Shandong, Jiangsu, etc.) and peripheral hinterworlds (Northwest, Southwest, South China, etc.) were in the face of being "captured" by other hubs. Meanwhile, Shanghai and Shenzhen's hinterworlds expanded rapidly, with their peripheral hinterworlds exhibiting substantial overlap. Shanghai's core hinterworlds rapidly expanded from the YRD to the northeast, the northwest, and the middle-and-upper reaches of the Yangtze River. Similarly, Shenzhen's core hinterworld grew significantly, and it was mainly concentrated in the south of the Yangtze River. Moreover, the core hinterworlds of Guangzhou and Shenzhen were largely overlapped and full of fierce competition because of their industrial, technological, and geographical proximities. Despite a larger overlap with Shenzhen's, Guangzhou's core hinterworlds were less influential and mostly concentrated in Guangdong, Hunan, and Guangxi Province.

5.2.4.3 *Evolutionary types of hinterworlds*

After analyzing the dynamics of hinterworlds of the major hubs, we can categorize three types of hinterworld growth: gradual expansion, leap expansion, and gradual contraction patterns (Table 5.8).

Table 5.8 Growth modes of the hinterworlds of technology transfer hubs.

Pattern	Hubs	The hinterworld of technology export				The hinterworld of technology import			
		2004–2008	2009–2013	2014–2018	Trend	2004–2008	2009–2013	2014–2018	Trend
Gradual expansion	Beijing	Surrounding +Periphery	Surrounding +Periphery	Surrounding +Periphery	Expansion	Surrounding +Periphery	Surrounding +Periphery	Surrounding +Periphery	Expansion
	Shanghai	Surrounding +Periphery	Surrounding +Periphery	Surrounding +Periphery	Expansion	Surrounding +Periphery	Surrounding +Periphery	Surrounding +Periphery	Expansion
	Shenzhen	Surrounding +Periphery	Surrounding +Periphery	Surrounding +Periphery	Expansion	Surrounding +Periphery	Surrounding +Periphery	Surrounding +Periphery	Expansion
	Guangzhou	Surrounding	Surrounding +Periphery	Surrounding +Periphery	Expansion	Surrounding +Periphery	Surrounding +Periphery	Surrounding +Periphery	Expansion
	Chengdu	Surrounding +Periphery	Surrounding +Periphery	Surrounding +Periphery	Expansion	Surrounding +Periphery	Surrounding +Periphery	Surrounding +Periphery	Expansion
	Suzhou	Surrounding +Periphery	Surrounding +Periphery	Surrounding +Periphery	Expansion	Surrounding +Periphery	Surrounding +Periphery	Surrounding +Periphery	Expansion
	Foshan	Surrounding +Periphery	Surrounding +Periphery	Surrounding +Periphery	Expansion	Surrounding +Periphery	Surrounding +Periphery	Surrounding +Periphery	Expansion
	Hangzhou	Surrounding +Periphery	Surrounding +Periphery	Surrounding +Periphery	Expansion	Surrounding +Periphery	Surrounding +Periphery	Surrounding +Periphery	Expansion
	Qingdao	Surrounding +Periphery	Surrounding +Periphery	Surrounding +Periphery	Expansion	Surrounding +Periphery	Surrounding +Periphery	Surrounding +Periphery	Expansion
	Tianjin	Surrounding +Periphery	Surrounding +Periphery	Surrounding +Periphery	Expansion	Surrounding +Periphery	Surrounding +Periphery	Surrounding +Periphery	Expansion
	Wuhan	**Periphery +Surrounding**	**Surrounding +Periphery**	**Surrounding +Periphery**	**Expansion**	**Surrounding +Periphery**	**Surrounding +Periphery**	**Surrounding +Periphery**	**Expansion**

(*Continued*)

Table 5.8 (*Continued*)

Pattern	Hubs	The hinterworld of technology export				The hinterworld of technology import			
		2004–2008	2009–2013	2014–2018	Trend	2004–2008	2009–2013	2014–2018	Trend
	Xi'an	Periphery +Surrounding	Surrounding +Periphery	Surrounding +Periphery	Expansion	Surrounding +Periphery	Surrounding +Periphery	Surrounding +Periphery	Expansion
	Chongqing	Surrounding +Periphery	Surrounding +Periphery	Surrounding +Periphery	Expansion	Surrounding +Periphery	Surrounding +Periphery	Surrounding +Periphery	Expansion
	Zhuhai	Surrounding	Surrounding +Periphery	Surrounding +Periphery	Expansion	Surrounding +Periphery	Surrounding +Periphery	Surrounding +Periphery	Expansion
	Xiamen	Surrounding	Surrounding +Periphery	Surrounding +Periphery	Expansion	Surrounding +Periphery	Surrounding +Periphery	Surrounding +Periphery	Expansion
	Zhongshan	Surrounding	Surrounding +Periphery	Surrounding +Periphery	Expansion	Surrounding	Surrounding +Periphery	Surrounding +Periphery	Expansion
	Dongguan	Surrounding +Periphery	Surrounding +Periphery	Surrounding +Periphery	Expansion	Surrounding	Surrounding +Periphery	Surrounding +Periphery	Expansion
	Ningbo	Surrounding	Surrounding	Surrounding +Periphery	Expansion	Surrounding +Periphery	Surrounding +Periphery	Surrounding +Periphery	Expansion
Leap expansion	Fuzhou	Surrounding +Periphery	Surrounding +Periphery	Surrounding +Periphery	Expansion	Surrounding +Periphery	Surrounding	Surrounding +Periphery	Expansion
	Nanjing	Surrounding +Periphery	Surrounding	Surrounding +Periphery	Expansion	Surrounding +Periphery	Surrounding +Periphery	Surrounding +Periphery	Expansion
	Changzhou	Periphery	Surrounding	Surrounding +Periphery	Expansion	Surrounding +Periphery	Surrounding +Periphery	Surrounding +Periphery	Expansion
	Changsha	Periphery	Surrounding +Periphery	Surrounding +Periphery	Expansion	Surrounding +Periphery	Surrounding+ Periphery	Surrounding +Periphery	Expansion
Gradual contraction	Shenyang	Periphery	Surrounding +Periphery	Surrounding +Periphery	Contraction	Surrounding +Periphery	Surrounding+ Periphery	Surrounding +Periphery	Contraction

The first is the gradual expansion pattern (from interior to exterior). It was the expansion of hinterworlds from the core to the peripheral, which is based on the preferential attachment mechanism and geographical proximity effect. This model dominates the growth of hinterworlds and is noticeably seen in all national and most regional technology transfer hubs, which is highly homogeneous with industrial gradient transfer.

The second is the leap expansion pattern (alternating interior and exterior). It indicates that the hinterworld distribution shows a frog-leaping diffusion pattern and breaks the constraint of geographical distance, which is represented by regional hubs such as Fuzhou, Nanjing, Changzhou, Changsha, etc. Their hinterworlds tend to expand in the opposite direction from far to near. Because of industrial transformation and upgradation in these cities and their surrounding areas, the size of technology transfer toward surrounding areas is gradually expanding, and evolving to a pattern of gradual expansion.

The third is the gradual contraction pattern (from exterior to interior). It was defined as the shrinking in scope of hinterworld from far to near, noticeably in Shenyang. This indicates a gradual contraction of the technology transfer hubs toward a gradual shrinkage pattern, as a result of the outflow of its technologies to the surrounding cities due to its weak industrial transfer capacity.

5.3 Evolution of Technology Categories

5.3.1 *The temporal dynamics*

By identifying the categories of domestic transferred patents, it is possible to know the differences in the fields of technology transferred between domestic cities. The transferred patents can be divided into five sectors and 35 technology fields, with each sector replaced by a different color when counting. It can be found that the size of transferred technologies in each sector continued to grow rapidly, and that the category preferences for domestic technology transfer differed from those of transnational ones over different time periods.

During the period, there was an obvious increment in the size of patents in all fields except in the number of transferred technology fields (Figure 5.15). In the period 2001–2006, almost all technology fields were transferred except microstructure and nanotechnology. Among them, pharmaceuticals topped the list with 347 transfers, followed by the fields of motor-instrument-energy, medical technology, and civil engineering, with about 200 transferred patents. From 2007 to 2012, patents were transferred in 35 technology fields (Figure 5.15). The number of transfers in different technology fields had increased considerably, noticeably in motor-instrument-energy, civil engineering, and pharmaceuticals with more than 2,000 patents. From 2013 to 2018, transferred patent covered all fields and sectors, and grown significantly in all fields (Figure 5.15). There was still an almost 10-fold increase over the previous period, with seven technology fields transferring more than 10,000, namely motor-instrument-energy, civil engineering, machine tools, chemical engineering, other special machinery, material handling, and mechanical measurement.

The share of the mechanical engineering and chemistry sectors was increasing in importance, while electrical engineering was decreasing. The technology field with the largest transfer volume shifted from pharmaceuticals to motor-instrument-energy, but the overall trend in the electrical engineering sector was downward. The period between 2001 and 2006 witnessed a conspicuous trend in domestic technology transfer, with the electrical engineering sector emerging as the clear frontrunner. This sector saw a substantial volume of transfers in computer technology, digital communications, and motor-instrument-energy. The chemical sector also demonstrated active technology transfer during this time, albeit to a lesser extent than electrical engineering. From 2007 to 2012, the chemistry and mechanical engineering sectors had increasing demand while the electrical engineering sector saw a decrease. From 2013 to 2018, the transfer of technologies in mechanical engineering and other sectors continued to increase, while electrical engineering experienced a relative decrease in demand though motor-instrument-energy remained at the top of the transfer fields. The possible

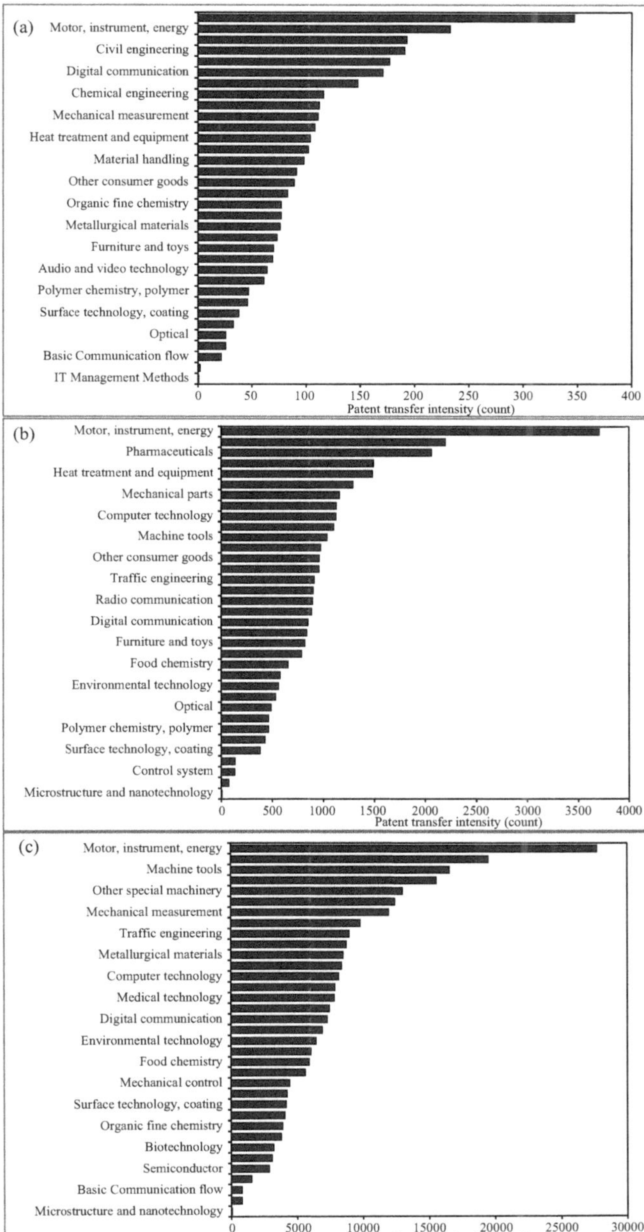

Figure 5.15 Categories of China's transferred technology (2001–2018). (a) 2001–2006; (b) 2007–2012 and (c) 2013–2018.

reasons for this were that the electrical engineering sector was more technologically advanced overall, and the computer and communications fields were generally concentrated in innovative cities with a high level of economic development and there was less demand in such fields, leading to a constant decrease in their circulation.

5.3.2 *The spatial evolution*

5.3.2.1 *Technology transfer-in*

The sectoral weight of domestic technology import was more evenly distributed than transnational import, with large differences in the degree of demand over time and a greater rate of growth in the demand for the mechanical engineering sector across cities.

In detail, the size of domestic patent import was relatively small, and concentrated in the three major cities of Beijing, Shanghai, and Shenzhen acting as the "pool" of technology flows from 2001 to 2006, (Figure 5.16). There was a regional division of functions among them, which is strongly related to their industrial structures. Shenzhen had the highest import from electrical engineering, Beijing was highly dependent on chemistry, and Shanghai greatly needed both electrical and mechanical engineering sectors.

From 2007 to 2012, the clustering effect of the three urban innovative agglomerations, centered around Beijing, Shanghai, and Shenzhen, had a significant impact on the national innovation system. It is evident that demand for electrical and chemical engineering was the highest. Compared to the GBA, the BTH and YRD urban agglomerations had shown a more even-distributed demand for technology across all sectors. Additionally, the YRD region exhibited higher demand for mechanical engineering technology than other regions, while the GBA was dominated by the electrical engineering sector for technology development.

From 2013 to 2018, there was a significant variation in the leading technology sector among different cities, which was greatly dependent on their industrial bases. Chengdu and Chongqing in the western region stood out with their notable performance in

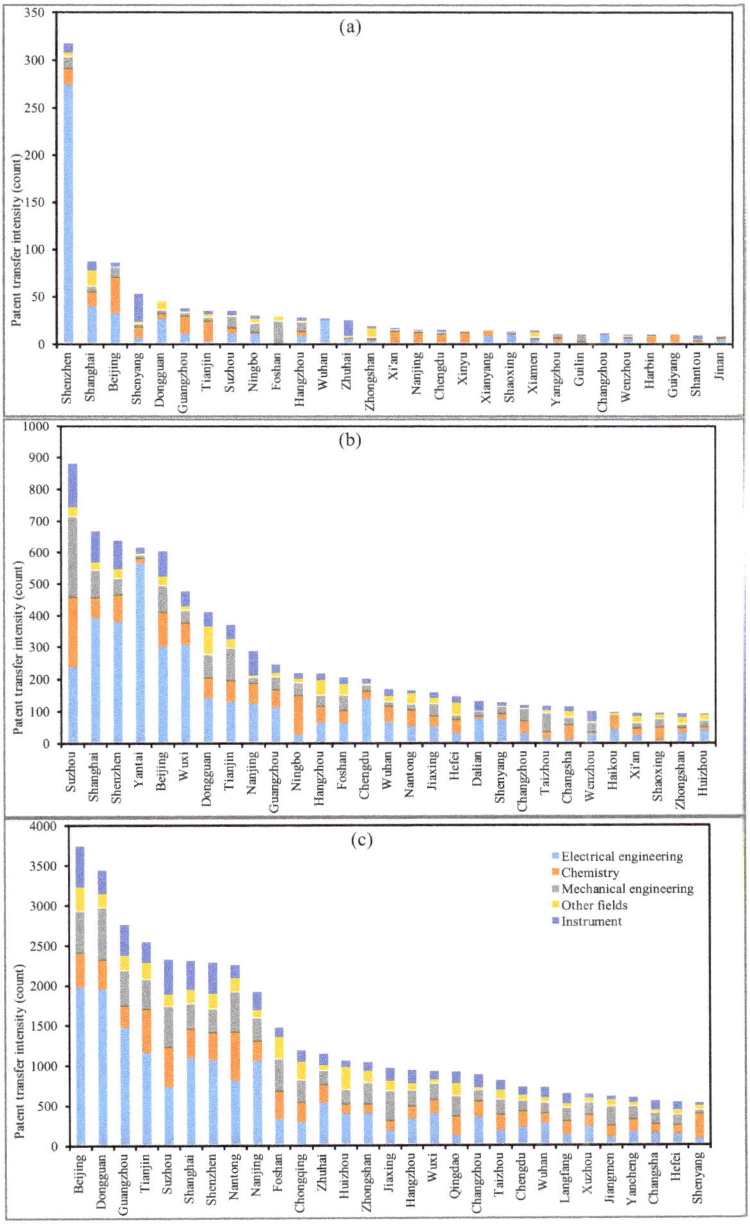

Figure 5.16 Top 30 cities of imported technology categories (2001–2018). (a) 2001–2006; (b) 2007–2012 and (c) 2013–2018.

absorbing domestic technologies in mechanical and chemical engineering. In contrast, the three major urban agglomerations experienced a decrease in the share of technology demand for electrical engineering during this period. The GBA, on the other hand, witnessed the fastest growth in demand for domestic technology in mechanical engineering, making it the most sought-after sector. The YRD maintained a relatively high level of technological demand for the mechanical engineering sector, while the BTH urban agglomeration continued to increase its demand for electrical engineering.

Moreover, the four main terminals exhibited varying demands for specific sectors. Beijing displayed a greater demand for chemistry and mechanical engineering, while Shanghai prioritized electrical engineering and chemical sectors. Shenzhen also had a high import for electrical engineering, while Guangzhou focused on chemistry. However, as time progressed, all four terminals shifted their focus mainly toward key technologies in the electrical and mechanical engineering sectors to support China's manufacturing power plan. It had been estimated that Beijing, Shanghai, Guangzhou, and Shenzhen represented 23–46% of the total technology import at different stages as can be seen in Figure 5.17.

Specifically, from 2001 to 2006, Beijing had the greatest demand for technologies of civil engineering, chemical engineering, and medicine; Shanghai was highly dependent on computer technology, medical technology, and biotechnology; Shenzhen also began to focus on digital communications and radio communications early; Guangzhou was mainly in the field of medicine and pharmaceuticals.

From 2007 to 2012, all four cities had primarily imported technologies of motor-instrument-energy, while Beijing and Shanghai still had a greater need for technologies of civil engineering, leading to some similarities between several cities in terms of directions of technological development.

From 2013 to 2018, technologies in the motor-instrument-energy remained the largest sector in the four cities, and there was no significant differentiation between the four core cities, which played a relatively similar role in the absorption and learning of domestic technologies.

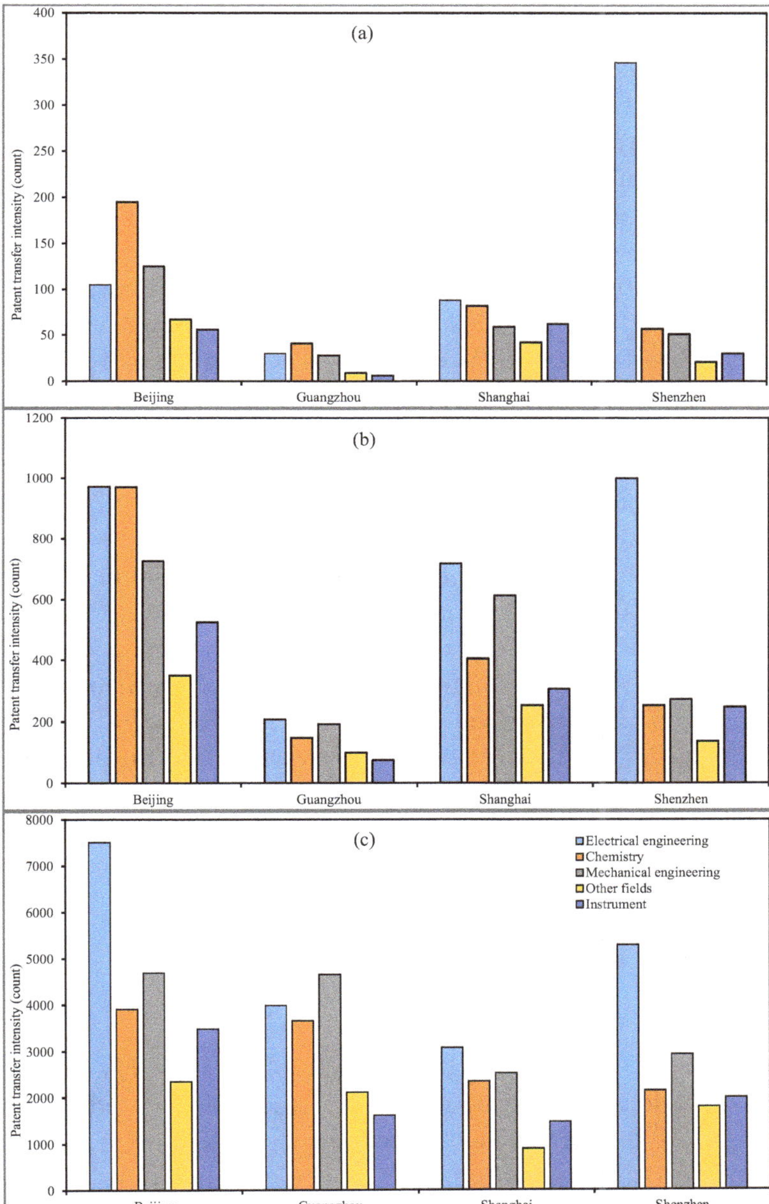

Figure 5.17 The distribution of imported technology sectors among the four hubs (2001–2018). (a) 2001–2006; (b) 2007–2012 and (c) 2013–2018.

5.3.2.2 *Technology transfer-out*

Similar to the technology transfer-in, the sectoral size of technology transfer-out was also more evenly distributed than the transnational one. There were larger differences in the degree of demand over time and greater rate of growth in demand for the mechanical engineering sector across cities.

From 2001 to 2006, domestic patent transfer-out was of a smaller size with the three major hubs of Beijing, Shanghai, and Shenzhen (Figure 5.18). During this period, the percentage of transferred-out patent in five sectors was more than 10%. From 2007 to 2012, the clustering effect of technologies in the three major hubs became increasingly evident. Within this period, the transferred-out technologies were led by electrical, chemical, and mechanical engineering, accounting for more than 3/4 of the total. Notably, the electrical engineering and chemical sectors were largely concentrated on Beijing, and Shanghai exhibited the greatest transfer-out in mechanical engineering. From 2013 to 2018, due to the effect of path dependence, the electrical, chemical, and mechanical engineering sectors still accounted for majority of transferred-out patent. Additionally, Chengdu and Chongqing had an outstanding growth in technology transfer size, owing to the clustering effect of manufacturing industries and the policy bonus of the Western Development Strategy.

5.4 Evolution of Technology Complexity

5.4.1 *Diversity*

Cities tend to adopt more diverse technologies after a period of intensive exploration, during which they may have focused on lower-related technologies. This exploratory phase can lead to an accumulation and recombination of pre-existing knowledge and ideas, resulting in radical innovation (Aghion & Howitt, 1992; Jones, 1995; Weitzman, 1998).

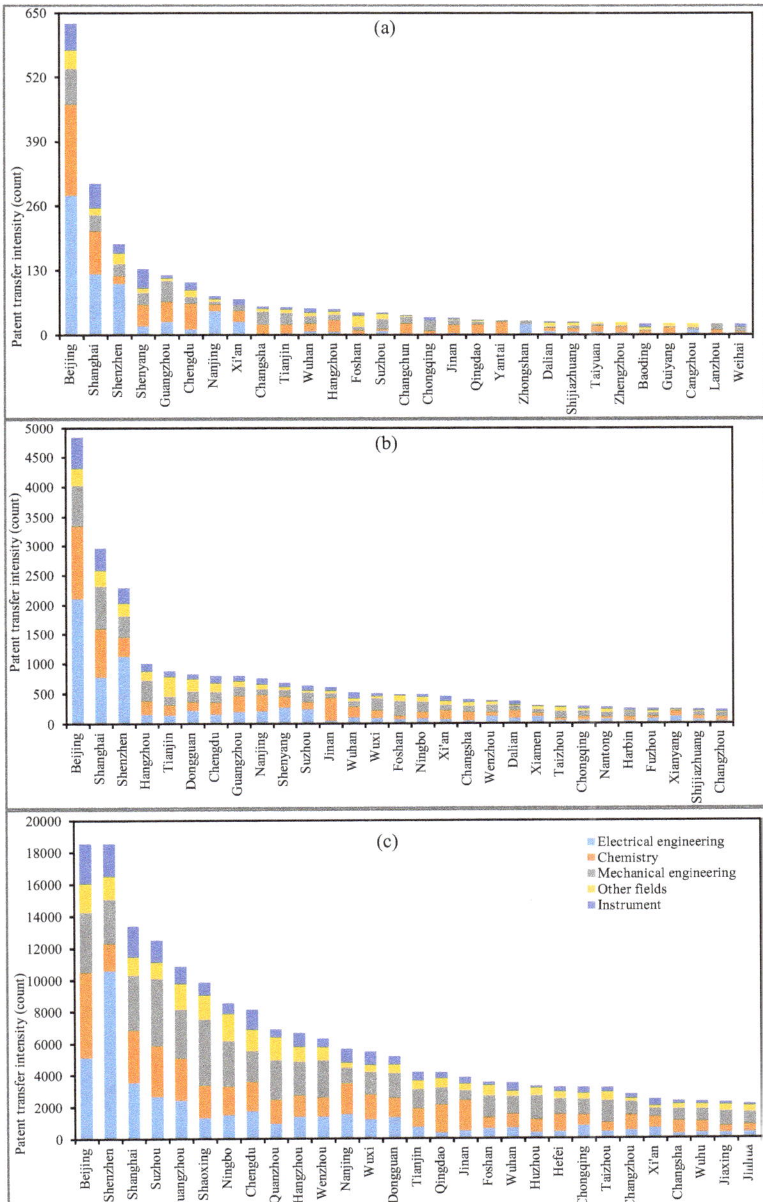

Figure 5.18 Top 30 cities of exported technology categories (2001–2018). (a) 2001–2006, (b) 2007–2012, and (c) 2013–2018.

The higher diversity of domestic imported technologies represents that these cities are more capable of learning and applying different types of technologies and show a greater capacity for technological innovation. As can be seen in Figure 5.19, the diversity of domestic import technologies was increasing across cities; there was spatial expansion of imported technologies from the eastern coastal region to the central and western regions, but the difference among cities was decreasing.

The diversity of domestic import technologies in cities had been increasing, owing to the enhancement of urban absorptive capacity. Furthermore, the cities with higher diversity of imported technologies had been expanding from the eastern coastal region to the central and western regions. However, there isn't a significant gap between these cities, and the dominance of core cities like Beijing, Shanghai, and Shenzhen has been weakened.

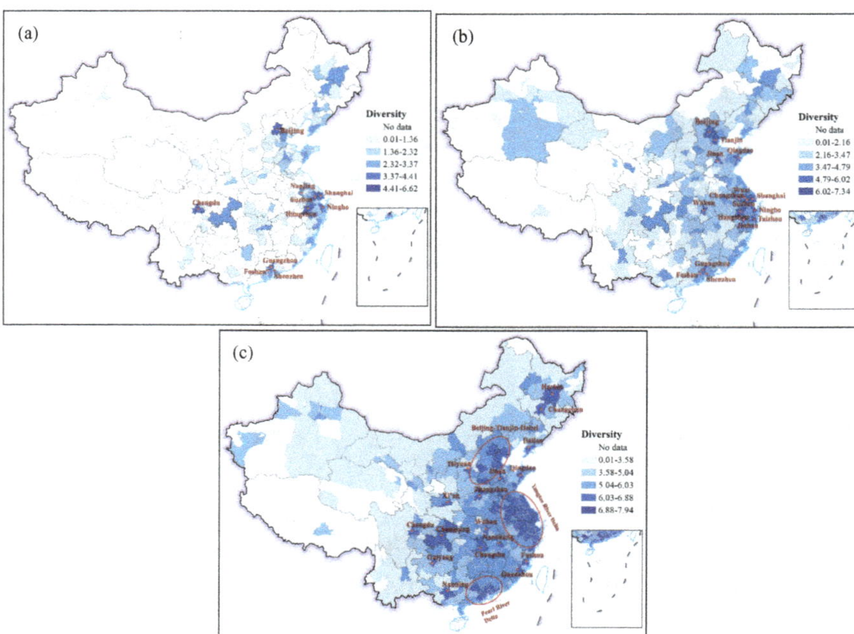

Figure 5.19 Spatial disparity of domestic importing technology diversity (2001–2018). (a) 2001–2006; (b) 2007–2012 and (c) 2013–2018.

From 2001 to 2006, Beijing, Shanghai, Guangzhou, Suzhou, and Hangzhou were among the cities with a higher diversity of imported technologies, surpassing a score of 5. Meanwhile, 152 cities displayed technological diversity, with more rich and diverse technologies than those sourced from international trade. From 2007 to 2012, there was an increase in the threshold range of diversity, which also witnessed an overall growth in the variability of imported technologies. A total of 302 cities displayed technological diversity during this phase, as there was a significant increment in the number of cities importing diverse technologies. From 2013 to 2018, the threshold range of diversity continued to expand. Most of the Chinese cities have imported diverse technologies from other domestic cities. Indeed, the overall technological strength of cities and their capacity to absorb external technologies had continued to grow.

5.4.2 *Relatedness*

The effectiveness of knowledge combination depends on the relatedness between imported and existing technologies, while a lower relatedness implies a more diverse combination. In other words, the specific types of knowledge involved before and after the combination are critical factors in determining the success of the innovation process (Boschma *et al.*, 2008).

The relatedness of transferred technologies can be calculated to understand the correlation between domestic technology absorbed by each city and its local technology bases. As can be seen from Figure 5.20, the number of cities importing related technologies showed an increasing trend. However, there was a spatial disparity in the values of specific relatedness, namely the imported technologies with higher relatedness were centralized in the three urban agglomerations of the BTH, YRD, and GBA, similar to the distribution of technology transfer intensity, network, and categories.

The variation in the relatedness between imported technologies and local technology bases was more diverse, but the number of cities that involved in the process showed an increasing trend.

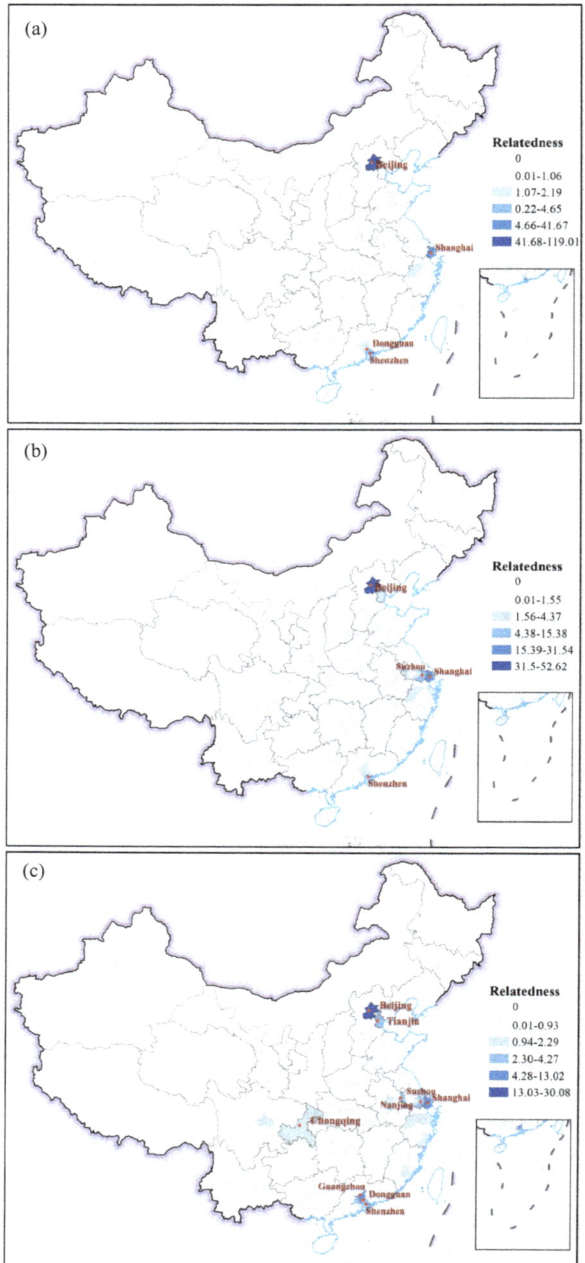

Figure 5.20 Spatial disparity of domestic importing technology relatedness (2001–2018). (a) 2001–2006; (b) 2007–2012 and (c) 2013–2018.

Meanwhile, the cities with a higher technology relatedness were concentrated in the three major urban agglomerations of the BTH, YRD, and GBA, so that their core position had been continuously highlighted (Figure 5.20). From 2001 to 2006, except for fewer innovative growthpoles such as Beijing, Shenzhen, and Shanghai, most of the cities had a relatedness value below 0.001, indicating a significant difference or similarity between their imported technology and local knowledge bases. From 2007 to 2012, the three core cities of Beijing, Shenzhen, and Shanghai demonstrated a narrowing threshold range and decreasing relatedness values. However, there was an increasing trend for most other cities. It further indicates that local technologies in some cities, particularly Beijing, Shanghai, and Shenzhen, were less correlated with their imported technologies, and instead, others were more likely to import similar technologies. Furthermore, it's worth noting that there was a significant increase of about three times over the first period, in terms of the number of cities importing related technologies. From 2013 to 2018, a narrowing threshold range and a decreasing degree of relatedness in a small number of cities could be seen. Despite this trend, a majority of cities demonstrated an increase in relatedness value during this period. In fact, a total of 320 cities imported related technologies that align with their local technology bases.

5.4.3 *Similarities*

Regions that have already patented technology in their local knowledge base stand to benefit from the influx of new knowledge, which can be harnessed in creative ways. The effect of such knowledge flow on a region's innovation performance is influenced by the similarity between its internal technology bases and external technology. In fact, research has shown that higher levels of technological similarity lead to greater innovation outcomes (Miguelez & Moreno, 2018).

The similarity of technologies represents the strength of a city's import of homogenous technologies. As can be seen in Figure 5.21, the similarity between external technologies and local knowledge bases was rising across cities, and the number of cities importing

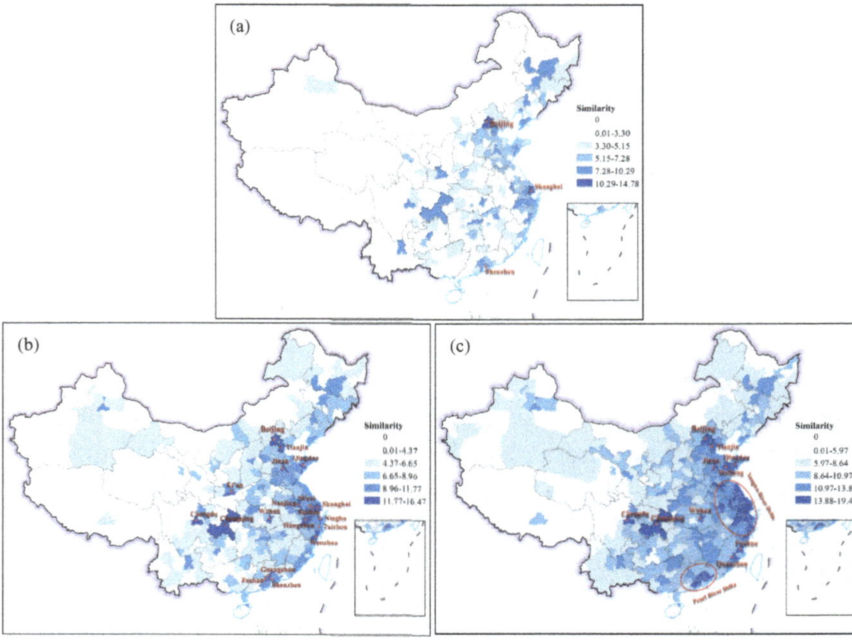

Figure 5.21 Spatial disparity of domestic importing technology similarity (2001–2018). (a) 2001–2006; (b) 2007–2012 and (c) 2013–2018.

more homogenous technologies was also increasing. The cities with a high-degree homogeneity of their imported technology had been expanding westward from the eastern coastal region toward YRD, GBA, and PRD, while the spatial differences in similarity values had been decreasing.

In particular, the technological similarity of cities showed an increasing trend, indicating increasing capacity to absorb homogenized technologies and foster incremental innovation among the cities. From 2001 to 2006, a total of 178 cities in China shared similar technologies, while the number of cities with technological similarity has reached 310 during the second period, with a growth rate of 74.16%, demonstrating the remarkable capacity of these cities to absorb homogenized technologies. From 2013 to 2018, there still was a noticeable trend toward increased technological similarity, and a total of 327 cities demonstrated technological similarity.

However, the similarity of imported technologies was not well distributed, and concentrated on a few megacities along the southeastern coast, particularly in the three urban agglomerations of BTH (i.e., Beijing and Tianjin), YRD (i.e., Shanghai, Suzhou, Nanjing, Wuxi, etc.), and PRD (i.e., Shenzhen, Guangzhou, and Dongguan), which recorded a similarity value exceeding 13.

5.4.4 *Path evolution*

As mentioned above, the dynamics of technological structures suggest that when cities adopt external technologies they tend to focus on areas related to their existing technology bases. This is attributed to the path-dependence effect, which reduces the cost and risk of learning new technologies, while also reinforcing local specialization (Jin & Zhu, 2021). On the other hand, the trajectory of regional industry is in a constant state, as the process of creative destruction (as first articulated by Schumpeter in 1947) continually redirects and redefines it.

Moreover, technological discontinuities can create new paths for regions by replacing outdated or inferior technologies with superior ones (as Baum & Dutton noted in 1996). As a result, regions may become less reliant on technological relatedness and venture into industries that are either less related or completely unrelated to their previous activities, which is called path breakthrough.

The paths of local innovation and the categories of external technologies are also mutually influenced. By analyzing the relatedness and similarity dynamics of domestic imported technologies, it is possible to determine the path dependence and path breakthrough of urban technological development under the influence of global pipelines. This book classifies the top 1% of cities from 2013 to 2018 as the base of cities with more obvious path breakthroughs (Zhang & Sun, 2016), and therefore uses the values Relatedness = 0.001 and Similarity = 12 as the dividing line to analyze the technological path evolution of each city in different periods. According to the coordinate, they can be divided into four modes, which are mixed path breakthrough and path dependence.

As shown in Figure 5.22, many cities were initially in the path exploratory stage. However, over time, a growing number of cities have transitioned to a path-dependence-based model. Core cities

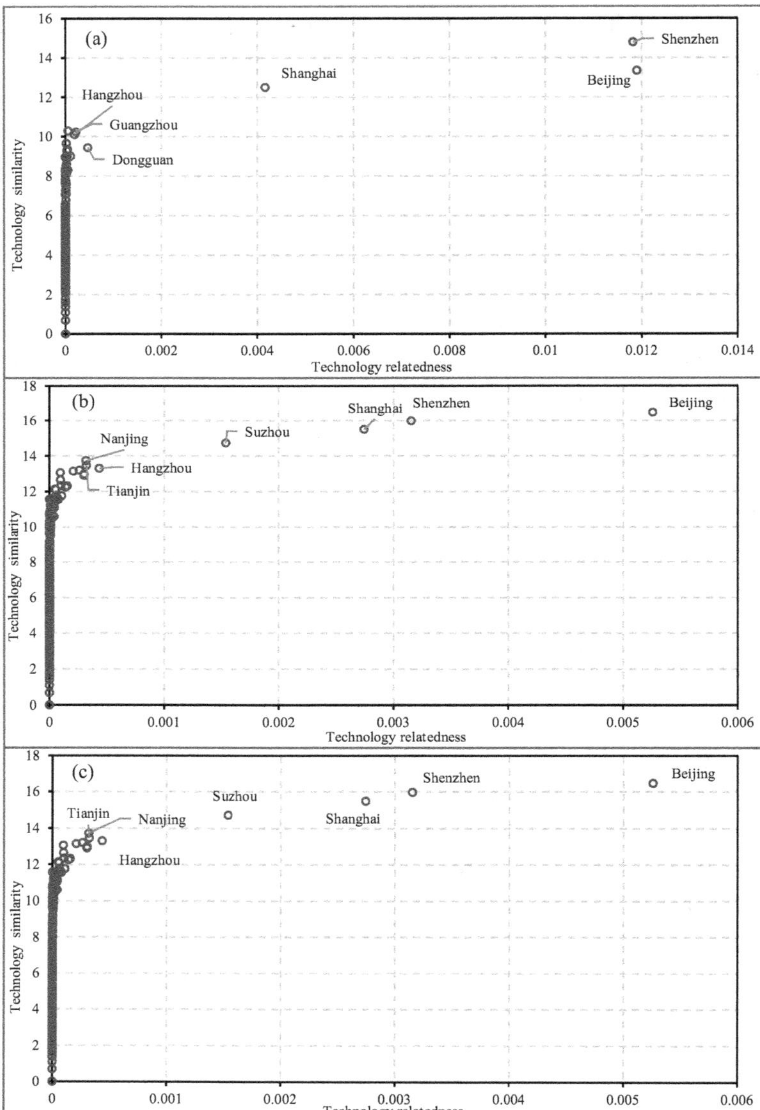

Figure 5.22 Evolutionary path of domestic technology transfer cities. (a) 2001–2006; (b) 2007–2012 and (c) 2013–2018.

such as Beijing, Shanghai, and Shenzhen have adopted a mixed model of path breakthrough and path dependence, while other economically advanced cities have embraced a path dependence model. Interestingly, no city has yet achieved a path breakthrough mode through the adoption of domestically imported technologies. These findings are explained in detail in different periods as follows:

From 2001 to 2006, the three innovative hubs such as Beijing, Shenzhen, and Shanghai imported a variety of domestic-related and similar technologies, indicating that they were not merely relying on existing local technologies, but were actively seeking out different technologies to achieve breakthrough innovation. However, it's worth noting that no city was solely dependent on a single path during this period, highlighting the cities' ability to absorb new technologies despite their weaker absorptive capacity of external technologies. Meanwhile, most other Chinese cities, such as Guangzhou, Hangzhou, and Dongguan were still in the early stages of exploring technological innovations and lacked imported technology spillover.

From 2007 to 2012, Suzhou grew up into the complex mode of both path dependence and path breakthrough, keeping up with the leaders — Beijing, Shenzhen, and Shanghai. These cities continuously imported technologies that set them apart from other cities, resulting in significant radical innovations. It demonstrated an increased ability for them to absorb domestic-related and similar technologies, which contributed to its momentum in technological innovation.

From 2013 to 2018, only Beijing, Shenzhen, and Guangzhou were in a state of mixed path dependence and breakthrough. It was apparent that Beijing and Shenzhen possessed the most advanced technological strength, as they continued to absorb domestic-related technologies and showed greater potential for achieving technological breakthroughs. Meanwhile, Guangzhou was on the rise, with an increasingly greater ability to absorb both related and similar technologies. In contrast, Shanghai regressed to a path-dependence mode preferring to absorb similar technologies, which ultimately reduced its potential for technological breakthroughs. Furthermore, numerous cities, such as Dongguan, Foshan, Tianjin, and Nanjing, also found themselves locked into the path dependence mode. This indicated that more cities had started to seek external technologies to support local innovation and had absorbed not just similar but also difficult and advanced external technologies. However, there were still many cities in the path exploration stage, which needed to break free from internal restrictions on technological development and promote local technological innovation with an open vision and actionable steps.

Overall, it was clear that more cities had advanced to this mode during this phase, and this trend signaled a growing awareness of the need to seek out new technological solutions beyond their current capabilities.

5.5 Summary

This chapter analyzes the dynamics of spatial heterogeneity on the intensity, network, category, and relatedness of intercity technology transfer in China. Some important findings are quantitatively revealed as follows:

The domestic technology transfer network has demonstrated scale-free characteristics, which have gradually been diminishing over time with the spatial differences becoming less pronounced. The rank–size distribution of interurban technology transfer activities has exhibited a clear power–law phenomenon. A few cities such as Shanghai, Beijing, Shenzhen, Suzhou, Tianjin, Qingdao, Hangzhou, Dongguan, Chongqing, and Wuxi are emerging as innovative growthpoles for technology transfer activities. However, it is worth noting that these activities are greatly centered around the four hubs — Beijing, Guangzhou, Shenzhen, and Shanghai, which have collectively imported less than 30% of the total. Furthermore, in the later stages of the network, the absolute value of the power exponent has decreased, resulting in lower network heterogeneity and a smaller gap in technology absorptive capacity among different cities.

Technology transfer activities have highly followed the clustering effect at macro-spatial level. Spatial centralization is particularly evident in technology productions and flows, resulting in the emergence of innovative clusters to facilitate collective learning and sharing of knowledge. It is not only the size of technological productions and technology transfers in the city and the connection of hubs of the technology transfer network that matters but also the concentration of external technological diversity and relatedness.

First, whether in terms of the intensity or centrality of technology flows, the three major cities of Beijing, Shanghai, Shenzhen, and their urban agglomerations of the BTH, the YRD, and the GBA, have been the hubs of interurban technology transfer network.

Second, interurban technology transfer activities have followed spatial heterogeneity including disparity, hierarchy, and clustering properties, which is consistent with a spatial distribution of urban innovation activities and capabilities.

Third, there is also a more pronounced agglomeration in the category structure of external technologies. The highest demand for intercity technologies has shifted from pharmaceuticals to motor-instruments-energy.

Fourth, there is still a more evident agglomeration effect in cities with high absorptive capacity for diverse and related technologies. A gradual diffusion from cores such as Beijing, Shanghai, and Shenzhen to provincial capitals or three urban agglomerations has been observed, which is dependent on their better economic levels and S&T infrastructures.

Cities are increasingly diversifying their demand for external technologies, with higher absorptive capacity for both different and similar technologies. Across all indicators related to technology categories and relatedness, cities have demonstrated an increase in their absorptive capacity. In addition, the technology fields covered by domestic patents are becoming more diverse, with greater complexity in the absorbed technologies. Meanwhile, the number of cities importing external technologies with varying levels of diversity, relatedness, and similarity has also grown, along with corresponding increases in the relevant indicators.

Technology transfer activities in spatial contexts exhibit characteristics of both path dependence and path breakthrough, with a trend toward greater complexity in technology absorption. The scope of intercity technology transfer activities has been expanding, and the number of transferring pipelines has increased to foster path creation. However, the triangular structure of domestic pipelines among Beijing, Shanghai, Guangzhou, and Shenzhen remains stable due to the "place dependence" effect of innovative activities. Most of the cities are currently at a stage where the intensity of both similar and related technology transfer is relatively low, but there is a faster growth rate for similar technology import, which reflects the learning of existing local technologies under the path dependence. On the other hand, core cities such as Beijing, Shanghai, and Shenzhen have imported relatively more related technologies, thereby enabling them to learn and combine these differential technologies, leading to a mixed pattern of both path breakthrough and path dependence.

China's technology transfer hubs exhibit triple characteristics: hierarchical organization, spatial polarization, and functional complementarity. These hubs represent high-level nodes to control large amounts of technology flows in the national innovation system, concentrated on the coastal cities such as Beijing, Shanghai, Shenzhen, and Guangzhou. Specifically, the hierarchy has developed into a stable pyramid-shaped structure, with stronger hierarchical consolidation, place dependency, and spatial stickiness. The hubs display a shift from the bipolar-led (Beijing and Shanghai) to a polycentric pattern with innovative clusters (the BTH, YRD, and GBA). In terms of functional organization, decentralization is predominant, with functional types evolving from dual-sectoral dominance to multisectoral fusion. It is concluded as three dynamic patterns: functional separation at the initial stage, functional integration at the developmental stage, and functional synergy at the mature stage. The hubs' hinterworlds exhibit significant dual properties with expansion and competition, displaying a two-way mode of spatially surrounding and peripheral expansions. It has formed three different hinterworld growth patterns: gradual expansion (from interior to exterior), frog-leaping expansion (alternating interior and exterior), and gradual contraction (from exterior to interior).

The spatial distributions of nodal degree and weighted degree in the domestic technology transfer network are significantly heterogeneous and hierarchical. The larger cities play the leading role in technology incubators, highly related to their economic levels, research and development (R&D) investment, and social filter conditions. The evolution of intercity technology linkages is driven by the networking mechanisms of preferential attachment and self-reinforcement effect, hierarchical and contagious diffusion, path dependence, and path breakthrough. Moreover, we found that geographical proximity and technology gaps are crucial determinants of intercity technology transfer intensity. As a result, it has been discovered that the network in the BTH urban agglomeration is organized in a tree-like network, while the YRD region features a polycentric network and the PRD has multistar characteristics. Beijing, Shanghai, and Shenzhen are emerging as innovative growth poles in the three urban agglomerations.

CHAPTER 6

SPATIAL EVOLUTION OF REGIONAL TECHNOLOGY TRANSFER NETWORK

This chapter delves into the spatial evolution of technology transfer networks (TTN) across typical regions in China, including the Beijing–Tianjin–Hebei (BTH) urban agglomeration, Yangtze River Delta (YRD) urban agglomeration, Pearl River Delta (PRD) urban agglomeration, and the three northeastern provinces (including Helongjiang, Jilin, and Liaoning). By utilizing social network analysis and negative binomial regression model, we aim to explore the spatial heterogeneity of TTNs in different regions in terms of nodal centrality and linkage intensity, as well as their influencing factors. Our findings reveal that these technology transfer networks exhibit a core–peripheral structure with hierarchies driven by multidimensional proximity mechanisms, which are common features across different regions. Furthermore, intercity technology flows in the three major urban agglomerations play a crucial role in sustaining their innovation vitalities and enhancing their innovative capacities (Liu *et al.*, 2019). However, we also identify a key issue hindering innovative performance in the three northeastern provinces, namely the strong nationalization and weak localization of their technology flows (Liu & Niu, 2019).

6.1 The Three Urban Agglomerations

6.1.1 *Evolution of urban centrality*

6.1.1.1 *Degree centrality of cities*

Table 6.1 lists the top five cities with the largest urban degree centrality and betweenness centrality within the three major urban agglomerations in 2008, 2012, and 2015. The four megacities including Beijing, Shanghai, Shenzhen, and Guangzhou are all highly central to the three TTNs, acting as their hubs through a hub-and-spoke networking organization.

Within their urban agglomerations, Beijing and Shanghai occupy dominant positions. In 2015, Beijing and Shanghai started to establish connections with other cities within the BTH and YRD regions, respectively. Meanwhile, Shenzhen and Guangzhou had the highest degree centrality in the PRD, indicating a trend for cities to link up with highly innovative and economically advanced urban areas. These findings further demonstrate that the spatial evolution of these TTNs was influenced by the preferential attachment principle or Matthew effect. Generally, degree centrality values were observed to increase within the three major agglomerations during the periods, but this increase was diminished in 2012 and 2015, compared to 2008, owing to the increase in external technology linkages outside the three regions.

6.1.1.2 *Betweenness centrality of cities*

Beijing, Shanghai, and Shenzhen were observed to dominate technology transfer in these networks. These cities acted as important bridges for technology flows within the networks. However, there was a considerable discrepancy among urban betweenness centrality in the three innovative clusters. Among them, the betweenness centrality for the BTH region increased from 2008 to 2015, while the YRD and PRD regions experienced a downward trend. As intercity linkages increased, the betweenness centralities of cities in the YRD and PRD regions tended to decrease for top-ranked cities in the hierarchy system. In other words, lower-ranked cities were less

Table 6.1 Top five cities with the largest degree centrality and betweenness centrality within the three urban agglomerations in 2008, 2012, and 2015.

Indicators	Degree centrality						Betweenness centrality					
Year	2008		2012		2015		2008		2012		2015	
The BTH region	Beijing	0.75	Beijing	0.92	Beijing	1.00	Beijing	0.43	Beijing	0.60	Beijing	0.66
	Tianjin	0.33	Baoding	0.54	Tianjin	0.67	Tianjin	0.12	Tianjin	0.05	Tianjin	0.06
	Baoding	0.17	Tianjin	0.46	Shijiazhuang	0.58	Tangshan	0.08	Shijiazhuang	0.03	Shijiazhuang	0.05
	Langfang	0.17	Cangzhou	0.38	Langfang	0.42	Langfang	0.00	Baoding	0.02	Hengshui	0.04
	Tangshan	0.17	Langfang	0.38	Xingtai	0.33	Qinghuangdao	0.00	Langfang	0.01	Langfang	0.04
The YRD region	Shanghai	0.44	Shanghai	0.88	Shanghai	1	Shanghai	0.33	Shanghai	0.29	Shanghai	0.19
	Hangzhou	0.32	Suzhou	0.64	Suzhou	0.88	Hefei	0.2	Suzhou	0.13	Nanjing	0.09
	Hefei	0.32	Hangzhou	0.6	Ningbo	0.8	Nanjing	0.2	Hangzhou	0.09	Suzhou	0.07
	Nanjing	0.28	Ningbo	0.56	Nanjing	0.8	Hangzhou	0.12	Nanjing	0.09	Hefei	0.07
	Ningbo	0.24	Nanjing	0.56	Hangzhou	0.72	Wuhu	0.06	Ningbo	0.06	Hangzhou	0.05
The PRD region	Shenzhen	0.62	Guangzhou	0.92	Shenzhen	0.92	Shenzhen	0.31	Guangzhou	0.28	Shenzhen	0.26
	Guangzhou	0.46	Shenzhen	0.85	Guangzhou	0.85	Foshan	0.19	Shenzhen	0.24	Guangzhou	0.14
	Foshan	0.38	Foshan	0.77	Foshan	0.85	Guangzhou	0.15	Foshan	0.15	Foshan	0.11
	Dongguan	0.31	Dongguan	0.46	Jiangmen	0.69	Dongguan	0.08	Dongguan	0.08	Jiangmen	0.02
	Zhongshan	0.31	Zhongshan	0.38	Dongguan	0.54	Zhongshan	0.00	Zhongshan	0.01	Dongguan	0.02

dependent on the top-ranked cities due to increased connectivity. However, there existed a different situation in the BTH region with a single-core network led by Beijing, namely most cities had stronger connections with Beijing and weaker connections with other cities in the region.

As mentioned earlier, the centrality of cities is based on their position in the network which in this chapter specifically refers to their technology transfer capacity. Urban technology transfer capacity is highly related to its social filter conditions such as the economic size, R&D investment, human capital, cultural milieu, etc. Thus, cities with higher centrality generally exhibit more developed economies, greater investment in R&D, and higher innovation performance, compared to those with lower centrality. Meanwhile, under the impetus of the preferential attachment mechanism, the cities with higher centrality hold a more significant position in the network either.

6.1.1.3 *Weight degree centrality of cities*

All cities in the three agglomerations are divided into four hierarchies according to natural breaks classification using ArcGIS 10.3, based on their degree centrality and weight degree centrality (Figure 6.1). Over time, cities experienced increases in both degree and weight degree, as well as a more positive patent transaction. At the same time, the scope of intercity technology transfer activities expanded to encompass the entire urban agglomeration (Figure 6.1).

According to Figure 6.1, there was a gap in technology transfer between Hebei Province and Beijing and Tianjin municipality. Beijing held the most prominent position in the BTH network with the highest degree and weight degree values. Tianjin also played an essential role as a subcore city in the network. However, most cities were locked in the peripheral position within the network, except for Shijiazhuang and Langfang in 2015. In the YRD, Shanghai served as the primary hub and spearheaded technology transfer activities within the network. Suzhou emerged as the secondary hub

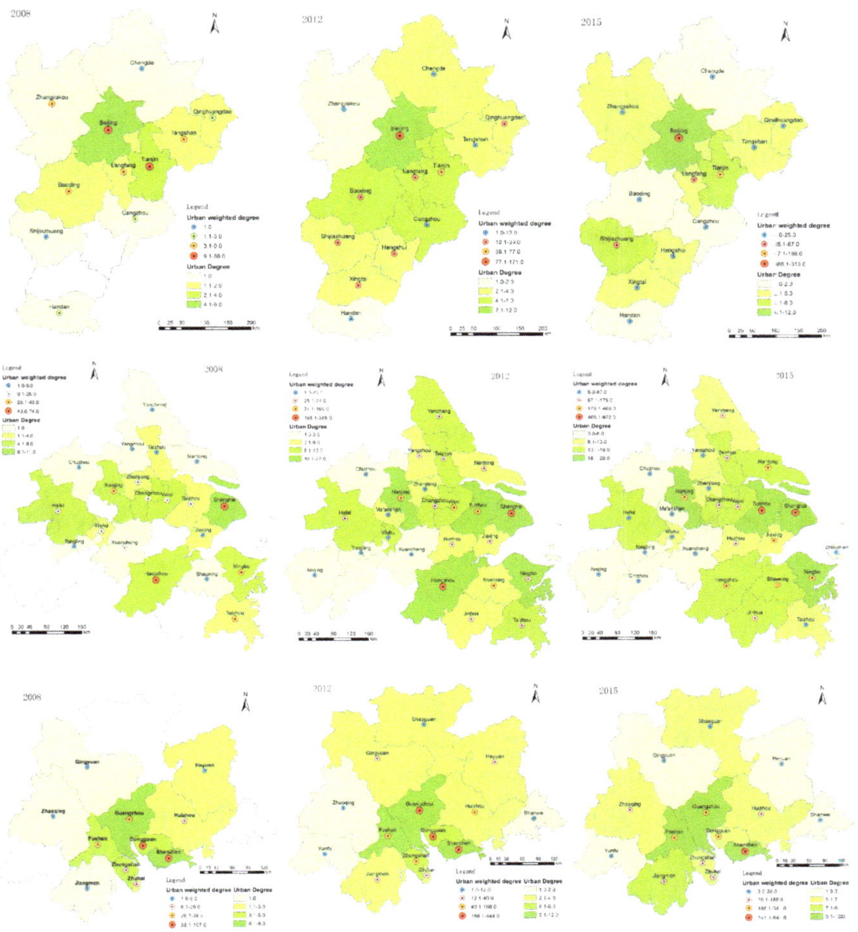

Figure 6.1 Degree and weight degree centralities of cities within the three urban agglomerations in 2008, 2012, and 2015.

Source: Liu *et al.* (2019).

in terms of weight degree values, overtaking Hangzhou in 2015. Notably, areas with higher weight degree gradually shifted from the central and southeastern toward the eastern areas. In the PRD, cities with larger degree and weight degree values were concentrated in the southcentral areas of the region. The weight degree value for

Shenzhen (648) was far higher than that of Guangzhou (341), despite their same hierarchies. This suggests that Shenzhen held a more influential position in the network compared to Guangzhou. It is worth noting that there was a discrepancy between the urban hierarchies based on degree and weight degree. In particular, Foshan had a higher degree centrality than Dongguan, whereas Foshan lagged far behind Dongguan in weight degree centrality. Thus, a high degree value does not mean that a city will have strong intercity linkages. Some cities may have a stable and strong technology flow with a few cities.

6.1.2 *Evolution of flow direction*

To analyze the direction of intercity technology transfers, we calculated the net flow in-degree (NFD) for all cities in the three major urban agglomerations. The NFD values for each city are listed in Table 6.2(a–c) with rankings based on weight degree within the respective networks. Notably, Beijing, Shanghai, and Shenzhen emerged as the most significant hubs due to their greater weight degree values in their respective networks.

Within the BTH, Beijing held a significant advantage due to its political and economic power, resulting in its overwhelming dominance. The city's role shifted from being a technology absorptive center to an innovative highland. While most of the surrounding cities had higher weighted in-degree values than out-degree values, Beijing's impact on them varied over time, ranging from an initial "siphon effect" to a diffusion effect or "trickle-down effect". In the earlier period, Beijing continuously absorbed technology from surrounding cities, thanks to its many enterprises with powerful transformation capabilities. This led to an economic gap between Beijing and the surrounding cities, indirectly contributing to uneven economic development in the BTH. During the later period, Beijing played a crucial role in disseminating technology to the neighboring cities. Facilitating technology transfer served as a pivotal strategy to foster integration within the BTH and alleviate population pressure as well as urban function congestion in Beijing.

Table 6.2 Urban weight degree (WD) and net flow in-degree (NFD) in the three urban agglomerations in 2008, 2012, and 2015.

(a) Beijing–Tianjin–Hebei region

2008			2012			2015		
City	WD	NFD	City	WD	NFD	City	WD	NFD
Beijing	69	31(50/19)	Beijing	171	15(93/78)	Beijing	313	−125(94/219)
Tianjin	62	−24(19/43)	Tianjin	77	21(49/28)	Tianjin	186	66(126/60)
Baoding	9	−5(2/7)	Shijiazhuang	39	−5(17/22)	Langfang	67	45(56/11)
Tangshan	7	5(6/1)	Baoding	31	−13(9/22)	Shijiazhuang	41	−7(17/24)
Langfang	6	−4(1/5)	Langfang	27	−15(6/21)	Tangshan	25	11(18/7)
Zhangjiakou	6	−6(0/6)	Xingtai	24	−4(10/14)	Cangzhou	19	−1(9/10)
Handan	3	3(3/0)	Hengshui	20	2(11/9)	Hengshui	14	−6(4/10)
Qinghuangdao	2	0(1/1)	Qinghuangdao	19	1(10/9)	Chengde	12	8(10/2)
Cangzhou	2	2(2/0)	Tangshan	12	−8(2/10)	Qinghuangdao	11	−7(2/9)
Shijiazhuang	1	−1(0/1)	Zhangjiakou	12	2(7/5)	Baoding	11	11(11/0)
Chengde	1	−1(0/1)	Cangzhou	11	−1(5/6)	Xingtai	9	5(7/2)
Xingtai	0	0(0/0)	Chengde	4	4(0/4)	Zhangjiakou	8	−2(3/5)
Hengshui	0	0(0/0)	Handan	1	1(1/0)	Handan	4	2(3/1)

(Continued)

Table 6. 2 *(Continued)*

(b) Yangtze River Delta region

2008			2012			2015		
City	WD	NFD	City	WD	NFD	City	WD	NFD
Shanghai	74	24(49/25)	Shanghai	355	−129(113/242)	Shanghai	972	−174(399/573)
Hangzhou	62	−12(25/37)	Hangzhou	257	−143(57/200)	Suzhou	647	−67(290/357)
Taizhou	43	−1(21/22)	Suzhou	195	39(117/78)	Hangzhou	469	−21(224/245)
Nanjing	35	1(18/17)	Nanjing	164	10(87/77)	Nantong	341	195(268/73)
Wuxi	34	−4(15/19)	Shaoxing	133	71(102/31)	Nanjing	331	−33(149/182)
Hefei	25	11(18/7)	Wuxi	114	−20(47/67)	Shaoxing	305	−147(79/226)
Ningbo	24	2(13/11)	Changzhou	74	26(50/24)	Ningbo	258	−48(105/153)
Wuhu	21	11(16/5)	Ningbo	63	−15(24/39)	Jiaxing	244	154(199/45)
Changzhou	19	−11(4/15)	Jiaxing	63	−1(31/32)	Changzhou	179	−29(75/104)
Suzhou	18	−12(3/15)	Hefei	53	−29(12/41)	Wuxi	176	−36(70/106)
Jiaxing	15	3(9/6)	Taizhou	51	35(43/8)	Jinhua	154	−78(38/116)
Zhenjiang	12	−12(0/12)	Nantong	50	6(28/22)	Huzhou	140	48(94/46)
Shaoxing	9	−1(4/5)	Yancheng	49	39(44/5)	Yancheng	100	74(87/13)
Nantong	4	2(3/1)	Huzhou	48	40(44/4)	Taizhou	95	59(77/18)
Taizhou	3	−3(0/3)	Yangzhou	37	27(32/5)	Hefei	87	53(70/17)
Tongling	3	3(3/0)	Jinhua	36	4(20/16)	Taizhou	71	−13(29/42)
Yancheng	2	0(1/1)	Taizhou	33	11(22/11)	Zhenjiang	62	−10(26/36)
Chuzhou	1	1(1/0)	Wuhu	25	15(20/5)	Yangzhou	61	33(47/14)
Xuancheng	1	−1(0/1)	Anqing	19	19(19/0)	Chuzhou	56	16(36/20)
Yangzhou	1	−1(0/1)	Ma'anshan	17	−7(5/12)	Wuhu	52	16(34/18)
Zhoushan	0	0(0/0)	Zhenjiang	16	−2(7/9)	Ma'anshan	27	5(16/11)
Chizhou	0	0(0/0)	Chuzhou	11	11(11/0)	Zhoushan	27	17(22/5)
Anqing	0	0(0/0)	Tongling	10	−6(2/8)	Xuancheng	19	−1(9/10)

Jinhua	0	0(0/0)	Xuancheng	5	−1(2/3)	Anqing	12	−6(3/9)
Huzhou	0	0(0/0)	Zhoushan	0	0(0/0)	Tongling	8	−8(0/8)
Ma'anshan	0	0(0/0)	Chizhou	0	0(0/0)	Chizhou	5	1(3/2)

(c) Pearl River Delta region

2008			2012			2015		
City	WD	NFD	City	WD	NFD	City	WD	NFD
Shenzhen	107	−3(52/55)	Shenzhen	444	−210(117/327)	Shenzhen	648	−262(193/455)
Dongguan	80	−10(35/45)	Dongguan	344	154(249/95)	Guangzhou	341	107(224/117)
Guangzhou	38	−14(12/26)	Guangzhou	294	−70(112/182)	Dongguan	252	70(161/91)
Foshan	32	4(18/14)	Foshan	168	2(85/83)	Foshan	250	12(131/119)
Zhongshan	26	12(19/7)	Zhongshan	93	51(72/21)	Huizhou	188	54(121/67)
Zhuhai	22	−14(4/18)	Huizhou	77	55(66/11)	Zhongshan	146	8(77/69)
Huizhou	22	16(19/3)	Zhuhai	40	−14(13/27)	Zhaoqing	128	−94(17/111)
Jiangmen	6	6(6/0)	Qingyuan	22	18(20/2)	Jiangmen	94	50(72/22)
Heyuan	3	3(3/0)	Heyuan	21	9(15/6)	Zhuhai	81	11(46/35)
Zhaoqing	1	−1(0/1)	Jiangmen	17	5(11/6)	Qingyuan	39	17(28/11)
Qingyuan	1	1(1/0)	Zhaoqing	12	0(6/6)	Shaoguan	31	13(22/9)
Shaoguan	0	0(0/0)	Shaoguan	5	1(3/2)	Heyuan	31	17(24/7)
Yunfu	0	0(0/0)	Shanwei	2	−2(0/2)	Shanwei	6	−6(0/6)
Shanwei	0	0(0/0)	Yunfu	1	1(1/0)	Yunfu	3	3(3/0)

Consequently, cultivating industrial growth and enhancing technological absorptive capacity in Hebei Province helped mitigate poverty in areas surrounding Beijing and Tianjin.

Within the PRD, Table 6.2(b) demonstrates that Shanghai, Hangzhou, Suzhou, and Nanjing held dominant positions within the network. Among these cities, Shanghai stood out as having a considerably more influential and central role compared to the others. Over time, all four cities transitioned from being technology-absorptive hubs to becoming diffusion centers. On the other hand, the cities of Nantong, Jiaxing, Hefei, Huzhou, and Yancheng were ranked as secondary in the network hierarchy and primarily acted as technology absorbers within the YRD.

Understanding why top-tier cities act as important technology transfer-out centers while second-tier cities are typically the destination for technology transfers in YRD is crucial. Firstly, this is due to the development strategies of highly developed cities such as Shanghai, Hangzhou, Suzhou, and Nanjing, which focus on new high-tech industries and invest heavily in research and development to foster innovation resources. This approach generates significant economic and social benefits, resulting in strong external technology spillovers and making them ideal technology transfer-out centers.

Secondly, the fact that second-tier cities are the primary absorption centers can be explained by the technology gap theory. When a technical gap exists between two regions, the less developed region will absorb and utilize the improved technology from the more developed region. The greater the technological gap, the harder it is for backward regions to imitate advanced technologies, making technology transfer more challenging. Therefore, second-tier cities are more likely to absorb technology from the top-tier cities in the network. Additionally, enterprises play a key role in transforming scientific breakthroughs into market-oriented products. Second-tier cities with superior technical absorptive capacity have more companies capable of doing so than lower-tier cities, making them the absorption center of the network.

Nevertheless, peripheral cities in the YRD network have smaller weight degree values because of their lower weighted in-degree and out-degree values, indicating that technology transfer activities are not positive for these peripheral cities.

In PRD, as shown in Table 6.2(c), Shenzhen and Guangzhou had the largest technology flows in the PRD during the study period, indicating that Shenzhen's innovation capacity exceeded that of the provincial capital, Guangzhou. This finding aligns with Pan *et al.* (2017) research on economic connections, which highlighted Shenzhen as the most influential city in the PRD. The top four cities with the largest technology flows during the study period were Shenzhen, Guangzhou, Dongguan, and Foshan. Among these cities, Shenzhen was also the biggest technology transfer-out. The remaining cities had a higher weighted in-degree than out-degree, highlighting Shenzhen's significance as the primary technology transfer hub in the region. Shenzhen has played a crucial role as a gateway to the world since China's opening up, and its independent innovation ability has been remarkable in recent years. Technological innovation has emerged as the key driver of economic development. In 2017, Shenzhen surpassed Guangzhou in terms of economic output for the first time. While Guangzhou previously served as a hub for technology transfer, it has since transformed into a city that actively absorbs new technologies to enhance its competitiveness. This shift highlights Guangzhou's ability to incorporate cutting-edge technologies, such as artificial intelligence, biomedicine, new energy, and new materials, within its technology industry. As a result, this has led to the rapid expansion of technology transfer-ins in the city.

It can be concluded that the top hierarchical cities within the urban agglomerations have prioritized the diffusion of technology over absorbing it, indicating that these cities are acting as technology transfer hubs and that innovation plays a significant role within the megalopolis. On the other hand, other cities within the same agglomerations seem to focus on adopting technology to enhance their productivity and competitiveness.

6.1.3 *Evolution of network hierarchy*

Figure 6.2(a–c) displays the evolution and structure of the intercity technology transfer network in the BTH, YRD, and PRD during the study period. The technology transfer network was visualized in ArcGIS 10.3 by employing linkage data between the original and destination city. The width and direction of the link represented the number of trades from origin to destination. The weighted connections were divided into four grades using natural breaks classification. The chord diagram plot is a useful instrument to investigate the bilateral flows by identifying the source, destination, direction, and volume. Thus, a chord diagram plot was used to visualize intercity technology transfer flows with the help of the circlize package in R programming. The chord diagram plots present the direction and volume of the flows within the three major urban agglomerations from 2008 to 2015 (Figure 6.2).

Generally, from a patent transaction perspective, the structural characteristics of technology transfer networks display some similarities. Over time, the connections for technology transfer between cities have continued to grow, resulting in an increased network density. This indicates that cities within urban agglomerations have become more closely connected. Furthermore, it is worth noting that the connectivity of technology transfer increased significantly over the study period. This can be attributed to the Chinese government's focus on promoting and supporting technology flows across different regions, which has led to a stronger network of technology transfer linkages.

The study found that intercity technology flows were unevenly distributed, leading to the emergence of a hierarchy of linkages. The analysis of spatial flow patterns revealed significant technology diffusion within urban agglomerations. This diffusion was mainly focused on neighborhood and hierarchical levels, with major cities

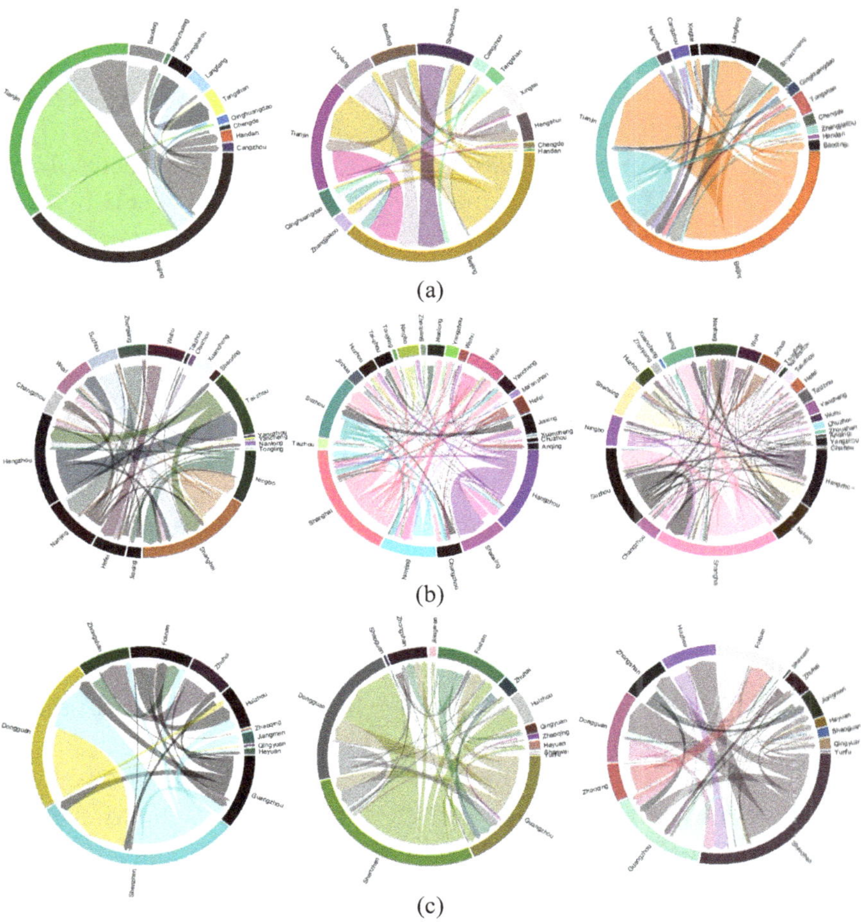

Figure 6.2 Intercity technology flows in the three urban agglomerations from 2008 to 2015. (a) Beijing–Tianjin–Hebei Region; (b) Yangtze River Delta Region and (c) Pearl River Delta Region.

Source: Liu *et al.* (2019).

acting as the source. For example, Shijiazhuang and Langfang became third-tier cities in 2015 due to hierarchical (jump diffusion) and contagious diffusion (neighboring diffusion). Similarly, Nanjing remained in the secondary tier due to hierarchical diffusion, while Nantong climbed the network hierarchy due to neighboring diffusion. Neighboring diffusion was also observed in the PRD, where Dongguan and Huizhou gained substantial technological innovations from Shenzhen (Figure 6.2).

There was also a significant path-dependence effect due to the embeddedness found in the technology transfer network. For instance, a lock-in effect was found in the intercity technology flows in the BTH, which was mainly concentrated between Beijing and Tianjin, as well as in Langfang and Shijiazhuang (Figure 6.2). The strength of connections among the three fixed city-dyads (between Beijing and the above-mentioned three cities) accounted for more than 50% of all connections during the study period. At the same time, the lock-in effect also existed in the YRD and PRD, as indicated by the sustainable growth of the weight value for Shanghai–Suzhou and Shaoxing–Hangzhou from 2008 to 2015 (Figure 6.2(b)) and Table 6.2). The strength of connections between Shenzhen–Dongguan, Shenzhen–Guangzhou, and Shenzhen–Huizhou also intensified over time in the PRD (Figure 6.2(c)). This indicated that the dynamic evolution of technology transfer network followed the retention mechanism of place dependence and maintained notable spatiotemporal inertia. In other words, the connections among fixed city-dyads maintained strong self-reinforcement.

However, it is worth noting that the path breakthrough effect was also significant in the YRD and PRD. This suggests that there was a variation mechanism that generated new technology transfer paths in the intercity technology transfer network. During the study period, several city pairs emerged as key players in the YRD, including Shanghai–Jiaxing, Nanjing–Nantong, Jinhua–Nantong, Shaoxing–Huzhou, and Jinhua–Shaoxing (Figure 6.2(b)). Similarly, the PRD also witnessed the emergence of new paths such as in Guangzhou–Huizhou, Guangzhou–Jiangmen, and Foshan–Jiangmen. Additionally, there were new linkages between peripheral and leading cities during the considered period (Figure 6.2(c)). These links expanded to new growth frameworks in technology transfer network and promoted network connectivity.

Specifically, in the BTH, Beijing possessed an overwhelming superiority within the network compared with other cities (Figure 6.3(a)).

Beijing was located at the center of the network in 2008 and was surrounded by other cities, forming a star network. This network was continuously extended over time and featured an increase in the number of connections. The star network gradually evolved into a tree-like network that was centralized in Beijing in 2015. The technology transfer network in the BTH maintained a monocentric structure that was based on dynamic connections. Among all the linkages between different cities, the connections between Beijing and Tianjin were consistently positioned at the top of the hierarchy. This was due to the high concentration of technology-intensive enterprises in both cities, which had a strong capability for transforming and applying new technologies. As a result, Beijing and Tianjin had a particularly dynamic exchange of technology within the region. Despite the active technology transfers between Beijing, Tianjin, and other cities within the region, there existed a significant economic disparity with Hebei Province. It is believed that cities within the province may have experienced a lower absorption of technology due to the lack of industrial development. However, upgrading urban industrial structures can stimulate technological demand and promote enterprise development. Therefore, it is important for cities within Hebei Province to actively absorb new technologies to facilitate their own growth. These findings suggest that technology transfers are influenced not only by the technological gap between cities but also by the active demand for such transfers from enterprises.

Figure 6.3(b) displays the spatial patterns for intercity technology flows in the YRD. The technology transfer network in the YRD evolved from a monocentricity centered on Shanghai to a polycentric structure during the 2008–2015 period. In 2008, Shanghai held a clear dominance over other cities in the region. However, by 2015, cities such as Suzhou, Nanjing, and Ningbo had risen to the top of the network hierarchy alongside Shanghai. The development of a tree-like structure within the YRD region indicated a complex and interconnected network, with most cities being actively connected to the technology transfer system. This suggests that even smaller cities within the YRD were likely involved in technology transactions.

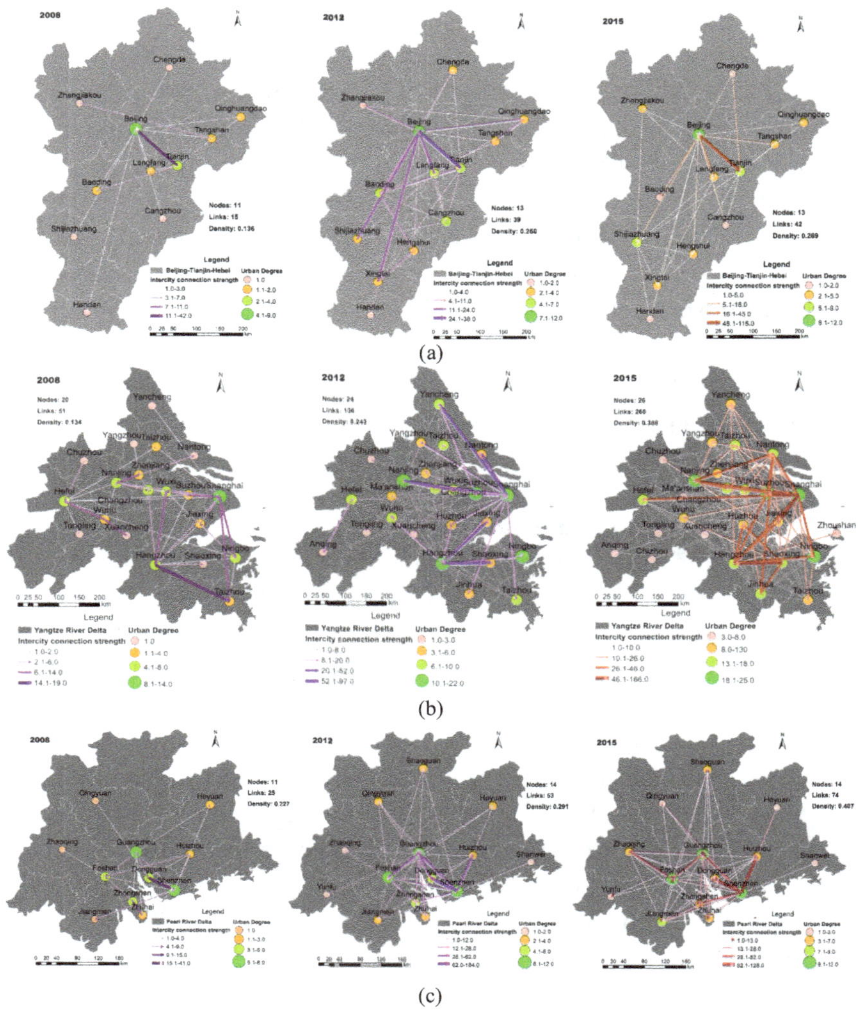

Figure 6.3 Intercity technology transfer network for the three urban agglomerations from 2008 to 2015. (a) Beijing–Tianjin–Hebei Region; (b) Yangtze River Delta Region and (c) Pearl River Delta Region.

In contrast to the YRD, technology transfers with smaller cities within the BTH region were primarily limited to Beijing. Additionally, intercity links were mostly generated by cities in the eastern part of the region, while there was less connectivity and lower weight values for cities in the western part. It is worth noting that most of the cities with the lowest urban degree tiers were located in Anhui Province, as these cities had weaker connections with other cities in the YRD — a pattern that was expected given their geographical location.

The technology transfer network in the PRD region underwent a significant evolution, transforming from a simple tree-like structure into a multistar network that was centered around Shenzhen and Guangzhou, with Foshan following closely behind (Figure 6.3(c)). The linkages in the network increased and strengthened during the 2008–2015 period. Shenzhen has the highest number of strength transfer links with other cities in the PRD region. Surprisingly, Dongguan has more links with Shenzhen than Guangzhou, despite Guangzhou's important position in the area. This explains why Dongguan has a higher weight degree than urban degree. In 2015, the connection between Shenzhen and Huizhou became the most significant link in the hierarchy. There are several possible reasons for this phenomenon. Firstly, Shenzhen is known as China's innovation capital, with over 30,000 high-tech enterprises and an average of 46 invention patents created daily. It has been compared to Silicon Valley by international media. Secondly, Dongguan and Huizhou are located near Shenzhen and likely benefit from the diffusion of technology in the region. Shenzhen's external technology transfer has led to a concurrent industrial shift to neighboring cities, including Dongguan and Huizhou, which has fostered economic growth. This shift has been observed in both traditional manufacturing and high-tech sectors, aided by improved transportation. Foshan is also a key player in this network and has strong links to Guangzhou. Urban integration has further facilitated technology transfer between these two interconnected cities.

6.2 The Three Northeast Provinces

6.2.1 *Evolution of urban centrality*

6.2.1.1 *At national scale*

The spatial differentiation of the weight degree to northeast intercity technology transfer network's weight degree from a national perspective is remarkable. It showed a "core–periphery" structure with hierarchies, and the core cities had been strengthening their control over the network.

Between 2005 and 2015, the coefficient of variation in degree among city nodes rose from 1.19 to 1.75. Meanwhile, the coefficient of variation in weight degree consistently exceeded 2.36, indicating a notable polarization in network linkage strength compared to the topology. As the number of cities increased, the "core–semiperipheral–peripheral" hierarchy of city weight degree gradually strengthened with Shenyang, Changchun, Harbin, and Dalian within the region, as well as Beijing and Shanghai outside the region playing a central role in the network (Figure 6.4). In addition, the percentage of cities with node degree above average decreased from 19.5% at the start to 12.24% by 2015. This illustrates further polarization and development of the network, with core cities gaining greater control and a consolidated dominant position. As a result, the evolution of the intercity technology transfer network in Northeast China exemplified a typical "Matthew effect".

From a national perspective, the intercity technology transfer network developed a geographical pattern of northeastern divergence and national convergence, that is, patented innovations have been spreading from the three northeastern provinces to the rest of the country at an increasing rate. Observations were made toward

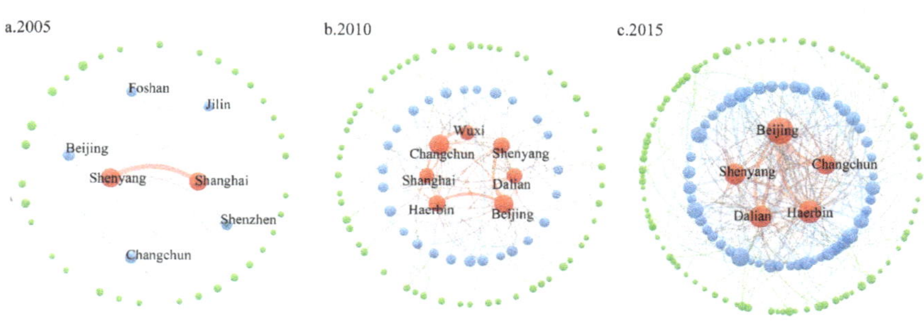

Figure 6.4 Core–peripheral structure of urban weight degree in the TTN between the northeast and the whole country in (a) 2005, (b) 2010, and (c) 2015.

Source: Liu & Niu (2019).

the weighted in-degree and out-degree as well as the degree centrality of the leading 10 cities in 2005, 2010, and 2015 (Table 6.3). The weighted in-degree of cities outside the region was typically greater than the weighted out-degree, so the net inflow was positive. On the other hand, the weighted in-degree of cities within the region was typically smaller than the weighted out-degree, so the net inflow was negative. Especially in 2015, external cities such as Beijing, Shenzhen, Shanghai, Guangzhou, and Qingdao had positive net inflows, while internal cities including Shenyang, Harbin, Dalian, and Changchun had negative net inflows. This suggested that the three northeastern provinces played a major role in technology transfer-out across the country and that the major cities in the northeastern provinces (Harbin, Changchun, Shenyang, and Dalian) were important technology diffusion centers in the network, whereas Beijing, Shanghai, and Shenzhen were both export and important convergence centers. In addition, the net inflow to cities in the three northeastern provinces continued to drop as a huge number of patents were absorbed and transformed by other cities across the nation, indicating a high concentration of patents in economically developed cities such as Beijing and Shanghai. This phenomenon is suggested by the traditional technology gap theory that economic and technological potential differences drive technology flows from developed to backward regions with a gradient-passing pattern of technology transfer. Table 6.3 shows that the northeast intercity technology transfer network developed as a typical counter-potential technology flow to developed innovation centers with higher levels of productivity development and stronger technologies.

The intercity technology transfer network in the northeast region had a higher density of nodal cities in the east along the coastal city belt, with fewer in the west (Figure 6.5). The cities for active intercity technology transfer in Northeast China are primarily centered around the Harbin–Dalian Metropolitan Corridor, which was well connected to the BTH, the YRD, and the PRD urban agglomerations. In general, the eastern coastal cities carried a higher weight degree in terms of technology transfer, with the three major urban agglomerations serving as the most active regions. These areas formed the core of the

Table 6.3 Top 10 urban degree centrality of interurban transfer network between the northeast and the whole country.

2005			2010			2015		
City	Degree centrality	In-flow	City	Degree centrality	In-flow	City	Degree centrality	In-flow
Shenyang	0.325	–33 (24/57)	Beijing	0.307	25 (77/52)	Harbin	0.459	–58(75/133)
Beijing	0.3	16 (21/5)	Changchun	0.25	30 (58/28)	Dalian	0.336	–74(56/130)
Harbin	0.175	2 (5/3)	Dalian	0.227	–16 (16/32)	Changchun	0.26	–20(48/68)
Changchun	0.175	–11 (7/18)	Shenyang	0.216	–26 (19/45)	Beijing	0.253	64(215/151)
Dalian	0.15	–2 (3/5)	Shanghai	0.17	–14 (16/30)	Shenyang	0.24	–4(73/77)
Shenzhen	0.15	4 (11/7)	Harbin	0.17	–50(9/59)	Shanghai	0.11	17(37/20)
Shanghai	0.125	21 (40/19)	Daqing	0.068	–5(6/11)	Shenzhen	0.089	27(45/18)
Daqing	0.075	3(4/1)	Wuhan	0.057	1(3/2)	Daqing	0.089	–24(8/32)
Panjin	0.05	–4(1/5)	Jilin	0.057	–5(5/10)	Qingdao	0.082	8(25/17)
Guangzhou	0.05	–2(0/2)	Tianjin	0.057	0(4/4)	Guangzhou	0.082	10(25/15)

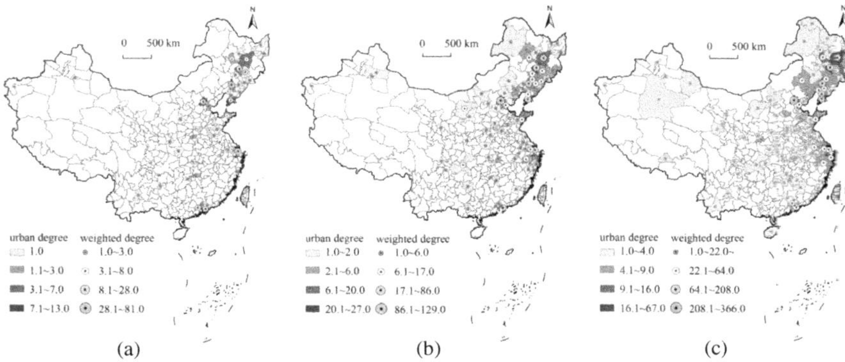

Figure 6.5 Spatial dynamics of urban degree and weight degree in the TTN between the northeast and the whole country.

technology transfer network while inland cities had a lower weight degree and were situated at the network's periphery. The technology transfer-out of cities in the three northeastern provinces followed a preference linkage mechanism, with a tendency to create relationships with innovation hubs on the eastern coast that had strong technical innovation capabilities and intense technology flows. Notwithstanding, it is worth noting that on a national scale, the northeast intercity technology transfer network has been expanding in terms of its breadth (number of cities) and intensity (volume of city patent transfers). As a result, the northeast's technology influence is gradually spreading from the eastern coast to the central and western regions of China. An increasing number of central and western cities are now part of the technology transaction flows with the three northeast provinces. Despite this expansion, a considerable gap still exists in technology transaction flows between the central and western with three northeast provinces and the western with the three northeast provinces.

6.2.1.2 *At regional scale*

The intercity technology transfer network in the three northeastern provinces from the local perspective had increased in agglomeration and exhibited centripetal contraction (Table 6.4). In 2005, the local technology transfer network consisted of 18 nodes and 17 edges, and only half of the cities participated in the regional technology transfer. Additionally, the number of network edges was smaller than the number of nodes, which meant there were a large number of isolated nodes, indicating low density and connectivity. The average degree and average weight degree of regional technology transfer network nodes increased between 2005 and 2015, as did the network's size, connectivity, and intercity technology interaction. By 2015, the technology transfer network had undergone further developments and improvements. The average degree and weight degree of network nodes increased, the network expanded in size, connectivity improved, and intercity technology interaction became closer. Additionally, the central potential of the network increased from 0.111 to 0.196, indicating that a preference linkage mechanism resulted in the self-organized growth of the network becoming concentrated on a few hub cities. This demonstrated a centripetal contraction of the network (Table 6.4).

The four sub-provincial cities of the three northeastern provinces were the hubs of the intercity technology transfer network, serving as "technology gatekeepers". By 2015, the peak cities (the first group) based on the city's degree centrality (>6) and weight degree (>28) were the subprovincial cities of Harbin, Changchun, Shenyang, and Dalian, and the administrative level of cities was spatially homogeneous with the network connectivity level. This was directly tied to the large concentration of innovation resources, industrial development, innovation investment, research, and

Table 6.4 Topological statistics of the northeast interurban technology transfer network.

Year	Node	Edge	Network density	Avg. degree	Avg. weight degree	Weight degree	Central potential
2005	18	17	0.056	1.889	1.111	40	0.111
2010	25	36	0.06	2.88	5.52	276	0.12
2015	28	74	0.098	5.286	8.857	496	0.196

education in subprovincial administrative centers. The technology gatekeepers emphasized the power of cities to absorb external technologies and diffuse them internally. In this way, the analysis determined that Shenyang, Changchun, Harbin, and Dalian were the core nodes and hubs of the regional technology transfer network. They acquired advanced technologies from other innovative cities across the country via a "global pipeline" that was tightly connected to the outside world and continuously diffused them to other local cities via "local buzz". This had established significant technological relationships with other cities nationally and locally, working as an intermediary and bridge in the national–regional technology transfer network and as a technology gatekeeper. Screening at the net inflow of local cities (Table 6.5), the difference in cities' roles in the regional technology transfer network was not evident in 2005 and 2010, and only became more evident in 2015, with Shenyang (–51), Harbin (–24), and Changchun (–27) all having a significantly negative net inflow. The four subprovincial core cities gradually played the role of technology exporters, and the small- and medium-sized cities primarily played the role of technology importers. This phenomenon was consistent with the classic gradient transfer theory, which holds that economic and technological potential inequalities promote the flow of technology from developed countries to lagging regions.

6.2.2 *Evolution of interurban connectivity*

6.2.2.1 *At national scale*

The northeast-domestic technology transfer network's spatial evolution was characterized by both path dependence and path creation. From the perspective of spatial structure (Figure 6.6), technology transfer in Northeast China deeply interacted with the three major urban agglomerations in China. It initially formed an urban technology flow with the core cities in the northeast as the sources of flow and pointed to the BTH, the YRD, and the PRD urban agglomerations with Beijing, Shanghai, and Shenzhen as the hubs. This highlights how certain regions tend to reinforce

Table 6.5 Top10 urban degree centrality in the northeast technology transfer network.

2005			2010			2015		
City	Degree centrality	In-flow	City	Degree centrality	In-flow	City	Degree centrality	In-flow
Harbin	0.294	2(4/2)	Shenyang	0.500	−2(13/15)	Shenyang	0.889	−51(17/68)
Changchun	0.235	0(2/2)	Harbin	0.375	10(19/9)	Harbin	0.630	−24(10/34)
Shenyang	0.235	−3(1/1)	Dalian	0.292	13(34/21)	Dalian	0.630	2(27/25)
Dalian	0.176	1(2/2)	Changchun	0.250	−4(4/8)	Changchun	0.481	−27(13/40)
Anshan	0.176	2(2/2)	Suihua	0.125	−3(0/3)	Tonghua	0.222	7(15/8)
Jiamusi	0.118	−1(1/1)	Mudanjiang	0.125	−15(1/16)	Daqing	0.185	2(12/14)
Liaoyang	0.118	−3(0/0)	Tonghua	0.125	0(2/2)	Jilin	0.185	7(12/5)
Benxi	0.059	1(1/1)	Daqing	0.125	1(2/1)	Dandong	0.185	2(6/4)
Baishan	0.059	−1(0/0)	Anshan	0.125	−39(2/41)	Benxi	0.185	1(7/8)
Tonghua	0.059	1(1/1)	Jiamusi	0.083	1(6/5)	Yingkou	0.185	1(4/3)

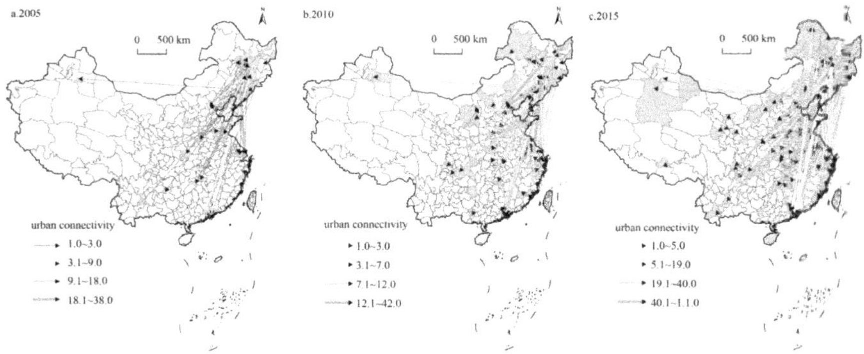

Figure 6.6 Spatial dynamics of urban connectivity in the TTN between the northeast and the whole country.

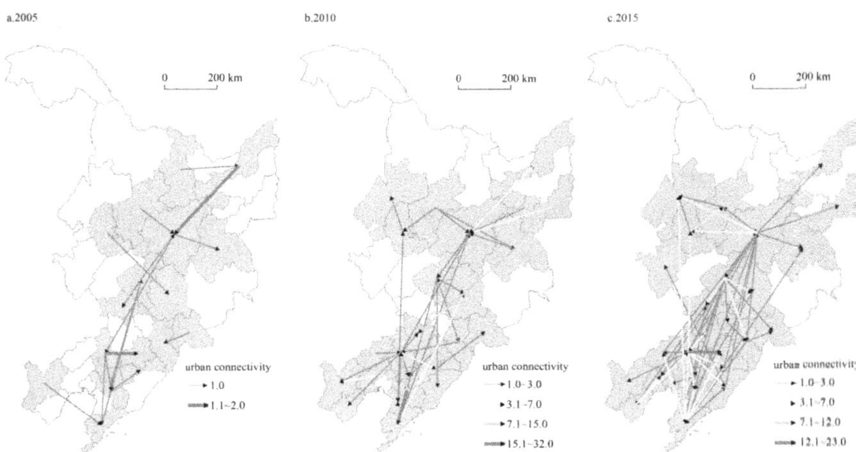

Figure 6.7 Spatial dynamics of urban connectivity in the northeast TTN.

Source: Liu & Niu (2019).

themselves and depend on each other for development, showing clear spatial stickiness and regional dependence. From the perspective of spatial expansion (Figure 6.7), technology transfer in the three northeastern provinces not only reinforced path dependence but also expanded to central and western regions, creating new paths of creation primarily directed toward core cities such as

Anhui, Hubei, Jiangsu, Sichuan, and Chongqing. The transfer was mostly from the central cities in the northeastern provinces to capital cities in the central and western provinces, indicating that the northeastern provinces were more economically advanced and had a stronger capacity for knowledge and technology production. From the aspects of technology categories and actors, the transfer of patents between the northeastern provinces and the rest of the country mainly consisted of invention patents with high technological standards and economic value, while utility model patents and design patents made up a minimal proportion. Enterprises had emerged as the primary actors of intercity technology transfer, with patent fields involving primarily medical equipment, automobile manufacturing, technology companies, biopharmaceutical compounds, new materials, equipment manufacturing, and other emerging manufacturing industries. The role of universities and research institutions in intercity technology transfer was found to be less significant in Northeast China and the rest of the country. This suggests that there is a weak collaboration between industry, academia, and research in the three northeastern provinces. To address this issue, it is important to enhance the transfer and transformation capacity of universities and research institutions for scientific and technological achievements.

The intensity of technology transfer between the three northeastern provinces and the rest of the country varied significantly in terms of hierarchy and spatial heterogeneity. It showed a significant "Matthew effect" and abrupt changes in the direction and intensity of city pairs (Table 6.6). From 2005 to 2015, a stable hierarchical structure with four levels was formed. First-tier city pairs were primarily concentrated in the core cities in the northeast and economically developed cities in the east, with technology flows dominated by spoke-and-hub diffusion from the northeast to the east. Second-tier city pairs overwhelmingly showed a convergence of technology flows from the three northeastern provinces to Beijing, and Beijing surpassed Shanghai and Shenzhen as the center of technology transfer. In the third-tier city pairs, technology flows from the three northeastern provinces to the eastern coastal urban

Table 6.6 Top 10 urban pairs of interurban transfer network between the northeast and the whole country.

2005			2010			2015		
Origination	Destination	Number	Origination	Destination	Number	Origination	Destination	Number
Shenyang	Shanghai	38	Wuxi	Changchun	42	Beijing	Yingkou	101
Shanghai	Shenyang	18	Harbin	Beijing	31	Dalian	Beijing	40
Shenyang	Shenzhen	10	Shenyang	Beijing	27	Harbin	Beijing	34
Changchun	Beijing	9	Shanghai	Fushun	12	Changchun	Beijing	31
Foshan	Jilin	7	Beijing	Shenyang	11	Suzhou	Shenyang	27
Panjin	Beijing	5	Dalian	Chongqing	9	Shenyang	Beijing	23
Changchun	Jining	5	Jilin	Zhuhai	8	Qiqihar	Beijing	22
Shenzhen	Changchun	5	Shanghai	Siping	7	Jinhua	Tieling	19
Shenyang	Wuxi	4	Daqing	Shanghai	7	Dongguan	Tieling	19
Xuzhou	Daqing	3	Harbin	Suzhou	7	Daqing	Shanghai	17

agglomerations were predominant, with the BTH, the PRD, and the YRD becoming the agglomeration areas for patent technology transfer from the three northeastern provinces. The fourth-tier city pairs dominated the northeast-domestic technology transfer linkage (over 60%). However, the level of technology transfer and interaction between them is generally low. More than half of these cities have a technology linkage intensity score of only 1, indicating that they mainly export technology to core cities located in the central and western regions of China, with limited interaction between themselves. Compared to the majority of central cities in the central and western provinces, the core cities in the three northeastern provinces had a greater technological gradient advantage and knowledge production capacity. External cities with a close technology transfer (first and second-tier city pairs) to the core cities of the three northeastern provinces were mainly concentrated in the three major urban agglomerations of the BTH, the PRD, and the YRD on the eastern coast. These external cities have a high concentration of science and technology innovation centers, such as Beijing, Shanghai, Suzhou, Shenzhen, and Guangzhou, with obvious spatial unevenness.

Moreover, between 2005 and 2015, the high-rank city pairs showed both stability and abrupt changes. First, the top 10 city pairs in terms of the large volume of technology transfer were highly locked between the capital cities of the three northeastern provinces (Harbin, Changchun, and Shenyang) and Beijing, maintaining both temporal and spatial inertia (or stickiness). Second, intercity technology transfer city pairs within the three northeastern provinces were consistently absent from the top 10 city pairs. This is due to weak technology interaction and knowledge spillover effects among these cities, as well as insufficient local knowledge and technology production for regional transformation and transfer. Third, despite the spatial stability of the flow and direction of the top 10 city pairs, there had been significant changes, with some prefecture-level cities in the three northeastern provinces (such as Panjin, Jilin, Daqing, and Qiqihar) rapidly emerging as radiation centers for northeast-national technology transfer. On the other hand, Fushun, Yingkou,

Siping, and Tieling have become spokes for external technology transfer, resulting in significant changes.

6.2.2.2 *At regional scale*

The three northeastern provinces' regional technology transfer was primarily centered around Harbin, Changchun, Shenyang, and Dalian. The pattern of technology diffusion displayed a combination of hierarchy, contact, and leap effects. From the standpoint of technology transfer city pairs and their directions (Table 6.5), a technology flow with Harbin, Changchun, Shenyang, and Dalian as agglomeration-and-diffusion hubs had been established by 2015, with intraprovincial technology transfer as the primary focus. In 2005, the intercity technology transfer flow was small sized and unidirectional, exhibiting a dendritic network formed by the lead diffusion pattern with Harbin, Shenyang, and other provincial capitals as axes. In 2010, the intercity technology transfer network had grown to its initial size, and the network showed a self-organized nature that appeared to have certain preferences for geographical proximity. For instance, Harbin focused primarily on technology transfer with nearby cities like Jiamusi, Jixi, Mudanjiang, and Suihua, while Shenyang primarily interacted with small- and medium-sized cities such as Fuxin, Benxi, Tieling, Liaoyang, and Anshan for technology transfer. The capital cities of each province played a pivotal role in the "local buzz" of technology transfer within the province, and the localization of technology transfer was significantly influenced by geographical proximity, with technology transfer tending to take place within the province and exhibiting certain contact technology diffusion patterns. In 2015, Harbin, Changchun, Shenyang, and Dalian began establishing themselves as first-tier core cities, while secondary core cities such as Jilin, Tonghua, Fushun, and Anshan continued to absorb patented technologies from first-tier core cities. At the same time, the technology diffusion to peripheral cities gradually strengthened, forming a typical hierarchical "hub-and-spoke" contact-hierarchical technology diffusion pattern. During the period from 2005 to 2015, the characteristics of medium and long-distance technology transfer across administrative boundaries were also noteworthy, resulting in the formation of leaping technology transfer city pairs such as Dalian–Mudanjiang, Fuxin–Harbin, Dalian–Qiqihar, Shenyang–Daqing, and Changchun–Huludao.

In terms of the categories and actors of technology transfer, enterprise-led invention and patent transactions predominated in the three northeastern provinces, just as they did for external technology transfer. During 2005–2015, there was a general trend of low proportions of utility model patents and designs, while the proportion of transferred invention patents increased steadily. At the beginning of the period, less than one-third of patents transferred were inventions, but by the end, this had risen to 54.49%. This suggests that regional patent transactions were becoming more technologically advanced over time. Large corporations played a significant role in transferring technology within their respective regions and had remarkable abilities to absorb patents. However, it would be unwise to underestimate the importance of individual entities that dominated patent exports. In fact, these individual entities surpassed corporate technology export capacity in both 2005 and 2010. On the other hand, universities and research institutions had a relatively lower proportion of technology transfer activities. Among these, Jilin University, Dalian University of Technology, Harbin Institute of Technology, Harbin Engineering University, and other universities of science and technology were dominant players in this field.

The intensity of intercity technology transfer within the three northeastern provinces was significantly influenced by the industrial transfer of large enterprises, with subprovincial cities having a greater breadth of transfer and the ability to create pathways. On the one hand, patent transactions in the three northeastern provinces were highly concentrated in large state-owned enterprises, and large-scale intercity technology transfer was led by the industrial transfer of large enterprises, manifesting as a technology flow under the headquarters–branch organizational structure. For example, Anshan exported a large number of patents to Dalian in 2010, primarily due to the export of patents from *Anshan Iron and Steel Group* (located in Anshan) to its joint venture in Dalian, and the technology transfer triggered by the relocation of *Zhongye Journai Automation Co., Ltd.* from Anshan to the *Dalian High-Tech Industrial Park*. In 2015, a large volume of patent rights sold from *Shenyang Hanking Industrial Group Corporation* to *Fushun Metallurgical Industry Branch* created the Shenyang–Fushun technology transfer path. On the other hand, the breadth of technology transfer in large cities was significantly greater than that in small- and medium-sized cities (Figure 6.8), with the four major central cities of Shenyang, Harbin, Changchun, and Dalian having a greater breadth of technology transfer with most cities in the regions. However, the breadth of transfer in small- and medium-sized cities was significantly smaller than that in the four major central cities, while they had greater intensity of transfer. In 2010, Anshan had the highest weight degree (patent transfer intensity) in the network, but only produced technology linkages with Dalian and Shenyang; Mudanjiang had a higher weight degree than Harbin and Changchun, but only had technology exchanges with Tieling, while Shenyang produced technology transactions with 12 other cities (nearly one half). Thus, it appeared that large cities had become the hubs with technology agglomeration and diffusion in regional technology transfer network, with a wider breadth of

Figure 6.8 Chord diagram of dominant technology flows in the northeast interurban technology transfer network in 2005, 2010, and 2015.

Source: Liu & Niu (2019).

external technology transfers and a greater capacity to create pathways from which new technology flows can be derived.

6.3 Summary

This chapter mainly targets the three major urban agglomerations and the northeast provinces at a regional scale. It aims at revealing the characteristics of network centrality and intercity technology flow. The main conclusion is as follows.

From a national perspective, the technology transfer network in the three northeastern provinces has a hierarchical and "core–peripheral" structure, which results in the dispersion of technology from the northeast to the national coast. On a regional level, there is a centripetal contraction where the four core cities of Harbin, Changchun, Shenyang, and Dalian act as gatekeepers for technology in the local network. The technology transfer process shows characteristics of being strongly nationalized but weakly localized.

Intercity technology flows in the three northeastern provinces are both path-dependent and path-creating. From a national perspective, technology transfer originates from the core cities in the three northeastern provinces and mostly flows to the BTH, YRD, and PRD urban agglomerations with Beijing, Shanghai, and Shenzhen serving as hubs, respectively. Local intercity technology transfer takes place through Harbin, Changchun, Shenyang, and Dalian as hubs, with intraprovincial transfers concentrating on a mixed spatial pattern of hierarchical, contact, and leap diffusion.

Several factors influence intercity technology transfer in the three northeastern provinces, including geographical proximity, industrial proximity, differences in economic level, similarity in innovation capacity, technology absorption capacity, and foreign direct investment.

PART IV
Spillover and Determinants of Technology Transfer Network

CHAPTER 7

EFFECTS OF TECHNOLOGY TRANSFER NETWORK ON LOCAL INNOVATION CAPACITY

Innovation is considered a crucial factor in accelerating economic growth (Hesse & Fornahl, 2020), and knowledge is widely acknowledged as the key to regional innovation (Tavassoli & Carbonara, 2014). As a function of complexity, the emergence of new knowledge is increasingly dependent on the integration and recombination of various kinds of knowledge from different sources (Bresnahan *et al.*, 2001; Breschi & Lenzi, 2015). The process of knowledge generation and combination is not only triggered through the interaction in the local knowledge base (Tavassoli & Carbonara, 2014), it can also be accomplished via establishing cross-regional linkages such as trade exchanges, foreign direct investment, technology alliances, incorporation into knowledge networks, and participation in global production networks among others (Boschma & Iammarino, 2009). Therefore, the creation of new knowledge can also be conceptualized as the outcome of integration between adjacent and distant interactions (Bathelt *et al.*, 2004). However, due to its character of high localization, knowledge communication between local actors may lose its value over time on account of knowledge redundancy and limited chances for recombination (Boschma & Frenken, 2010; Breschi & Lenzi, 2015). As a consequence, the entry of external knowledge is crucial to knowledge creation and innovation capacity development within the region (Tavassoli & Carbonara, 2014; Lee & Sohn, 2019), and knowledge flow in turn interconnects with innovation in distant places in the background of current globalized knowledge economy (Corpataux *et al.*, 2009; Binz & Truffer, 2017).

In this chapter, we develop the "buzz-and-pipeline" theory to construct a framework of "dual pipelines" toward the imported technology. Our contribution is two-fold. First, patent transfer, as a key to technology commercialization, can not only represent the economic value of patents but also measure the knowledge flows between regions. This section explores the impact of technology introduction on local innovation capacity from the perspective of patent transfer. Second, we analyze whether factors such as intensity, technological relatedness, and technological similarity contribute to the development of local innovation capacity of both domestic and transnational linkages. At the same time, we will also examine the potential trade-offs and joint impacts of these two types of linkages. To test these theoretical claims, we will use a database containing information on imported technology between Chinese prefecture-level cities, as well as between domestic cities and foreign cities from 2001 to 2018.

7.1 Conceptual Background and Hypotheses

Against the backdrop of economic globalization, innovation activities are constantly breaking through geographical boundaries, and the cross-regional flow of innovation elements has become the primary character of regional opening innovation. The imported external knowledge and the interaction between external enterprises and local enterprises are crucial in realizing technology diversification and local innovation (Miguélez & Moreno, 2018).

The previous literature has demonstrated a well-established fact that the integration and recombination of unrelated thoughts may lead to knowledge generation and technology innovation (Aghion & Howitt, 1992; Weitzman, 1998; Miguélez & Moreno, 2018). This kind of recombination innovation can be achieved not only through interaction within the local knowledge base but also by participating in the global innovation network (Binz *et al.*, 2014). However, current research has concentrated more on how local capacity affects regional innovation, while little is known about the function of cross-regional knowledge flows in regional innovation (Miguélez & Moreno, 2018). Most of the literature usually implicitly assume that innovation-producing activities come primarily from geographically localized knowledge sources but neglect the linkages across regions (Audretsch & Feldman, 2004). In doing so, it ignores the fact that the integration and recombination of knowledge which has the same source at a certain time may make local knowledge redundant and valueless, leading to negative lock-in (David, 1993; Boschma, 2005). Therefore, it is necessary to source knowledge and technology from outside the region by creating extra-regional linkages in order to avoid path lock-in (Bathelt *et al.*, 2004; Bergman & Maier, 2009; Boschma & Iammarino, 2009; Miguélez & Moreno, 2013).

A lot of studies also have a consensus that integrating both internal and external knowledge can enhance urban innovation capacity (Cao *et al.*, 2022). Moreover, scholars usually agree that the interaction between global and local knowledge flow is critical to innovation (Moodysson, 2008; Trippl *et al.*, 2009), which is in accordance with the basic concept of global–local interaction. Bathelt *et al.* (2004) proposed the "global pipelines-local buzz" theory to be a classic framework for illustrating the process of global–local interaction. It revealed new knowledge not only can be created through the integration and recombination of localized knowledge but also can be acquired from the outside when it is not readily available locally through non-local linkages at various scales, so that the global–local cross-regional knowledge and technology interaction can be achieved (Bathelt *et al.*, 2004; Swyngedouw, 1997, 2004).

The terms "global" and "local" in the "buzz-and-pipeline" model are not fixed scales but two terminals of a series of complex interactions (Dicken, 1994). "Local buzz" can be the one-on-one interaction and also the distant and virtual interaction and therefore can be both local and global (Asheim *et al.*, 2007b; Jones *et al.*, 2010). "Global pipeline" represents not only the transnational connections but also the domestic connections across the region and cluster (Cao *et al.*, 2022). In this study, the purpose is to construct a framework called "dual pipelines" of imported technology on the basis of "local buzz" and "global pipeline". The dual pipelines of imported technology can be discoursed as domestic imported technology pipeline and transnational imported pipeline. Domestic technology imported pipeline implies that technology transfer activity is happening between cities within the country, and transnational technology imported pipeline refers to the introduction of technology in Chinese cities from foreign cities (Figure 7.1).

On a global scale, numerous empirical researches have already proved the impact of imported international technology on local innovation capacity (Laursen & Salter, 2006; Awate *et al.*, 2015;

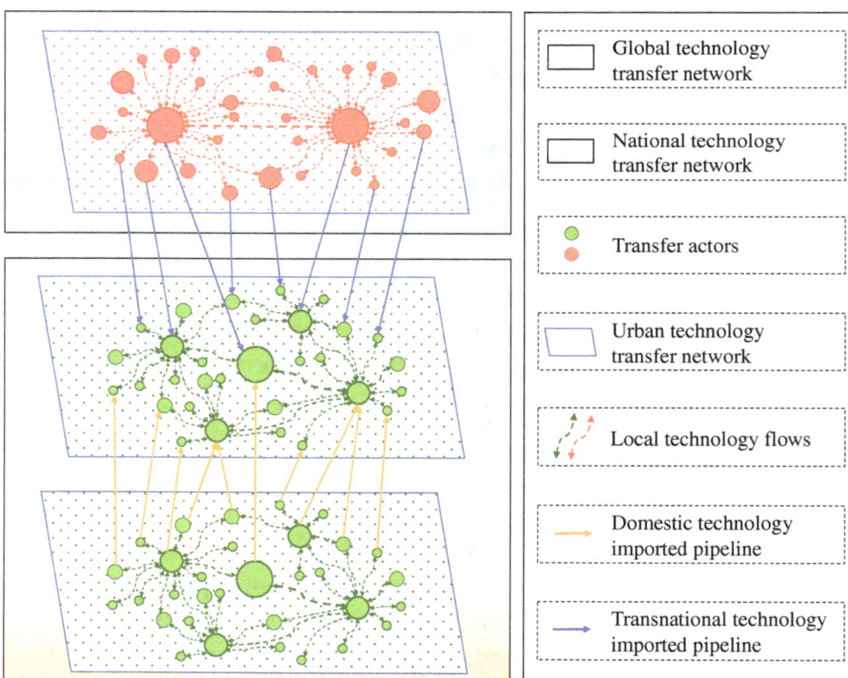

Figure 7.1 The dual-pipeline structure of TTN.

Source: Wang *et al.* (2023).

Yu *et al.*, 2019). Late-developing advantage theory considers, in this way, that the late-developing countries can leverage the technology spillover and diffusion from the developed countries to improve their local technology capacity (Yu *et al.*, 2019). On the one hand, the locals can introduce and learn new technology from other countries to accelerate urban development and regeneration (Yu *et al.*, 2019). On the other hand, innovation can be carried out using imported technology to quickly catch up with the current technology (Awate *et al.*, 2015; Yu *et al.*, 2019). However, it has also been confirmed that developing countries cannot apply the advanced technology from developed countries sophisticatedly due to the lack of knowledge base and absorbing capacity (Fernández & Gavilanes, 2017). Therefore, dedicating itself to importing external knowledge and technology may also have a negative effect on local innovation capacity (Laursen & Salter, 2006), that is, the reliance on technology import may lead to the unbalance of innovation mechanism and the vicious circle of "import-backward-reimport" (Yu *et al.*, 2019). Therefore, we raise the following hypotheses:

Hypothesis 1a: The intensity of imported transnational technology will have an inverted U-shaped relationship with local innovation capacity.

As mentioned above, not all types of imported knowledge can successfully integrate with the existing knowledge, even though communication with the external knowledge is crucial. The degree of integration rests on the knowledge relatedness between the imported and the existing one (Boschma & Iammarino, 2009). According to Nooteboom's (2000b) theory, the two knowledge bases will be considered related if they overlap to some extent and develop through similar capacities. Therefore, only the imported external technology is related to local existing technology cognitively, that is, presenting technological relatedness, can the innovation be promoted (Boschma & Iammarino, 2009; Miguélez & Moreno, 2018; Hesse & Fornahl, 2020). By easing learning and integrating pertinent technologies, technological relatedness significantly enhances technological diversification and local innovation capability (Breschi *et al.*, 2003; Boschma, 2017; D'Adda *et al.*, 2020; Li *et al.*, 2020). At the same time, the imported external knowledge should also present differences with the local knowledge base (Boschma *et al.*, 2014), which means related but not the same technology can greatly promote innovation generation and technology progress (Boschma & Iammarino, 2009). Previous research discovered that the entry of similar external technology can enhance local innovation because the external knowledge has also presented the differences and novelty though the similarities with local knowledge and technology (Boschma & Iammarino, 2009; Miguélez & Moreno, 2018), so local enterprises and technicians can rapidly learn the corresponding knowledge and conduct innovation. In this way, the introduction of similar technologies from abroad can have strong spillover effects on local innovation (Miguélez & Moreno, 2018). Given the aforementioned justification, we propose the group of hypotheses listed in the following:

Hypothesis 1b: The relatedness of imported transnational technology will have a positive effect on local innovation capacity.

Hypothesis 1c: The similarity of imported transnational technology will have a positive effect on local innovation capacity.

Another crucial pathway for technology innovation is domestic knowledge communication. Some scholars have paid attention to the effect of external knowledge inflows with a country or region on local innovation capacity (Bottazzi & Peri, 2003; Miguélez & Moreno, 2018; Hesse & Fornahl, 2020). Within a country, due to the similar culture and language (Lawson & Lorenz, 1999), it is easier to build trustworthy relationships among the actors, thus acquiring and absorbing the new knowledge from other cities more effectively (Morrison *et al.*, 2013).

However, opportunities for regrouping of different but complementary knowledge fragments are exhausted and excessive external linkages may render knowledge redundant over time, with communication and repeated interactions with external knowledge becoming less valuable (Boschma & Frenken, 2010; Breschi & Lenzi, 2015). Therefore, we propose the following hypothesis:

Hypothesis 2a: The intensity of imported domestic technology will have an inverted U-shaped relationship with local innovation capacity.

In addition, as what is mentioned in the previous paragraph, if the imported external knowledge is distinct from but related to the local knowledge base, it can greatly promote knowledge generation and innovation (Boschma & Iammarino, 2009; Miguélez & Moreno, 2018). On the one hand, the related external technology can be easier for the locals to understand and absorb so as to enrich the local knowledge base, which can promote local innovation capacities toward knowledge generation and technology development (Boschma & Iammarino, 2009). On the other hand, the knowledge flowing into a region that has been already patented in the local technology base would appear to have many opportunities to be used in creative ways, and the higher the similarity of technological composition between local knowledge and cross-regional flowed knowledge, the greater the impact on the region's innovation performance (Miguélez & Moreno, 2018). Therefore, we also put out the following group of theories in this regard:

Hypothesis 2b: The relatedness of imported domestic technology will have a positive effect on local innovation capacity.

Hypothesis 2c: The similarity of imported domestic technology will have a positive effect on local innovation capacity.

Excessive internal orientation may reduce the creative drive, but excessive external exposure may lead to technological dependence and depletion of the endogenous capacity for autonomous innovation (Breschi & Lenzi, 2015; Evangelista *et al.*, 2002). There is an inherent trade-off between excessive internal and external linkage networks in facilitating knowledge creation (Bathelt *et al.*, 2004; Bathelt, 2007). In this way, local innovation capacity is fully promoted when there is a certain balance between the two types of internal (domestic) and external (transnational) linkages. In other words, internal and external linkages are complementary and mutually reinforcing, rather than substitutes (Bathelt *et al.*, 2004; Graf, 2011). Zhang *et al.* (2016) found empirically that in China, provinces with more transnational technology introduction also received more patents from other domestic provinces, suggesting a complementary relationship between transnational and domestic technologies. Therefore, we propose the following hypothesis.

Hypothesis 3: Transnational and domestic technology imports complement each other in promoting local innovation capacity.

7.2 The Model and Measurements

This section provides a detailed description of dependent variables, explanatory variables, control variables, and empirical models. The definition of the variables is illustrated in Table 3.6.

7.2.1 *The dependent variables*

Despite the difficulties in quantifying innovation in any direct manner (Foray, 2004), a large amount of literature has tried to measure innovation outcomes by the patent application (Pavitt, 1985; Griliches, 1998; Bottazzi and Peri, 2003; D'Adda, *et al.*, 2020; Wagner *et al.*, 2020). Patent statistics are considered to be the primary choice to evaluate the different units' innovation capacity at different scales in quantitative studies (e.g., Patel & Pavitt (1997) for large firms and Zucker *et al.* (2007) for regions). In this way, our dependent variable, innovation capacity referring to the previous research, will be quantified by the number of patent applications.

However, the usage of patent data to measure the level of technology has been widely criticized as it only reflects a limited range of innovations (Pavitt, 1985; Griliches, 1998; Nathan & Rosso, 2022). Firstly, patents are codified knowledge that cannot comprehensively represent knowledge generation as it neglects a large amount of tacit knowledge produced in universities, firms, and research institutes. Secondly, many knowledge products are also not recorded in the library of patent applications (Pavitt, 1985) because not all enterprises and departments rest on the patent to protect their innovation idea (Graham *et al.*, 2009). Nevertheless, we still consider the patent application as the representative during the research due to its effective way of recording technology that is relatively consistent and homogenous.

7.2.2 *Explanatory variables*

As mentioned in the conceptual framework, the explanatory variables in our research are grouped into three indicators on the basis of the underlying theories: the intensity, relatedness, and similarity of transnational and domestic introduction of technology in a city. Among those, the *intensity* of transnational and domestic introduction of technology is measured by the number of imported patents from abroad (including Hong Kong, Macau, and Taiwan in China) and from Chinese mainland cities. The indicator of *relatedness* is used to analyze the imported knowledge related to but not the same as the local knowledge base. If a city has a larger value of external technology relatedness, its external technology is more related to the local technology base. *Similarity* is the degree of similarity between external knowledge and local knowledge base. When a city focuses on a specific subclass of technology that can be easily imported and absorbed from external sources, its *similarity* value is at its highest.

In addition, the double-sided effect of individual variables and the joint effect of different variables which are mentioned above are represented by their squared terms and interaction terms. Before the regression, the input variables are centered to decrease multicollinearity (Cao *et al.*, 2022).

7.2.3 *Control variables*

Inspired by previous research works, we have controlled the potential drivers of regional innovation capacity (Fritsch & Storey, 2014).

We then introduce the reasons for the variable choice. First, research and development (R&D) expenditure can primarily explain the differences in regional growth and development (Crescenzi *et al.*, 2007; Gilsing *et al.*, 2008). Second, industrial enterprises, as important subjects of technological innovation, play an important role in innovation. Finally, we consider the *variety* of the technology as a large number of studies consider both related and unrelated variety as important for regional innovation (Boschma & Iammarino, 2009; Tavassoli & Carbonara, 2014; Miguélez & Moreno, 2018; Hesse & Fornahl, 2020).

7.3 Empirical Results

7.3.1 *Local innovation capacity*

Figure 7.2 depicts the distribution of patent applications across China from 2001 to 2018. The number of patent applications in China increased from 83,158 in 2002 to 329,317 in 2018. The innovation output of Chinese cities demonstrates spatial persistence and spatial heterogeneity. In terms of the spatial distribution of patent applications, the hierarchical scale distribution of innovation output remained stable over the 18-year period, with high innovation output cities mainly concentrated in three major city clusters — the BTH, YRD, and PRD. These clusters comprise provincial capitals and regional centers, where Beijing, Shanghai, and Shenzhen are the most innovative cities due to their high levels of economic development. While the proportion of innovation output in cities located in China's central and western regions has been increasing steadily over the years, certain cities, such as Hefei and Wuhan in the central region, and Chengdu, Chongqing, and Kunming in the western region, have all surpassed 100,000 patent applications between 2014 and 2019. Despite this progress, the eastern coastal region continues to dominate in terms of innovation output, with its total innovation

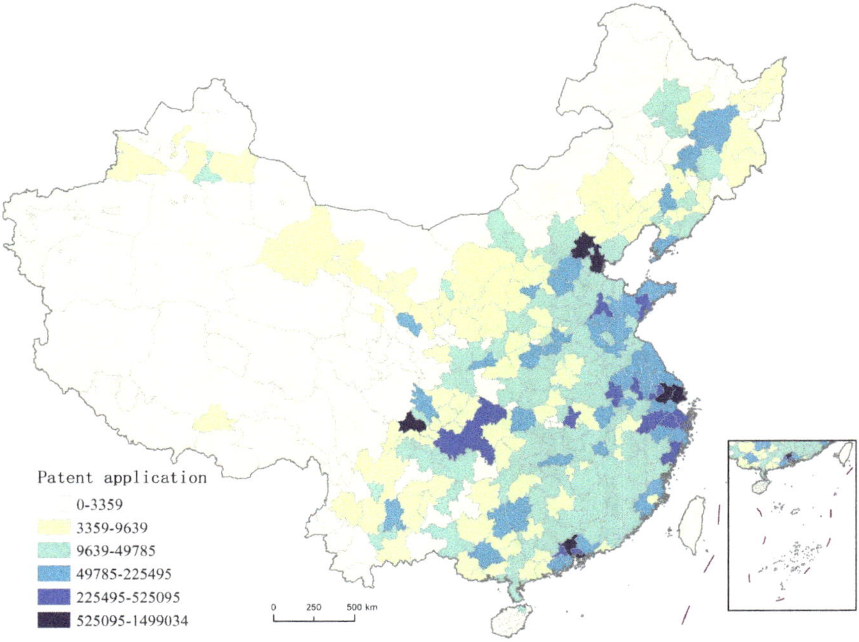

Figure 7.2 Spatial disparity of urban patent application in China from 2001 to 2018.

output and number of patent applications accounting for 70–80% of the total. This suggests that regional innovation in China may be subject to the "Matthew effect", meaning that the strong continue to grow stronger while the weak fall behind (Zhang *et al.*, 2016).

Table 7.1 describes the variance inflation factors (VIFs) and correlation coefficients for each variable. There is also no serious multicollinearity issue according to correlation coefficients and VIF tests. By the way, to avoid the potential of reverse causation in the empirical section, all explanatory and control variables are lagged by one year (Castaldi *et al.*, 2015; Hesse & Fornahl, 2020).

7.3.2 *Spillover effect of technology import network*

The results of OLS estimation over the period of 2000–2018 are shown in Table 7.2. Model 1 represents the baseline model with

Table 7.1 Descriptive statistics and correlation matrix.

Variable	Observations	Minimum	Maximum	Mean	SD	1	2	3	4	5	6	7	8	9	10	11
1 Patents	4,291	2	250,115	5,443	16,784	1										
2 Intensity$_d$	4,291	0	5513	68.97	292.0	0.828***	1									
3 Relatedness$_d$	**4,291**	**0**	**18.42**	**0.121**	**0.546**	**0.367*****	**0.364*** **	**1**								
4 Similarity$_d$	**4,291**	**0**	**16.19**	**3.183**	**3.692**	**0.528*** **	**0.426*** **	**0.350*** **	**1**							
5 Intensity$_f$	4,291	0	477	2.429	15.31	0.737***	0.618***	0.302***	0.316***	1						
6 Relatedness$_f$	4,291	0	16.86	0.126	0.807	0.414***	0.297***	0.517***	0.269***	0.606***	1					
7 Similarity$_f$	4,291	0	9.216	0.495	1.743	0.669***	0.530***	0.398***	0.574***	0.534***	0.466***	1				
8 Rel. variety	4,291	0.446	3.391	2.307	0.723	0.299***	0.229***	0.164***	0.625***	0.133***	0.112***	0.264***	1			
9 Unrel. variety	4,291	0	2.312	2.063	0.181	0.135***	0.092***	0.096***	0.258***	0.047***	0.056***	0.120***	0.390***	1		
10 R&D	4,291	0.130	9,600,000	168,118	554,193	0.836***	0.654***	0.329***	0.453***	0.716***	0.407***	0.587***	0.247***	0.134***	1	
11 Enterprise	4,291	0	303	8.638	22.38	0.828***	0.647***	0.542***	0.472***	0.745***	0.585***	0.626***	0.251***	0.141***	0.818***	1

Note: *Significant at the 0.05 level, **Significant at the 0.01 level, ***Significant at the 0.001 level.

Table 7.2 Results of least squares method.

	Model 1	Model 2	Model 3	Model 4	Model 5	Model 6
$Intensity_f$		0.004***				0.001***
		(0.001)				(0.000)
$Intensity_f^2$		−0.003***				
		(0.001)				
$Relatedness_f$			0.006			
			(0.011)			
$Similarity_f$			0.032***			
			(0.005)			
$Intensity_d$				0.001***		0.003***
				(0.000)		(0.001)
$Intensity_d^2$				−0.001***		
				(0.000)		
$Relatedness_d$					0.012***	
					(0.016)	
$Similarity_d$					0.028***	
					(0.001)	
$Intensity_f$*$Intensity_d$						−0.002***
						(0.001)
Rel. variety	0.536***	0.548***	0.565***	0.572***	0.545***	0.564***
	(0.022)	(0.022)	(0.022)	(0.022)	(0.022)	(0.022)
Unrel. variety	−0.227***	−0.225***	−0.217***	−0.225***	−0.214***	−0.223***
	(0.044)	(0.044)	(0.044)	(0.044)	(0.044)	(0.044)
R&D(log)	0.149***	0.152***	0.152***	0.155***	0.150***	0.154***
	(0.012)	(0.012)	(0.012)	(0.012)	(0.012)	(0.012)
Enterprise(log)	0.289***	0.270***	0.251***	0.235***	0.240***	0.246***
	(0.026)	(0.027)	(0.027)	(0.028)	(0.027)	(0.028)
Time fixed effects	Yes	Yes	Yes	Yes	Yes	Yes
Region fixed effects	Yes	Yes	Yes	Yes	Yes	Yes
Constant	3.335***	3.315***	3.292***	3.294***	3.333***	3.298***
	(0.126)	(0.126)	(0.126)	(0.126)	(0.125)	(0.126)
Observations	4,051	4,051	4,051	4,051	4,051	4,051
Adjusted R^2	0.9616	0.9621	0.9620	0.9619	0.9624	0.9619
F-statistic	26.811***	24.403***	23.014***	22.566***	20.835***	23.092***

Note: Robust standard errors are shown in parentheses; *Significant at the 0.05 level, **Significant at the 0.01 level, ***Significant at the 0.001 level.

only the control variables. We discover that related variety is positively significant, which is partly consistent with prior research on the contribution of knowledge variety to regional innovation (Tavassoli & Carbonara, 2014). The negative coefficient of unrelated variety indicates that cities with higher unrelated variety have more variability in knowledge, thus making it more difficult to diffuse and learn the knowledge and technology, which to some extent hinders the local innovation capacity of the city. Besides, our research has revealed that cities with a concentration of industrial enterprises and high R&D investment are positively correlated with local innovation capacity.

Model 2 examines the impact of imported intensity of transnational technology on local innovation capacity. The positive sign on the linear term and the negative sign on the quadratic term indicate the more transnational technologies are imported to the region, the better for the innovation capacity of the city. However, when transnational imported technologies in the region increase to a threshold, the cost of knowledge absorption will increase dramatically and it will become too expensive to establish and maintain formal and informal linkages. At the same time, the reliance on technology introduction may lead to an imbalance in the innovation mechanism (Yu *et al.*, 2019), which ultimately may have a negative impact on local innovation capacity. This discovery supports Hypothesis 2a, that is, the intensity of transnational imported technology will have an inverted U-shaped relationship with local innovation capacity.

Model 3 refers to the impact of the relatedness between transnational imported technology and local knowledge base on local innovation capacity. The results revealed in model 3 confirm that transnational technology has a complex effect on local innovation capacity, and transnational knowledge benefits knowledge generation while transnational knowledge is not essential. Firstly, model 3 shows the coefficient of relatedness of transnational technology is not significant enough to reach the statistical standard, and thus cannot support Hypothesis 1b. The transnational technology related to existing local technology is generally advanced which has a large gap with the local technology base. This kind of technology gap leading to local innovation promotion is not obvious. However, transnational technology still overall has a great contribution to the generation of new technology that can be confirmed by the positive value of $Similarity_f$ and significance coefficient, which supports Hypothesis 1c.

According to the results reported in model 4, the intensity of imported domestic technology will have an inverted U-shaped relationship with local innovation capacity, thus supporting our Hypothesis 2a. The potential reason for this result is that once imported domestic technologies in the region increase to a threshold, opportunities for novelty and idea regrouping will be more limited, and later diminishing returns stand out. Model 5 refers to the impact of the relatedness between domestic technology and local knowledge base on local innovation capacity. The results of model 5 suggest that the relatedness between domestic technology and the local knowledge base can also promote the local innovation capacity, which confirms our Hypothesis 2b. The coefficients in model 5 have the expected sign and are statistically significant, which supports our Hypothesis 2c. The results show that the more similarity the domestic technology has with the local knowledge base, the more effective it is in the promotion of local innovation capacity, which is in line with Miguélez and Moreno (2018). In models 3 and 5, it can be seen that the local innovation capacity is influenced by the combination of the domestic and transnational technology pipeline.

The result of model 6 on the interaction term shows a positive significant effect, which confirms our Hypothesis 3, that is, transnational and domestic technology introductions complement each other in promoting local innovation capacity. On the one hand, the introduction of advanced foreign technologies provides novel technologies and regrouping opportunities for the local knowledge base, while the introduction of domestic technologies helps to better decipher and absorb externally acquired technologies. On the other hand, this may be because many Chinese technologies are subsidiary technologies that complement foreign technologies, so cities with more foreign technologies may acquire more domestic technologies too, and the reverse engineering process of foreign technologies may also enhance local innovation capacity (Zhang *et al.*, 2016). Thus, domestic and foreign technology transfer is complementary.

7.4 Modeling Specification and Robustness Test

By using the dependent variable replacement approach and the number of granted patents for discoveries with a one-year lag as a proxy variable for urban innovation capacity, we were able to examine the robustness of the study's findings. The robustness of the regression results is shown by models 7–11 in Table 7.3 which demonstrates that the technology dual pipelines still have a favorable impact on urban innovation capacity when the dependent variable has been replaced.

Another robustness test is to use a subsample that excludes the top 10 GDP-producing cities (Guo *et al.*, 2021), which account for only 3% of the number of cities in the country and 23% of the country's economic output. These cities are more likely to have large innovation resources and stronger innovation potential due to their economic position and polarizing effects. Models 12–16 in Table 7.3 produce results that exclude the 10 cities with the highest GDP in 2018, which are consistent with the earlier findings. This confirms that the dual pipelines of technology import are effective for all cities and not just economically developed ones.

7.5 Summary

The "buzz-and-pipeline" concept has had a significant and pervasive impact on the study of the geography of innovation (Bathelt *et al.*, 2004; Bathelt, 2007). It is difficult to find empirical investigations that combine these two various technological pipelines at various geographic scales into a single analytical framework. With the performance of ordinary least squares, this chapter has investigated the impact of extra-regional technology at various sizes on local innovation in a sample of Chinese cities between 2001 and 2018.

Table 7.3 The results of robustness test.

	Model 7	Model 8	Model 9	Model 10	Model 11	Model 12	Model 13	Model 14	Model 15	Model 16
$Intensity_f$	0.159^{***}				0.006^{***}	0.014^{***}				0.007^{***}
	(0.019)				(0.001)	(0.003)				(0.002)
$Intensity_f^2$	-0.024^{**}					-0.000^{***}				
	(0.011)					(0.000)				
$Relatedness_f$		0.020					0.001			
		(0.014)					(0.019)			
$Similarity_f$		0.040^{***}					0.037^{***}			
		(0.007)					(0.007)			
$Intensity_d$			0.001^{***}		0.000^{***}			0.001^{***}		0.000^{***}
			(0.000)		(0.000)			(0.000)		(0.000)
$Intensity_d^2$			-0.000^{***}					-0.000^{***}		
			(0.000)					(0.000)		
$Relatedness_d$				0.481^{***}					0.162^{***}	
				(0.073)					(0.046)	
$Similarity_d$				0.107^{***}					0.026^{***}	
				(0.014)					(0.003)	
$Intensity_f$*$Intensity_d$					-0.000^{***}					-0.000^{***}
					(0.000)					(0.000)
Rel. variety	0.281^{***}	0.262^{***}	0.281^{***}	0.199^{***}	0.267^{***}	0.576^{***}	0.579^{***}	0.605^{***}	0.569^{***}	0.590^{***}
	(0.028)	(0.028)	(0.028)	(0.027)	(0.028)	(0.023)	(0.023)	(0.023)	(0.022)	(0.023)

(*Continued*)

Table 7.3 (*Continued*).

	Model 7	Model 8	Model 9	Model 10	Model 11	Model 12	Model 13	Model 14	Model 15	Model 16
Unrel. variety	−0.488***	−0.491***	−0.500***	−0.497***	−0.494***	−0.220***	−0.217***	−0.216***	−0.212***	−0.218***
	(0.055)	(0.056)	(0.056)	(0.055)	(0.056)	(0.045)	(0.045)	(0.045)	(0.045)	(0.045)
R&D(log)	0.072***	0.074***	0.078***	0.065***	0.076***	0.148***	0.152***	0.150***	0.152***	0.149***
	(0.015)	**(0.015)**	**(0.015)**	**(0.015)**	**(0.015)**	**(0.012)**	**(0.012)**	**(0.012)**	**(0.012)**	**(0.012)**
Enterprise(log)	**0.619***	**0.638***	**0.603***	**0.627***	**0.624***	**0.255***	**0.252***	**0.201***	**0.223***	**0.225***
	(0.034)	(0.034)	(0.035)	(0.033)	(0.035)	(0.028)	(0.028)	(0.029)	(0.028)	(0.029)
Time fixed effects	Yes	Yes	Yes	Yes	Yes	Yes	Yes	Yes	Yes	Yes
Region fixed effects	Yes	Yes	Yes	Yes	Yes	Yes	Yes	Yes	Yes	Yes
Constant	1.453***	1.413***	1.409***	1.576***	1.413***	3.232***	3.198***	3.216***	3.227***	3.230***
	(0.159)	(0.158)	(0.158)	(0.157)	(0.158)	(0.127)	(0.127)	(0.126)	(0.126)	(0.127)
Observations	4051	4051	4051	4051	4051	3,881	3,881	3,881	3,881	3,881
Adjusted R^2	0.9421	0.9415	0.9418	0.9428	0.9416	0.9563	0.9563	0.9567	0.9569	0.9565
F-statistic	20.608***	20.399***	20.290***	19.579***	20.605***	22.658***	22.106***	20.657***	18.981***	21.473***

Note: Robust standard errors are shown in parentheses; *Significant at the 0.05 level, **Significant at the 0.01 level, ***Significant at the 0.001 level.

It investigates a macroscopic mechanism of the impact of the intensity, relatedness, and similarity of imported technologies on local innovation capacity at different geographical scales.

The findings demonstrate that technology may also be imported into an area from the outside, and technological innovation gains from the use of trans-local pipelines for introduction, which strengthens the local innovation capacity. Indeed, more externally imported technology is not better, and the intensity of transnational and domestic patent introduction both have an inverted U-shaped relationship with innovation capacity. China is currently making the shift from focusing only on quantity to focusing on both innovation quantity and quality, so it should balance technology introduction and independent innovation in urban development to form a virtuous cycle of "transfer, absorption, diffusion and re-innovation" (Yu *et al.*, 2019). Most significantly, our findings provide novel insight into the impact of external linkages and show that domestic and transnational technology introductions are equally vital for cities, that is, if a city relies on only one technology introduction channel, its innovation capacity and potential cannot be fully developed in the long term, while domestic technology pipeline is more crucial in fostering innovation capacity. These findings support the favorable impact of knowledge externalities that have been demonstrated. However, it offers stylized viewpoints with which to formalize an investigation of the contribution of external technology "dual-pipelines" dynamics.

Furthermore, research suggests that regional innovation benefits from the extra-regional technological relatedness and similarity. This concept has been discussed extensively in recent literature (Miguélez & Moreno, 2018) and has shown to have a positive effect on local inventive capacity. In actuality, new knowledge that is progressively built upon existing cognitive structures across related technologies will be produced whether it comes from relatedness or similar technologies. However, what this research has found different from the vast literature of previous studies is the insignificant role of technology relatedness in the transnational pipeline. The findings contradict our prediction and indicate that the greater the similarity between extra-regional knowledge and the local knowledge base, the greater the rise in regional innovation capacity. Therefore, decision-makers should thus determine the kind of local knowledge base, focus on relatedness and similarity technology, and build connections with the domestic regions which have advanced technology.

Finally, this study also has some restrictions which need further analysis. During the research, we distinguished the impact of imported domestic and transnational technology on local innovation capacity, but failed to investigate the difference between them. Even if the study still has certain flaws, the analysis offers a unique viewpoint and produces intriguing results that open up new research avenues. It has discovered the significance of dual pipelines of technology import (domestic and transnational pipelines) on local innovation capacity and raised the debate on how knowledge relatedness and similarity work in the local knowledge base.

In future studies, we will further distinguish between different types of knowledge and technology sources (e.g., universities, enterprises, research institutes, etc.) and analyze whether imported domestic and transnational technologies have an impact on improving local innovation quality and innovation performance. In addition, due to the restriction of patent data, we will also consider other types of data such as trademarks and products in the future. In order to extend our research framework, it is necessary to supplement the case studies at the regional level. Regional case studies could clarify the ways in which external technology influences the diversification of regional technology over time and the methods by which relatedness and similarity technology is transferred.

CHAPTER 8

DETERMINANTS OF TECHNOLOGY TRANSFER NETWORK EVOLUTION

As China's economy transitions into a new era, it must confront the challenges of transformation and upgradation of traditional as well as emerging new industries. To promote the transfer and transformation of scientific and technological achievements, there is an urgent need for a new and powerful impetus. Against this background, expediting technology transfer emerges as an imperative choice to facilitate China's economic restructuring and transition toward an innovation-driven development path. As a post-developmental economy, China displays conspicuous non-heterogeneity in its regional innovation resources and industrial structure. Consequently, interregional technology transfer has assumed a critical role in enabling the flow and dissemination of innovative resources, thereby advancing technological progress and propelling economic growth in underserved areas.

However, there are relatively few studies on the pattern and mechanism of technology transfer in China (Liu *et al.*, 2018a), and the commonly employed regression analysis methods are incapable of revealing the network construction mechanism entirely. In this chapter, we use exponential random graph model (ERGM) as a method of complex network measurement to discover the dependence between relationships. ERGM integrates the influence of network structure, actor attributes, and proximity, which has been gradually applied to innovation linkage research (Wang *et al.*, 2022).

China has greatly benefited from the technological exchange between the global and local communities. By actively integrating into global innovation networks, China has emerged as a key player in the global economy and a center for technology innovation. Its extensive involvement in globalization has enabled it to evolve into a dynamic global economic and technological entity. Unfortunately, existing studies of global–local perspectives are relatively fixed in scale and inevitably neglect the complex technological links that they encompass between international and domestic regions. Therefore, the mechanism of action between global access and local embeddedness remains a riddle (Liu & Yan, 2022).

This chapter focuses on patented innovative technologies. Firstly, we examine the factors that promote or inhibit the integrated flow of technology. Then, an econometric model is constructed to examine the dynamics of the generation and evolution of China's urban technology innovation system, as well as the factors influencing the scale of domestic and international patent rights trade, which is based on the cross-regional technology transfer pipelines.

8.1 Conceptual Background and Hypotheses

The evolutionary path of the growth of innovation networks does not simply involve the accumulation of nodes; rather, it involves the interaction and fusion of knowledge that accompanies the process of replicating, changing, and renewing nodes. As many micro-rules repeatedly interact, fundamental changes in network structure occur (Wang *et al.*, 2022). We identify three growth mechanisms for technology transfer networks (TTNs) in China: other-organization, proximity, and self-organization factors. The detailed illustration can be found in Chapter 2.

Technology transfer linkages have a multiscale coupling effect (Zhang *et al.*, 2013). At the city scale, such linkages are resolved into transnational technology pipelines and domestic technology pipelines. There is a dual-way interaction effect between transnational and domestic intercity technology pipelines, forming a complex mechanism of competition-complementarity in terms of the intensity, breadth, distance, and control capacity of technology pipelines (Liu & Yan, 2022). The local technology innovation milieu continuously influences the external technology pipeline network through local embeddedness, while the domestic intercity technology pipeline enhances the local technology innovation capacity through the technology transmission effect, which in turn acts on the global nexus of patented technologies. In this way, we put forward the following potential determinants of the transnational TTN:

Hypothesis 1: Local economic development has a positive effect on TTN;

Hypothesis 2: Local innovation strength has a positive effect on TTN;

Hypothesis 3: The development of domestic TTN has a positive effect on transnational TNN.

8.2 Determinants of Transnational Technology Transfer

8.2.1 *The determinants of technology transfer-in*

We measure the impact of several domestic technology pipeline expansion capacities on the import of transnational technology by controlling for the socio-economic attributes of the city, respectively (Models 1–5). The regression models analyze the degree to which the local technology innovation milieu impacts the situation (Models 6–8).

8.2.1.1 *Expansion of domestic pipelines*

Models 1–5 in Table 8.1 show that the expansion of domestic innovation networks significantly contributes to the integration of

Table 8.1 Results of panel negative binomial regression for transnational technology transfer-in.

	Dependent variable: T_in							
Variables	Model_1	Model_2	Model_3	Model_4	Model_5	Model_6	Model_7	Model_8
Indegree	0.25^{**}							
WIndegree		0.30^{***}						
Outdegree			0.26^{**}					
WOutdegree				0.27^{***}				
Closeness					**3.34^{***}**			
Betweenness					**−2E-05**			
GDP_per_ca	1.14^{***}	0.69^{***}	1.08^{***}	0.75^{***}	0.70^{***}	1.46^{***}		
Exp	0.34^{***}	0.26^{**}	0.34^{**}	0.29^{**}	0.31^{***}	0.42^{***}		0.32^{***}
Imp	-0.25^{***}	-0.16^{*}	-0.24^{**}	-0.18^{**}	−0.14	-0.31^{***}		-0.11^{*}
FDI	0.12	0.13	0.13	0.13	0.10	0.09		0.22^{***}
Tech_Work	0.72^{***}	0.67^{***}	0.71^{***}	0.69^{***}	0.65^{**}	0.74^{***}	0.36^{**}	
College	-0.46^{**}	-0.46^{**}	-0.46^{**}	-0.46^{**}	-0.39^{**}	-0.39^{**}	-0.42^{**}	
List_Com	−0.06	−0.04	−0.08	−0.11	−0.07	−0.01	0.19	
Ent_Tech	−0.04	−0.07	−0.05	−0.05	−0.08	−0.001	0.39^{***}	

Note: Significant at: $^{***}p < 0.001$, $^{**}p < 0.01$, $^{*}p < 0.005$.

innovative cities into global innovation networks as well as the introduction and absorption of transnational technologies.

The breadth of domestic intercity technology transfer has a significant positive effect on the absorption of transnational technology. The results of Models 1 and 3 show that the in-degree and out-degree values of domestic technology transfer (reflecting the scope and breadth of its technological linkages) have a significant positive contribution to the intensity of transnational technology import. This indicates that the more domestic technology transfer pipelines a city has, the stronger its ability to diffuse or absorb transnational technologies, which in turn is conducive to the development of its transnational technology import pipelines.

The intensity of domestic technology transfer has a significant positive effect on the import of transnational technologies, and this effect is stronger than the promotion effect of the breadth. Combining the results of Models 2 and 4, the weighted in-degree and weighted out-degree values of domestic technology transfer (which portray the intensity of technology transfer) in China have a positive effect on the import of transnational technology. This suggests that domestic technology transfer plays a vital role in enhancing a city's technological absorption and innovation capacity, which strengthens its ability to imitate and innovate external technologies. Furthermore, this process also expands the pipeline for importing transnational technology, thus validating previous studies on the driving mechanism of regional technology transfer (Yang & Liu, 2014).

The closeness of domestic technology transfer is conducive to the introduction of transnational technology. Model 5 regression results show that the closeness centrality of cities in domestic TTN is positively related to their transnational patent import, that is, the stronger a city's linkage capacity in TTN, the shorter the access or diffusion path to patented technologies in other domestic cities. This not only facilitates local innovation but also promotes the diffusion of transnational technology. It further confirms the mechanism of the role of technology network structure in innovation performance (Cheng & Wang, 2013; Zeng & Wang, 2019).

Model 6 shows that there is no significant association between the control ability of domestic technology transfer and transnational technology import.

8.2.1.2 *Local innovation milieu*

Models 6–8 in Table 8.1 show that the technological innovation milieu of the city itself has a significant impact on its integration into global innovation networks and the import of transnational technologies.

The level of the local economy significantly affects the absorption of transnational technologies. The results of regression Models 1–6 show that the city's economic development level (GDP per capita) has a significant impact on the absorption of transnational technologies.

The regression Models 1–6 indicate a strong positive correlation between the level of urban economic development (as measured by GDP per capita) and the import of transnational technologies. This suggests that cities with higher levels of economic development have a greater demand for technology due to their industrial activity, which in turn drives the import and absorption of external technologies. This finding was reported by Liu *et al.* (2018a).

Local technological strength has multiple effects on the import of transnational technologies. Models 1–6 show that to enhance the intensity of external technology import in a city, it is effective to increase its number of science and technology employees and establish national-level enterprise technology centers. However, the number of universities in the city may inhibit its transnational technology import. This is because enterprises are the main actor of transnational technology transfer, and cities with high concentrations of enterprise technicians or national-level enterprise technology centers have stronger technological reserves and R&D strength, which enables them to better absorb advanced technology from abroad. In contrast, universities are less involved in the process of transnational technology transfer and make limited contributions to technology transfer. Therefore, focusing on building strong enterprise technology centers and increasing the number of science and technology employees can help cities optimize their strategy for importing external technology. The coupling between the extent of higher education in urban areas and their technological production capacity tends to be imbalanced (Xiao & Liu, 2018), which can impede the innovation potential of businesses (Xu, 2013). Therefore, a surplus of universities may actually hinder the establishment of transnational technology import pipelines.

The degree of open economic orientation has multiple effects on the intensity of transnational technology import. The results of Models 1–6 and 8 show that the amount of goods exported by cities can promote the import of transnational patents, while the amount of goods imported is negatively correlated. Exports of goods are an important way for cities' external economic ties, leading to more information spillovers and the formation of stable intercity cooperation relationships, thus facilitating the import and absorption of transnational technologies. Goods imports, on the other hand, are not conducive to transnational technology import, which differs from Zhang & Sun (2011). The reason may lie in the rising size of consumer goods in imports, growing at an average annual rate of over 10%. It has been discovered that dependence on imported consumer goods can hinder local innovation capacity (Zhou, 2019), posing challenges to achieving technological learning and absorption. In addition, an increase in actual foreign investment utilized by cities would promote economic cooperation and technology exchange between domestic cities and foreign firms, thus positively contributing to the expansion of transnational technology import pipelines, but the regression results are not robust enough.

8.2.2 *The determinants of technology transfer-out*

Consistent with the studies of transnational technology transfer-in, we measure the impact of domestic technology pipeline expansion capacity on transnational technology transfer-out separately by controlling for the socioeconomic attributes of the city (Models 9–13).

We have also investigated the influence of the local technological innovation milieu on this impact (Models 14–16).

8.2.2.1 *Expansion of domestic pipelines*

From Models 9–13 in Table 8.2, it is clear that the expansion of the domestic innovation network significantly contributes to the transnational export and global networking of patented technologies in innovative cities.

The breadth of domestic technology transfer has a significant contribution to the export of transnational technologies. The results of Models 9 and 11 show that the in-degree and out-degree values of domestic technology transfer have a significant positive effect on the intensity of transnational technology transfer-out. According to Cheng & Wang (2013), cities with more extensive domestic technology transfer pipelines tend to have greater openness in intercity technology import and export. This, in turn, enhances their own technological innovation capacity and ability to participate in international technological competition and cooperation, thus facilitating the development of transnational technology transfer-out pipelines.

The strength of domestic technology pipelines also has a significant contribution to local embeddedness in the global innovation network. The findings from regression Models 10 and 12 indicate that both the intensity and breadth of domestic technology transfer in cities have a positive correlation with transnational technology transfer-out. Thus, it suggests that having a diverse range of output locations for domestic technology transfer can contribute significantly to improving innovation performance and optimizing global innovation networks (Gui *et al.*, 2018a, 2018b).

The closeness of domestic intercity technology transfer is also beneficial to local technology transfer-out transnationally. Model 13 shows that there is a significant positive correlation between the closeness centrality of a city in the domestic TTN and the intensity of its transnational technology transfer-out. That is, an increase in the closeness of technology transfer links between cities can

Table 8.2 Results of panel negative binomial regression for transnational technology transfer-out.

	Dependent variable: *T_out*							
Variables	Model_9	Model_10	Model_11	Model_12	Model_13	Model_14	Model_15	Model_16
Indegree	0.35**							
WIndegree		0.38***						
Outdegree			0.35**					
WOutdegree				0.35***				
Closeness					5.44***			
Betweenness					–2E-05			
GDP_per_ca	0.81*	0.40	0.73	0.36	0.13	1.17***		
Exp	0.10	0.10	0.12	0.11	0.15	0.19		0.35***
Imp	0.04	0.08	0.07	0.09	0.14	–0.03		–0.06
FDI	–0.27	–0.32*	–0.29	–0.27	–0.29	–0.32		0.04
Tech_Work	0.21	0.18	0.17	0.18	0.19	0.22	0.06	
College	0.003	–0.02	0.02	–0.007	–0.04	0.04	–0.17	
List_Com	–0.04	–0.09	–0.08	–0.12	–0.05	–0.03	0.23	
Ent_Tech	0.10	0.20	0.12	0.12	0.10	0.15	0.41***	

Note: Significant at: $^{***}p < 0.01$, $^{**}p < 0.05$, $^{*}p < 0.1$.

significantly enhance the exchange between industrial clusters, multicluster networks, and the spillover of new technologies (Zhou *et al.*, 2019).

Model 13 shows that there is no significant association between the control ability of domestic intercity technology transfer and transnational technology transfer-out.

8.2.2.2 *Local innovation milieu*

As can be seen from Models 14–16 in Table 8.2, the technological innovation milieu of the city itself has a positive but poorly robust effect on the intensity of transnational technology transfer-out.

The local economic level has a positive effect on the intensity of transnational technology transfer-out. Models 9–14 appear that there is a positive correlation between a city's level of economic development (as measured by GDP per capita) and its productivity, technological investment, and R&D capacity. This relationship ultimately results in a stronger capacity for technological export.

Local technological strength and openness have some influence on the intensity of their transnational technological exports, but the robustness is low. Models 9–16 show that when only innovation capacity and economic openness are selected as independent variables respectively, the coefficients of the two indicators, which are the number of national-level enterprise technology centers and the amount of export trade in goods, are positive. When including multiple indicators in a regression model, it has been observed that most of these indicators lose their significance. However, research suggests that an increase in the number of national-level enterprise technology centers is positively correlated with the technological innovation capacity and international competitiveness of a city. This also leads to a higher intensity of transnational technological exports. Cities with larger export trade in goods tend to have closer transnational economic ties and technological exchanges and are required to meet more stringent international standards for export goods. This necessitates the mastery of more advanced technologies by relevant enterprises, which ultimately promotes their technological innovation capacity (Li *et al.*, 2016).

8.3 Determinants of National Technology Transfer

There are two causes for the failure of the model to converge. The first is multicollinearity, which is caused by the large number of indicators and the fact that the measured dimensions overlap to some extent. For the model to accommodate the greatest number of variables and to identify the greatest number of influential factors while still meeting the colinearity requirement, variables were added and removed in a step-by-step manner. After running the model iteratively using the variables in Tables 8.3 and 8.4, it was discovered that the inclusion of human capital investment, technological intermediation capacity, similarity of industrial structure, and economic development potential difference led to multiple cointegration problems in both periods, generating more information redundancy, and were therefore excluded from the model in both periods. Furthermore, in the first period, homogeneity of administrative hierarchy led to multiple cointegration problems and was excluded from the model in the first period. The second reason is that the introduction of an excessive number of endogenous structural variables tends to degrade the model, particularly for complex configurations such as triangular structures, resulting in poor fitting and failure to converge (Milo *et al.*, 2002). By iteratively adding and removing endogenous structural variables, it was discovered that, in both stages, the model converged only when scale control variables with reciprocity were included, indicating that TTN is probably not susceptible to local structures other than reciprocity.

8.3.1 *Influence mechanism during the first period (2007–2012)*

The EGRM for intercity technology transfer is a model based on edge dependencies, with weighted edges always among the ensemble of independent variables, reflecting the underlying tendency for transfer

Table 8.3 The descriptive statistic of city attribute.

Indicators	Number of cases	Average	Median	Std. Dev.	Min.	Max.
2007–2012						
GDP_per_ca	297	3.167	2.581	2.100	0.530	13.880
Sci_Pub	297	0.014	0.010	0.011	0.002	0.064
Sci_Workforce	297	0.008	0.006	0.007	0.001	0.056
Dec_Index	297	0.365	0.291	0.254	0.063	1.671
Mar_Index	297	8.836	8.816	1.708	4.557	12.721
Adm	297	1.199	1.000	0.585	1.000	4.000
New_Hitech	297	0.005	0.001	0.0208	0.000	0.280
ITS	297	1.593	0.000	16.826	0.000	285.000
Rel_Div	297	3.563	3.557	1.602	0.000	7.341
Non_Rel_Div	297	1.524	1.693	0.631	0.000	2.270
Tech_Simi	297	7.073	6.583	3.013	0.000	16.468
2013–2018						
GDP_per_ca	297	5.356	4.530	3.265	1.102	21.929
R&D_GDP	297	0.372	0.010	6.156	0.000	106.107
R&D_Workforce	297	0.007	0.005	0.007	0.000	0.060
Dec_Index	297	0.368	0.305	0.241	0.098	1.676
Mar_Index	297	11.885	11.773	1.770	7.650	16.447
Adm	297	1.199	1.000	0.585	1.000	4.000
New_Hitech	297	1.132	0.140	4.477	0.000	59.890
ITS	297	6.394	0.000	49.735	0.000	824.000
Rel_Div	297	1.834	1.866	0.732	0.000	3.344
Non_Rel_Div	297	1.978	2.059	0.333	0.000	2.317
Tech_Simi	297	6.676	6.390	3.388	0.000	16.468

relationships to occur. As shown in Table 8.5, Models 17–19 include nodal attributes, proximity, and reciprocity relationship, respectively, to examine the impact of urban innovation attributes, intercity relationship attributes, and endogenous network structure on intercity technology transfer. Model 20 then includes all three types of variables which includes all dimensional explanatory variables. In terms of AIC and BIC, Model 20 has the highest degree of fitting among the four models.

Table 8.4 The descriptive statistic of city pair's dualism.

Indicators	Number of Cases	Average	Median	Std. Dev.	Min.	Max.
2007–2012						
Pro_Geo	43956	2.439	2.333	0.678	1.000	6.528
Pro_Tec	43956	0.411	0.415	0.160	0.000	0.967
Diff_Inno	43956	2.923	2.553	1.561	1.000	13.593
2013–2018						
Pro_Geo	43956	2.439	2.333	0.678	1.000	6.523
Pro_Tec	43956	0.765	0.809	0.176	0.059	0.999
Pro_Ins	43956	0.534	0.562	0.189	0.000	0.976
Diff_Inno	43956	3.061	2.520	2.030	1.000	14.936
Diff_Eco	43956	2.076	1.872	0.865	1.000	7.199

Table 8.5 ERGM for network of technology transfer during the period 2007–2012, with isotropy of city node properties.

Variables	Model_17	Model_18	Model_19	Model_20
GDP_per_ca	0.090***			0.122***
Sci_Pub	35.460***			22.480***
Dec_Index	1.159***			1.164***
Mar_Index	0.094***			0.027***
New_Hitech	0.017***			0.061***
Pro_Geo		0.847***		0.917***
Pro_Tec		3.732***		2.382***
Diff_Inno		0.253***		-0.092
Mutual			2.799***	0.692***
Edges	-8.785***	-8.841***	-4.549***	-10.820***
AIC	-322938	-320844	-320956	-326361
BIC	-322882	-320806	-320938	-326247

Note: Significant at: $^{***}p < 0.01$, $^{**}p < 0.05$, $^{*}p < 0.1$.

In terms of the role of city attributes in the generation of TTN, Models 17–20 default to homogeneity. However, the explanation of the dependent variable based on the nature of city characteristics may not be consistent when two distinct directions of technology

Table 8.6 ERGM results for TTN during the period 2007–2012.

Variables	Model_21		Model_22	
GDP_per_ca_in	0.111	***	0.039	***
GDP_per_ca_out	0.132	***	0.078	***
Sci_Pub_in	27.680	***	7.671	***
Sci_Pub_out	17.140	***	−5.050	*
Dec_Index_in	1.061	***	0.573	***
Dec_Index_out	1.259	***	0.833	***
Mar_Index_in	0.027	**	−0.014	
Mar_Index_out	0.028	**	0.003	
New_Hitech_in	0.043	***	0.035	***
New_Hitech_out	0.078	***	0.090	***
Rel_Div_out			0.434	***
Tech_Simi_in			0.234	***
Pro_Geo	0.916	***	0.957	***
Pro_Tec	2.377	***	1.534	***
Diff_Inno	−0.092	***	−0.034	**
Mutual	0.723	***	0.531	***
Edge	−10.820	***	−12.330	***
AIC	−326463		−327618	
BIC	−326323		−327503	

Note: Significant at: $^{***}p < 0.01$, $^{**}p < 0.05$, $^{*}p < 0.1$.

flows, transfer-in and transfer-out, are considered. To obtain the heterogeneous effects cities may exhibit when taking on the role of technology transfer in versus transfer out, we further expand the model (shown in Table 8.6). Compared with Model 20, Model 21 divides the city attribute effects into heterogeneous associations, i.e., the effects of the transfer-in and transfer-out city attributes on the respective technology transfer-in and transfer-out. 283 of the 297 sample cities are involved in technology import and 284 are involved in technology transfer-in, whereas only 15 cities are involved in only one of the technology transfer-in or transfer-out roles. Therefore, the transfer-in and transfer-out cities in the network are highly overlapping, and an examination of the effects of transfer-in and transfer-out city attributes is nearly equivalent to an inferential statistic for the sample as a whole. Although city attributes only express the direction of linkage, all variables involve binary linkages, so the model retains the entire spectrum of network structure variables. By examining the regression coefficients, we can differentiate between the effects of city node characteristics on technology transfer-in and transfer-out activities. Model 22 adds the technological attributes variable to model 21 because the technological attributes of a node influence the innovative capacity of a city's technology (Balland & Ridby, 2017), which in turn shapes the various potentials in the TTN.

8.3.1.1 *The effect of local innovation milieu*

The results of Model 17 indicate that economic development level, innovation capital investment, decentralization level, marketization level, and technological innovation capacity all have a positive impact on technology transfer. The significance of the corresponding coefficients remains unchanged before and after the introduction of other dimensional variables, with only minor changes in magnitude. These findings suggest that the effects of these factors are robust and consistent. We will use Model 20 to analyze the determinants influencing technology transfer and the direction and strength of their effects.

The coefficient of the level of economic development is 0.122, i.e., for every 10,000 RMB increase in the GDP per capita of a city in the network, its probability of technology transfer to the outside world is approximately 1.13 times higher than before ($e^{0.122}$). It can be deduced that there is a strong correlation between the level of economic development and the evolution of the TTN. Specifically, with an average annual GDP per capita increment of RMB 0.385 million from 2007 to 2012, it is estimated that the density of the TTN could increase by approximately 4.8% per year.

When comparing the ratio of science spending to public finance spending in Chinese cities, a higher coefficient of innovation capital investment at 22.480 results in an increase of approximately 25% in the probability of technology transfer volume growth for every 1% increase in this ratio. This reflects the significant contribution of science investment in cities in promoting technology linkages across regions. Science investment is mainly government funding in the field of science and technology, such as basic research, applied research, marketing and promotion, and S&T training. This significant effect reflects the great advantage of the public system of S&T development in enabling the enhancement of scientific and technological capabilities and regional synergy of innovation.

With a decentralization level coefficient of 1.164, Chinese cities are 3.2 ($e^{1.164}$) times more likely to experience technology transfer for every 1 increase in decentralization level than before. Combined with the standard deviation of decentralization in the network (0.254), it is calculated that for every 0.5 standard deviation increase in the decentralization index (0.127), the average linkage intensity of the network increases by a factor of 0.159.

With a coefficient of 0.027 for the level of marketization, for every 1 increase in the level of marketization in Chinese cities, their probability of technology transfer is 1.02 ($e^{0.027}$) times higher than before. Combined with the standard deviation of marketization in the network (1.708), this calculates that for every 0.5 standard deviation increase in the marketization index (0.854), the average linkage intensity of the network increases by a factor of 0.023.

While previous literature has acknowledged the influence of local institutions on technology transfer, their impact has not been quantified much. This study demonstrates that the level of structural reforms in a region's fiscal and economic systems plays a crucial role in determining the balance between supply and demand for patented technologies at the institutional level. As a result, it contributes to the flow of innovation information between different cities and facilitates interaction among various actors. After taking other variables into account, this research found that both public financial resources and market competitiveness are essential in dominating innovation resources. However, cities with stronger financial power are more likely to emerge as hubs for TTNs.

According to the technology innovation capacity coefficient, which stands at 0.061, for every 100 new high-tech enterprises established within a Chinese city, the likelihood of technology transfer increases by 6.26% compared to the previous rate. Among various actors involved in market opportunity capture and technological innovation, firms exert the most significant influence. As part of their business strategy, firms may need to acquire new technologies that are beyond their existing knowledge base. In such cases, they may resort to outsourcing or collaborative R&D to obtain the required technology. In addition, technological innovation has a certain degree of spillover, with firms considering whether to license or transfer ownership of a technology that has already received intellectual property protection when it can bring economic benefits. This study confirms that the growth in the number of high-tech enterprises can provide the dynamism for technological innovation and provide the impetus for increased interactor technology exchange across regions.

8.3.1.2 *The effect of intercity relationship*

Models 18 and 20 show that both geographical and technological proximity have a consistently positive effect on technology transfer, even after controlling for city node variables. On the other hand, while the difference in STI potential changes from significantly positive to insignificant, its effect overlaps with the effects of city attributes. To fully comprehend the direction and magnitude of the impacts of geographical and technological proximity, we will focus on Model 20.

The coefficient of geographical proximity in Model 19 is 0.917. A one-unit increase in the standardized index of geographical distance between two cities in the network increases the probability of the city developing technology transfer links by approximately 2.5 times ($e^{0.917}$). Combined with the standard deviation of geographical proximity in the network (0.678), each 0.1 standard deviation increase in geographical proximity (0.068) is expected to increase the average linkage intensity of the network by a factor of 0.064, proving that there is a clear trend toward cooperation between neighboring cities and that technology transfer is currently limited by spatial frictions. Technology transfer, as an interactor market transaction behavior, generally requires face-to-face negotiations and contracts between the two parties to be successfully realized, and shorter geographical distances can reduce the temporal and spatial costs of transactions (Liu *et al.*, 2018a). In some ways, the exchange and absorption of tacit knowledge are synchronized with the trading of explicit knowledge such as patents, reflecting the spillover effects brought about by geographical proximity (Maskell & Malmberg, 1999a, 1999b). Thus, reduced geographical distance can provide more frequent face-to-face opportunities for both parties to technology transactions, increasing mutual understanding between firms and facilitating the establishment of trusting relationships for long-term cooperation (Heringa *et al.*, 2014). At the city level, technology transfer between neighboring cities is encouraged by the governments of neighboring or higher administrative regions as it helps to activate technology markets, enhance innovation, stimulate industrial development, and promote interregional synergies. This is similar to the findings of Burhop & Wolf (2013) on inter-regional technology transfer in Germany.

The coefficient of technological proximity in Model 19 is 2.382, and a one-unit increase in the technological proximity index between two cities in the network is 10.8 times more likely to develop technology transfer links than before ($e^{2.382}$). Combined with the standard deviation of technological proximity in the network (0.160), an increase of 0.1 standard deviation (0.016) in technological proximity is expected to increase the average linkage intensity of the network by a factor of 0.04. In order to establish themselves as innovators, companies must engage in technology exchange with other companies possessing similar technological expertise. This facilitates the effective absorption and integration of knowledge on both ends of the transaction, resulting in a more seamless alignment between technological supply and demand. At the city level, cities that share technological proximity exhibit a collective learning orientation, market goals, and industrial structures that are more homogeneous. This leads to lower costs associated with innovation linkages and significant economic benefits attributed to technology transfer.

Although the coefficient of technological proximity is larger than that of geographical proximity, the contribution of geographical proximity to domestic technology transfer linkages is greater relative to technological proximity when considered in conjunction with the statistical distribution of proximity for all cities in China (Wang & Wang, 2018). What is revealed here is a general pattern of technology transfer in China, and some studies have also found that industry characteristics bias the effect of proximity on technology transfer. For example, for the IC industry, which is more intraconglomerate and transnationally traded, geographical proximity has no effect on technology transfer, while the need for technological proximity is greater (Liu *et al.*, 2022), and cluster externalities make geographical proximity become non-essential (Lin & Lan, 2016). Regions with well-integrated economic and transportation infrastructure will be less limited by physical distance within their borders, and innovation will be able to flow more freely across geographical boundaries (Wang, 2022).

8.3.1.3 *The effect of network self-organization*

In Models 17–20, the estimated parameters of edges are always significantly negative, indicating that there are costs associated with establishing and maintaining cooperative relationships between cities that are not generated by stochastic processes (Zhang *et al.*, 2020). At the same time, the mutual coefficient is always significantly positive, indicating that two-way interactions between cities promote intercity technology transfer. When controlling for all variables in Model 20, the mutual coefficient remains high at 0.692. This indicates that the likelihood of intercity technology transfer within the network increases exponentially ($e^{0.692}$–1) with every additional city pair, highlighting a reciprocal tendency for technology transfer among Chinese cities. In this way, the overall pattern of technology transfer is dominated by the "counter-flow" and "back-flow" of innovation linkages (Wang *et al.*, 2022), which revealed a trend toward interregional equilibrium development and generation capacity of technological innovation.

8.3.1.4 *The effect of city attributes*

Model 21 introduces the anisotropic effect of city attributes, and Model 22 adds technology attribute variables based on Model 21. Comparing Model 21 with 22, we can see that the significance of the coefficients and the relative magnitude of the coefficients in both directions do not change for each city node attribute variable in Model 21, except for the innovation capital investment and the level of marketization. We aim to examine the evolution of TTNs taking into account the anisotropic effect of city attributes. We will achieve this by closely examining the coefficients of the variables in Model 21. In addition, we seek to explain how a specific variable (innovation capital input) impacts the network when switching from Model 21 to Model 22, leading to a sudden change in the relative magnitude of the anisotropic effect.

The coefficients for the level of economic development of the transfer-in city and transfer-out city are 0.111 and 0.132, respectively. For every 10,000 RMB increase in the GDP per capita of the technology-transferring city, the probability of increasing the number of patents flowing into that city will increase by approximately 11.7% ($e^{0.111}$–1). An equivalent sizeable increase in the GDP per capita of the technology-transferring city brings about a 14.1% ($e^{0.132}$–1) increase in the probability of an increase in technology transferring out. This is evidence that a stronger level of economic development stimulates the export of technology and the demand for its import. From a national scale, as China's economy grows at a high rate, TTN in cities will tend to be dense, at the same time, there is a tendency for supply to outstrip demand, and this is particularly evident in economically strong cities.

The coefficients of innovation capital investment in transfer-in cities and transfer-out cities are 7.671 and −5.05, respectively, with the effect of innovation capital investment on technology transfer-out and transfer-in aspects of cities being opposite. China's technology transfer market is at a preliminary stage of development, and technology transfer is generally at the stage of technology import and absorption. Information asymmetry, the absence of intermediary pipelines, and imperfect regional collaborative innovation platforms all contribute to conservative technology transfer-out practices across regions. As a result, certain barriers exist that hinder effective technology transfer-out processes.

The coefficients of the level of decentralization in transfer-in cities and transfer-out cities are 1.061 and 1.259, respectively, and there is no significant difference between the role of decentralization in the transfer-out and transfer-in of cities.

The coefficients on the level of marketization for both transfer-in and transfer-out cities are significantly positive. However, when the city's technology attribute variable is introduced, as shown in Model 22, the coefficients on the level of marketization in both transfer-in and transfer-out cities become insignificant. It is possible that the facilitative effect of the level of marketization on technology transfer is manifested after masking the technology attribute.

The coefficient of technology absorption capacity in the transferring city is 0.043, and the coefficient of technology transfer-out capacity in the transferring city is 0.078, indicating that the promotion effect of the number of new high-tech enterprises on technology transfer-out is greater than that on technology transfer-in.

8.3.1.5 *The effect of technology diversity*

There is an increasing debate on the influence of technology diversity on technology transfer (Lian *et al.*, 2021). Not all scholars agree that high diversity will play a positive role in technology transfer (Sampson, 2007). Accordingly, we analyze the effects of technological diversity through Model 22. The diversity of technologies owned by a city is to some extent representative of the technological innovation capacity of the city. Unrelated diversity is not included in the model because of the colinearity problem. The divergent patterns of technology-specific drivers of technology transfer are not yet evident, and regional bifurcation paths represented by related and unrelated diversity have not yet been distinguished. The similarity of imported technologies acts positively on technology transfer, demonstrating that technology import is a self-organizing behavior that tends to benefit, with cities tending to import technologies that are similar to their own technology bases.

8.3.2 *Influence mechanism during the second period (2013–2018)*

We use a similar modeling approach to the first stage, with weighted edges always in the ensemble of independent variables. As shown in Table 8.7, Models 23–25 add node attributes, proximity, and reciprocity, respectively, to examine the attributes of urban innovation, the attributes of intercity relationships, and the impact of the endogenous structure of the network on domestic technology transfer. Model 26 then incorporates all three types of variables. Model 26, which includes all dimensional explanatory variables, has the highest degree of fitting among the four models in terms of AIC and BIC.

Models 23–25 default to homogeneity in the role of city attributes toward the generation of TTN. However, the interpretation of the dependent variable by the nature of city attributes may not be consistent when two different directions of technology flows, transfer-in and transfer-out, are undertaken. To obtain the heterogeneous effects that cities may exhibit when assuming the role of technology transfer-in versus transfer-out, we extend the model further (shown in Table 8.8). Model 27, compared to Model 26, employs a more detailed approach by splitting the effects of city attributes into distinct links for technology transfer-in and transfer-out. This involves considering the unique impact of transfer-in and transfer-out city attributes on their respective technology transfers. There is a high degree of overlap between transfer-in and transfer-out cities in the network with all 297 sample cities involved in technology transfer-in and 293 of them involved in technology transfer-out. The

Table 8.7 Results of ERGM for TTN during the period 2013–2018.

Variables	Model_23	Model_24	Model_25	Model_26
GDP_per_ca	0.085***			0.091***
R&D_GDP	0.010***			0.009***
Dec_Index	0.496***			0.597***
Mar_Index	0.082***			0.053***
Homo_Adm_1	−0.475***			−0.342***
Homo_Adm_2	0.420***			0.102
Homo_Adm_3	1.261***			0.715*
Homo_Adm_4	150.800*			−5.852*
New_Hitech	0.033***			0.036***
Pro_Geo		0.575***		0.641***
Pro_Tec		2.453***		1.522***
Diff_Inno		0.208***		−0.003
Mutual			1.614***	0.705***
Edges	−5.643***	−6.062***	−3.067***	−7.854***
AIC	−221928	−217497	−218186	−228151
BIC	−221834	−217459	−218167	−228020

Notes: Significance level: '***'(0.001), '**'(0.01), '*'(0.05); *: *p* value (significance) could not be calculated due to the small sample size.

examination of attribute effects for transfer-in and transfer-out cities is almost equivalent to an inferential statistic for the sample as a whole. In addition, as with Models 21 and 22, Models 27 and 28 retain the full range of network structure variables. Model 28 adds technology attribute variables on the basis of Model 27.

8.3.2.1 *The effect of local innovation milieu*

As can be seen from Model 23, the effects of economic development level, innovation capital investment, level of decentralization, level of marketization, and technological innovation capacity on technology transfer all show positive effects. The significance of the

Table 8.8 Results of ERGM for TTN during the period 2013–2018.

Variables	Model_27	Model_28
GDP_per_ca_in	0.000***	0.000**
GDP_per_ca_out	0.000***	0.000***
R&D_GDP_in	0.009***	0.008***
R&D_GDP_out	0.009***	0.006***
Dec_Index_in	0.664***	0.406***
Dec_Index_out	0.632***	0.411***
Mar_Index_in	0.052***	0.024***
Mar_Index_out	0.055***	−0.012*
Homo_Adm	−0.311***	−0.065**
New_Hitech_in	0.000***	0.000***
New_Hitech_out	0.000***	0.000***
Div_out		1.087***
Non_Rel_Div_out		0.582***
Tech_Simi_in		0.212***
Pro_Geo	0.635***	0.565***
Pro_Tec	1.561***	0.805***
Diff_Inno	−0.003***	0.046***
Mutual	0.736***	0.313***
Edge	−7.924***	−10.540***
AIC	−221838	−235350
BIC	−221726	−235171

Notes: Significance level: '***'(0.001), '**'(0.01), '*'(0.05); *: *p*-value (significance) could not be calculated due to the small sample size.

coefficients stays consistent even with the introduction of other dimensional variables, although there may be minor variations in their magnitude, indicating that their effects remain robust. To investigate the impact of the administrative stage of a city, we created four rank categories. This allowed us to determine if two cities in the network belonged to any of these categories (represented by four dummy variables: rank 1 match, rank 2 match, rank 3 match, and rank 4 match). However, as only four cities had an administrative rank of 4, there were only 16 samples that matched this rank. Consequently, it was impossible to accurately estimate the significance of the coefficients. On the other hand, for the remaining three homogeneous variables, the coefficients rank 1 match and rank 3 match remained unchanged both before and after introducing additional dimensional variables. We will use Model 20 to analyze the determinants of technology transfer and the direction and intensity of their effects.

The coefficient of the level of economic development is 0.091, i.e., for every 10,000 RMB increase in the GDP per capita of a city in the network, its probability of technology transfer to the outside world is 1.095 times higher than before ($e^{0.091}$). It can be deduced that economic development has a significant impact on the evolution of the TTN. Specifically, with an average annual GDP per capita growth rate of RMB 0.42 million from 2013 to 2018, the density of the TTN has the potential to increase by approximately 4% per year. Compared with the first stage (Model 20), the driving effect of GDP per capita on technology transfer has weakened.

With an innovation capital investment coefficient of 0.009, the probability of increasing the volume of technology transfer in Chinese cities increases by approximately 0.08% for every 10% increase in the ratio of internal expenditure on R&D to GDP. Based on an average annual increment of 3.8% in the ratio of internal expenditure on R&D to GDP from 2013 to 2018, the density of TTNs has the potential to increase by about 0.03% per year.

When comparing the impact of economic development and R&D investment on technology transfer, it becomes clear that economic development has a stronger influence in stimulating technology transfer, while internal R&D activities have a weaker effect. This suggests that there exists a reciprocal feedback mechanism between economic benefits and technology transfer, which is driving the marketization of technology transactions. The supply of R&D activities however has little catalytic effect on technology transfer, which may be related to a lack of high-quality innovation or a low conversion rate.

The decentralization level coefficient is 0.597, and for every 1 increase in decentralization in Chinese cities, their probability of technology transfer is 1.817 ($e^{0.597}$) times higher than before. Combined with the standard deviation of decentralization in the network (0.241), each 0.5 standard deviation increase in the decentralization index (0.121) increases the average linkage intensity of the network by a factor of 0.075.

With a coefficient of 0.053 for the level of marketization, for every 1 increase in the level of marketization in Chinese cities, their probability of technology transfer is 1.054 ($e^{0.053}$) times higher than before. Combined with the standard deviation of marketization in the network (1.770), this calculates that for every 0.5 standard deviation increase in the marketization index (0.885), the average linkage intensity of the network increases by a factor of 0.048.

Comparison with Model 20 shows that the impact of both decentralization and market institutional forces on technology transfer is diminishing. Meanwhile, the gap between the two is decreasing, proving that economic market reforms are beginning to bear fruit and are gradually becoming one of the institutional dominant factors in technology trade.

When two cities with the same administrative rank of 1 engage in technology transfer, there is a high probability of success ($e^{-0.342}$). However, when one party does not have an administrative rank of 1, the probability of successful technology transfer decreases. Lower-ranking cities tend to form connections with higher-ranking cities or political power centers, often resulting in cross-ranking transfers known as "leapfrogging" diffusion. These transfers play a significant role in linking different areas. With a coefficient of 0.715 for a rank 3 match, a technology transfer between two cities with the same administrative rank of 3 is two ($e^{0.715}$) times more likely to occur than between two cities where at least one city does not have an administrative rank of 3. This implies that planned cities with considerable political power are more closely related to each other and are an important part of the core of the network (the other part is the higher-ranked municipalities, a fact that is not difficult to detect through statistical observations, although the regression analysis cannot confirm it). The analysis of administrative hierarchy homogeneity shows that the "hub-and-spoke" links of technology transfer are homogeneous with the administrative hierarchy of Chinese cities, with high-ranking nodes forming strong innovation links, while low-ranked nodes are attracted to high-ranking nodes. As a consequence, priority will be given to establishing technology transfer linkages with administrative centers that have mature technology trade markets and a siphon ability of innovative resources.

With a technology innovation capacity coefficient of 0.036, for every 100 new high-tech enterprises in a Chinese city, the probability of technology transfer in that city is 3.65% higher than before. The role of high-tech enterprises suffers a marginal diminution, and it is possible that those actors that have not been the focus of this study are more representative of urban innovation dynamism at this stage, such as established giants of the industry, emerging unicorns, and system integration platforms for information.

8.3.2.2 *The effect of intercity relationship*

Models 24 and 26 show that geographical proximity and technological proximity consistently have a positive effect on technology transfer regardless of whether city node variables are controlled for. However, the impact of STI potential difference changes from significantly positive to insignificant, and its effects overlap with the effects of city attributes. We will analyze the direction and strength of the effects of geographical proximity and technological proximity through model 26.

The coefficient of geographical proximity in Model 19 is 0.641. A one-unit increase in the standardized index of geographical distance between two cities in the network increases the probability of the city developing technology transfer links by approximately two times ($e^{0.641}$). Combined with the standard deviation of geographical proximity in the network (0.678), each 0.1 standard deviation increase in geographical proximity (0.068) is expected to increase the average linkage intensity of the network by a factor of 0.045. However, compared with Model 20, the hindering effect of geographical distance decreases, on the one hand, as the market for technology trading becomes more open and mature, the stickiness of knowledge decreases. On the other hand, the rapid development of ICT changes the shape or manner of various activities such as knowledge search, willingness exchange, cooperation negotiation, and long-term relationship maintenance for technology trading, making it less dependent on geographical proximity.

The coefficient of technological proximity in Model 19 is 1.554, and a one-unit increase in the technological proximity index between two cities in the network is 4.581 times more likely to develop technology transfer links than before ($e^{1.552}$). Combined with the standard deviation of technological proximity in the network (0.176), an increase of 0.1 standard deviation (0.018) in technological proximity is expected to increase the average linkage intensity of the network by a factor of 0.028. In comparison with Model 20, the role of technological proximity decreases. Intercity knowledge transfer in China becomes weaker and more diverse by the limitations of cognitive structures, which is consistent with the process of knowledge diversification and regional bifurcation in cities.

8.3.2.3 *The effect of network self-organization*

In Models 23–26, the estimated parameters of edges are always significantly negative, indicating that there are costs associated with establishing and maintaining cooperative relationships between cities that are not generated by stochastic processes (Zhang *et al.*, 2020). The mutual coefficient is always significantly positive, indicating that two-way interactions between cities promote intercity technology transfer. And in Model 26, where all variables are included, the mutual coefficient is 0.709, indicating that for every city pair in the network that generates technology mutual giving, the probability of generating technology transfer between cities in the network increases by a factor of 1.032 ($e^{0.709}$–1). The propensity for reciprocity in intercity technology transfer in China is stronger than it was in the previous period.

Combined with the declining role of geography and technological proximity, it can be assumed that the tendency for intercity technology transfer is to be based on familiarity-based linkages (social ties similar to "acquaintances"). It is interesting to see whether a pattern of path-dependent association-based technology transfer will develop in the future.

8.3.2.4 *The effect of city attributes*

Model 27 inputs the anisotropic effect of city attributes and Model 28 adds the technology attribute variable based on Model 27. The comparison between Models 28 and 27 shows that the relative magnitudes of the coefficients of city node attribute variables in both directions in Model 21 do not change significantly in Model 28, except for the level of marketization. We will analyze the dynamics of TTN evolution under the anisotropic effect of city attributes mainly by observing the coefficients of the variables in Model 27, and explain the effect of the variables that change abruptly when the variable system is switched from Model 27 to Model 28 (level of marketization).

The coefficients for the level of economic development of the transfer-in city and the transfer-out city are 0.074 and 0.107, respectively. In this way, for every 10,000 RMB increase in the GDP per capita of the technology transfer-in city, the probability of increasing the number of patents flowing into that city will increase by approximately 7.7% ($e^{0.074}$–1). An equivalent sizable increase in the GDP per capita of the technology transfer-out city brings about an 11.3% ($e^{0.107}$–1) increase in the probability of an increase in technology transferring out. Evidence that stronger levels of economic development stimulate technology transfer-out, but drive their demand for technology transfer-in is relatively slight. From the national perspective, the increase in economically strong cities exacerbates the tendency for supply to outstrip demand in the trade of technology patent rights, and their reliance on endowments to enhance their STI strength creates a situation where other relatively weak cities are fed and radiated by technology.

The coefficient of innovation capital investment in both transfer-in and transfer-out cities is 0.009, implying that a 0.1 increase in R&D intensity will stimulate the creation of innovation linkages and increase the probability of increased technology inflow and output by about 0.09% ($e^{9E\text{-}4}-1$), respectively. Therefore, it is a very ineffective way to promote the evolution of TTNs, considering the difficulty of increasing R&D intensity.

The coefficients for the level of decentralization in transfer-in cities and transfer-out cities are 0.664 and 0.632, respectively, with no significant difference in the effect of decentralization on the technology transfer-out versus transfer-in cities.

Similarly, in addition to decentralization, the coefficients related to the level of marketization for both transfer-in and transfer-out cities show no significant difference. However, when the city's technology attribute variable is introduced, as shown in Model 28, the coefficients on the level of marketization in the transfer-in city and the transfer-out city are 0.024 and −0.012, respectively, which differ largely from the level of decentralization. The impact of marketization on technology transfer has been contrasting for transfer-in and transfer-out cities. While the marketization of economic structure has increased the demand for technology, it has also discouraged the export of technology. As stated above, the level of economic development boosts technology supply more than it does technology demand, but the degree of market economy development only stimulates technology demand. This reflects a misaligned structure in the economic development model. As a consequence, cities with highly developed economies are configured with a strong science and technology innovation force and are suppliers of innovation networks, while cities with active private economies and a high degree of liberalization are in greater demand for a large number of innovations to support industrial development.

The coefficient of technology absorption capacity of transfer-in cities is 0.029, and the coefficient of technology transfer-out cities is 0.041. It demonstrates the promotion effect of the number of new high-tech enterprises on technology transfer-out is greater than the promotion effect on technology transfer-in.

8.3.2.5 *The effect of technology structure*

We analyze the effect of technological characteristics through Model 28 and also report the change in the proximity effect due to the introduction of the technological eigenvectors — the difference in STI potential between Models 27 and 28. Firstly, the diversity of technologies owned by a city is to some extent representative of its technological capabilities, and the coefficients on both related and unrelated diversity are significantly positive, confirming the hypothesis that technological endowment positively affects technological output capabilities. Of these, the effect of related diversity is greater, suggesting that specialization in narrow areas is conducive to the development of mature technological systems that dominate innovation linkages at the regional and even national levels. The similarity of imported technologies acts positively on technology transfer, proving that technology import is a self-organizing behavior that tends to benefit, with cities tending to import technologies that are similar to their own technology bases. Secondly, after the introduction of the technology characteristics variable, the potential difference in technology innovation shows a positive effect, which to some extent coincides with the effect of homogeneity. It also demonstrates that the "rich club" and "Matthew effect" in the TTN are not obvious. In the TTN, the diffusion pipelines of technology flows are mainly established between cities across innovation levels.

8.4 Summary

As an important way to achieve the flow and sharing of innovation resources, the influencing factors of technology transfer still remain a riddle. There are some points that we need to understand: firstly, the dynamics of the generation and evolution of complex networks dependent on global relationships, and secondly, the role of international cross-regional technology linkages from a global–local perspective. We use an ERGM and a negative binomial panel regression analysis to explore the influencing mechanisms of China's domestic TTNs and the determinants of China's transnational technology transfer, respectively. For the study of domestic intercity

TTNs, a phased approach is adopted, with the first phase covering the period 2007–2012 and the second phase covering the period 2013–2018. For the study of the determinants of transnational technology transfer in China, the data for the three time periods 2004–2008, 2009–2013, and 2014–2018 are aggregated to construct a short panel. The main findings we discovered are as follows.

Firstly, the innovation milieu of cities has a significant impact on the evolution of domestic TTN. The level of economic development pulls the evolution of TTNs, but its role diminishes in the second phase compared to the first. Innovative financial inputs promote cross-regional technology linkages. Increased levels of decentralization and marketization will facilitate outward technology transfer. Technological innovation capacity positively affects technology transfer. Administrative hierarchy matching reveals differential effects. The "hub-and-spoke" organization of technology transfer is isomorphic to the administrative hierarchy of Chinese cities, with high-ranking nodes forming strong innovation links with each other, while lower-ranking nodes are attracted to higher ranks and will prioritize technology transfer links with administrative centers that have mature technology trade markets and have a siphoning power of innovation resources.

Secondly, intercity proximity has a significant effect on the evolution of domestic TTN. A decrease in the geographical distance between two cities in the network increases the probability that both cities will be interested in developing technology transfer links. Increased technological proximity between two cities in the network will also increase their probability of developing technology transfer links, which suggests that the lower the cost of generating innovation links between cities with greater technological proximity, the higher the economic benefits of technology transfer are likely to be. Finally, with the introduction of heterogeneous urban attributes, the STI potential differential positively affects technology transfer, confirming that the "rich club" and "Matthew effect" in TTNs are nót significant.

Thirdly, the domestic TTN has the character of self-organizing, that is, there are costs associated with establishing and maintaining cooperative relationships between cities that are not generated by random processes. At the same time, the two-way interactions between cities promote intercity technology transfer, demonstrating the reciprocal tendency of intercity technology transfer in China.

When it comes to the determinants of transnational TTN, first of all, the local innovation milieu has multiple effects on transnational technology transfer. The level of the local economy, the increase in the number of science and technology employees, and the establishment of national enterprise technology centers can effectively increase the intensity of foreign technology import, while the number of universities in a city can inhibit its transnational technology import. There is also a coupling dissonance between a city's level of higher education and its technological production capacity. The amount of goods exported by a city can promote the introduction of multinational patents, while the amount of goods imported is negatively correlated.

The expansion of domestic technology pipelines boosts the intensity of transnational technology transfer-out. The breadth of domestic intercity technology transfer has a significant positive contribution to transnational technology export. The intensity of domestic technology transfer has a significant positive effect on the export of transnational technology, and the diversity of cities' output locations for intercity technology transfer plays a significant role in enhancing innovation performance and optimizing innovation networks. The intensity of domestic intercity technology transfer facilitates local technology export to foreign countries.

The local innovation milieu has multiple effects on transnational technology export. The level of the local economy significantly affects the export of transnational technologies. The greater the number of national enterprise technology centers in a city, the greater its technological innovation capacity and international competitiveness, and the larger its transnational technology exports. The greater the value of a city's export trade in goods, the stronger the transnational economic ties and technology exchange.

CHAPTER 9

IMPACTS OF TRANSPORTATION NETWORKS ON TECHNOLOGY TRANSFER NETWORK

In the era of knowledge economy, innovation happens to be the primary element and key support for national and regional economic growth (Mankiw, 1992; Dunning, 2002). Cities are facing a fresh round of scientific and technical revolution, that is, actively taking part in the global innovation network to improve their global competitiveness. This phenomenon recognizes "global pipeline" as an important path to improving the regional innovation capacity for absorbing external advanced technology. From the perspective of "Space of Flows" (Castells, 1996), cities can enhance their innovation performance through knowledge flow at different scales, and their innovation development increasingly relies on the physical and virtual networks they involve in. The urban transportation can then alleviate the restriction that the geographical distance has on knowledge spillover of the innovation actors (Basile *et al.*, 2012), which is a significant variable to influence the position of the cities in the global innovation network. In this way, the positive externality of transportation can be understood as a basis of innovation promotion in the cities due to its indirect effect on the spillover of technology innovation. Therefore, the research on the mechanism of how transportation affects technology transfer can not only enrich the theory of how physical space influences innovation space under the background of innovation geography but also help to further understand the basic law of innovation geography and the effect of transportation.

Patent transfer is regarded as the spatial flow and diffusion of a certain technology or knowledge which has clear geographical characteristics and can be used to measure the technology transfer directly (Liu & Guan, 2018). The development of aviation and high-speed railway provides opportunities for face-to-face communication between the technology suppliers and demanders, making it easier for technology spread and diffusion (Luo *et al.*, 2022). In this way, this chapter will discuss the effect of transport network on technology transfer network with the case studies of aviation and high-speed rail (HSR) (Figure 9.1). The methods used here are negative binomial regression (NBR), differences-in-differences (DID), and propensity score matching (PSM). This chapter has two proposals. On the one hand, it aims at making clarity of the complex relationship between transportation and technology transfer, so as to further research the relationship between physical space and virtual space on the basis of innovation geography. On the other hand, the outcome of this chapter can also provide a reference for rational allocation of national transportation, orderly flow of innovation elements, and efficient allocation of innovation resources.

9.1 The Effect of Aviation Network

9.1.1 *Conceptual background and hypotheses*

With the great development of the airline industry and relaxing of global aviation regulations, air transportation is playing an important role in individual journeys and business travel, which is bridging the spatiotemporal gap generated by regional borders and geographical distance (Jiao *et al.*, 2014). The rapid development of civil aviation also promotes the socioeconomic element flows between cities, and then further drives knowledge diffusion and technology spillover. In addition, the accessibility of aviation will accelerate the concentration of human capital and financial investment, at the same time, increase the communication probability between innovation elements and actors (e.g., business talents, technical talents,

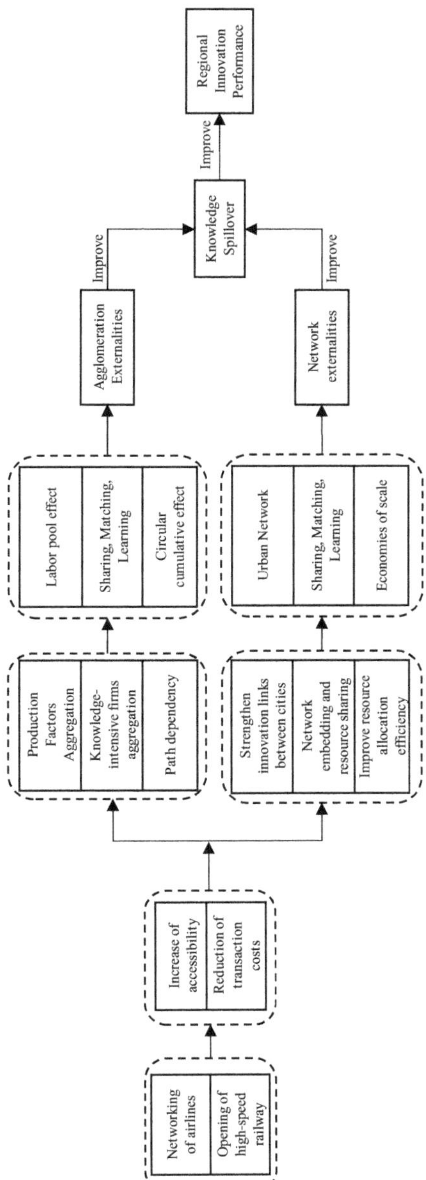

Figure 9.1 The research framework of this chapter.

scientists, and entrepreneurs) through business conference, academic conference, and international fair (Simmie, 2002; Andersen & Dalgaard, 2011; Catalini *et al.*, 2016; Yi, 2021). It has been indicated that a long-term common relation and fixed space may do little help to the knowledge producing and spillover (Gallié, 2009), while the primary knowledge is achieved through temporary meet or occasional face-to-face communication such as trade exhibitions, short-term business visits, academic conferences, and business negotiations (Bathelt & Schuldt, 2008). In this way, the air transportation can significantly decrease the cost of knowledge interaction between innovation actors to effectively promote the regional flow and transfer of tacit knowledge (Polanyi, 1966; Vonhippel, 1994; Storper & Venables, 2004; Boschma, 2005; Basile *et al.*, 2012). On the basis of the above analysis, we then raise the following two hypotheses:

Hypothesis 1: Air transportation may benefit technology transfer of a city and its relative cities.

Hypothesis 2: Air transportation may also promote technology transfer within the city pair once they have air contact.

The development of aviation network has greatly improved the experience of long journeys and geographical accessibility and alleviates the restriction of physical distance, which means the innovation actors can have technology interaction at remote places. In addition, different modes of transportation may have various adaptability to the distance. Taking the high-speed railway and air transport as examples, 80% of their volume is concentrated in the interval from 200 km to 1,200 km and from 800 km to 2,600 km, respectively (Ding, 2013; Wang, 2019). Passengers will be more eager to choose high-speed railway if their journey is under 700 km while choosing air transport if their journey is above 700 km. At the same time, the commercial activities which are sensible to the distance also influence the distance adaptability of transportation, such as the USA risk investment's "one-hour rule" (Lerner, 1995) and Chinese risk investment's "300 km rule" (Long, 2016). Technology transfer is regarded as a type of knowledge spillover with the basis of market behavior, thus greatly influenced by geographical distance. Therefore, with the integration of the above theories, we also raise the third hypothesis:

Hypothesis 3: The radiation range of air transport to technology innovation transfer is nonlinear and has a certain optimal distance.

9.1.2 *Empirical results*

Before conducting the regression analysis, it is advisable to perform a descriptive statistical analysis on the dependent variable, core explanatory variables, and control variables. Table 9.1 presents the descriptive statistics of the variables at the city node level, while Table 9.2 shows the descriptive statistics of the variables at the city-pair level. Then, to test for multicollinearity in the variables associated with the model, the VIF test between variables was used to analyze whether there was multicollinearity between variables. At the city node level, the maximum VIF value for the equation regression model was 2.31, with a mean value of 1.39; at the city-pair

Table 9.1 Descriptive statistics of nodal variables.

Variables	Sample	Average	Std. Dev.	Min.	Max.
The volume of city node's transfer	2935	300.1	1097	1	21396
City node's flights	2935	16241	48874	0	487762
City node's economic development level	2935	3.884	4.208	0.149	49.31
City node's innovation investment	2935	4.057	7.726	0.0100	59.80
City node's research and education resources	2935	173.8	226.5	0.00205	1294
City node's road accessibility	2935	9235	15030	0	554203
Whether city node has high-speed rail	2935	0.323	0.468	0	1
City node's administrative level	2935	0.126	0.332	0	1

Table 9.2 Descriptive statistics of city pair variables.

Variables	Sample	Average	Std. Dev.	Min.	Max.
The volume of city pair's transfer	14594	6.230	18.90	1	924
City pair's flights	14594	548.7	1575	0	18272
City pair's economic development level	14594	8.965	5.397	0.499	45.59
The geographical distance between cities in a city pair	14594	798708	601021	33200	3.600e+06
City pair's industrial development level	14594	91.13	15.57	26.78	155.5
City pair's technology similarity	14594	64.78	12.40	14.94	68.4
City pair's academia resources	14594	375.9	236.5	9.084	1235
City pair's innovation investment	14594	13.36	16.78	0.210	89.80
City pair's road accessibility	14594	34034	21022	2140	579128
Whether city pair has high-speed rail	14594	0.817	0.386	0	1

level, the maximum VIF value was 2.41, with a mean value of 1.43. The VIF values for both models were less than 5, indicating that there was no multicollinearity between the variables.

9.1.2.1 *The effect of aviation accessibility at nodal level*

To ensure the robustness of the regression results, a stepwise regression was used, starting with a univariate regression analysis of urban air accessibility and technology innovation transfer. The results of Model 1 showed that the regression coefficient of the univariate variable was 0.0357 and was significant at the 1% level of significance. Incorporating control variables improved the regression coefficients of air accessibility variables in Models 2–5, yielding a significant positive correlation. The most explanatory model was Model 5 (Table 9.3). In terms of the control variables, in line with the previous empirical results, the regression coefficients of urban economic development level, industrial development level, and city level are all significantly positive, with a positive effect on intraurban technology transfer. The regression coefficients for research and education resources were not significant, indicating that the role of research and education resources of cities in the industrial transformation of technological innovation transfer is limited. Regression

Table 9.3 Results of regression analysis at the city node level.

Variables	Model_1	Model_2	Model_3	Model_4	Model_5
Flight frequency	0.0357***	0.0245***	0.0173***	0.0134***	0.0133***
	(0.00409)	(0.00606)	(0.00626)	(0.00625)	(0.00621)
Economic development level	0.0213***	0.0182***	0.0203***	0.0221***	0.0229***
	(0.00559)	(0.00595)	(0.00576)	(0.00584)	(0.00582)
Research and education resource		0.000686***	0.000498***	0.000207	0.000186
		(0.000113)	(0.000119)	(0.000142)	(0.000142)
Innovation input		0.00206	0.00274	0.00292	0.0340***
		(0.00479)	(0.00473)	(0.00453)	(0.00450)
Industrial development level			0.0136***	0.0111***	0.0102***
			(0.00311)	(0.00317)	(0.00317)
Road accessibility				1.27e-06*	1.32e-06*
				(6.81e-07)	(6.87e-07)
Whether has high-speed rail				0.459***	0.121***
				(0.118)	(0.0377)
City level					0.480***
					(0.118)
Intercept term	−0.715***	−0.815***	−1.305***	−1.278***	−1.247***
	(0.0846)	(0.122)	(0.166)	(0.163)	(0.153)
Time fixed effects	YES	YES	YES	YES	YES
City fixed effects	YES	YES	YES	YES	YES
Sample	2935	2935	2935	2935	2935

Note: Significant at ***$p < 0.01$, **$p < 0.05$, *$p < 0.1$.

coefficient in Model 5 for the air frequency variable is 0.0133, and it has passed the significance level test at 1%. This suggests that when a city has a higher air frequency it tends to have better external air transport links, more frequent knowledge interaction, and is more likely to facilitate the transfer of patented technology between itself and other cities.

9.1.2.2 *The effect of aviation accessibility at city-pair level*

Consistent with the above, a stepwise regression was used for Equation (3.22) to ensure the robustness of the regression results. Firstly, the relationship between the airline frequency and technology transfer of univariate city pairs was analyzed and the regression coefficient in Model 1 was 0.0273 which passed the 1% significance level test. The regression coefficients of Models 2–6 were found to be significantly positive when control variables were gradually added, and Model 6 was finally chosen as the explanatory model. From the perspective of control variables, the level of economic development, industrial structure, and technological similarity at the city-pair level had a positive effect on urban technology transfer, while the role of geographical distance remained negative. In terms of the core explanatory variables, the results of Model 6 show a regression coefficient of 0.0171, which is significant at the 5% level of significance (Table 9.4). It indicates that the larger the city flight frequency, the more beneficial it is to promote technological innovation interactions between cities.

9.1.2.3 *The effect of aviation accessibility under different thresholds*

We performed a regression analysis to examine how aviation accessibility affects the transfer of technological innovation at varying distances. Our results highlight the impact of aviation accessibility on technological innovation transfer across different distance ranges. The regression results show that the effect of air transport on the transfer of technological innovation in cities is not significant

Table 9.4 Results of regression analysis at the city-pair level.

Variables	Model_1	Model_2	Model_3	Model_4	Model_5	Model_6
Flight frequency	0.0273***	0.0173**	0.0171**	0.0164**	0.0189**	0.0171**
	(0.00717)	(0.00755)	(0.00821)	(0.00824)	(0.00827)	(0.00853)
Economic development level		0.0151***	0.0165***	0.0163***	0.0148***	0.0143***
		(0.00333)	(0.00345)	(0.00346)	(0.00349)	(0.00348)
Innovation input		0.00201**	0.00300***	0.00312***	0.00309***	0.00276***
		(0.000822)	(0.00101)	(0.00102)	(0.00102)	(0.00103)
Research and education resource			0.000245***	0.000246***	0.000242***	0.000246***
			(7.36e-05)	(7.36e-05)	(7.38e-05)	(7.45e-05)
Industrial development level			−0.00227	−0.00234	−0.00160	0.00692***
			(0.00147)	(0.00147)	(0.00152)	(0.00154)
Technological similarity				0.0124***	0.0125***	0.0105***
				(0.00118)	(0.00117)	(0.00117)
Geographical distance					−6.16e-08**	−5.73e-08*
					(3.06e-08)	(3.07e-08)
Administrative level					0.0359*	0.0356*
					(0.0195)	(0.0195)
Road accessibility						2.41e-07
						(6.24e-07)
Whether has high-speed rail						-0.0164
						(0.0362)
Intercept term	−0.471***	−0.561***	−0.459***	−0.539***	−0.571***	−0.632***
	(0.0725)	(0.0753)	(0.128)	(0.148)	(0.149)	(0.150)
Time fixed effects	YES	YES	YES	YES	YES	YES
City fixed effects	YES	YES	YES	YES	YES	YES
Sample	**14594**	**14594**	**14594**	**14594**	**14594**	**14594**

Note: **Significant at ***$p < 0.01$, **$p < 0.05$, *$p < 0.1$.**

Table 9.5 Regression result for different distance ranges.

	Model_1	Model_2	Model_3	Model_4	Model_5	Model_6
Variables	Below 600 km	600–900 km	900–1200 km	1,200–1,500 km	1,500–1,800 km	Above 1,800 km
Flight frequency	−0.0386	0.00838	0.0275**	0.0234	0.204***	0.157***
	(0.0248)	(0.0250)	(0.0136)	(0.0295)	(0.0371)	(0.0302)
Economic development level	0.0250***	0.0320***	0.00742	0.00706	0.0158	0.00192
	(0.00506)	(0.0103)	(0.00764)	(0.0161)	(0.0191)	(0.0159)
Innovation input	0.000334	0.000506	0.0115***	0.00882**	0.0121***	0.00619
	(0.00159)	(0.00309)	(0.00248)	(0.00364)	(0.00461)	(0.00453)
Research and education resource	6.51e-05	0.000507**	0.000252	0.000706**	0.00119**	0.000434
	(9.78e-05)	(0.000216)	(0.000185)	(0.000317)	(0.000519)	(0.000573)
Industrial development level	0.00534**	−0.00941*	−0.00504	−0.0199***	−0.0396***	−0.0129**
	(0.00212)	(0.00491)	(0.00407)	(0.00636)	(0.00732)	(0.00588)
Technological similarity	0.000275	0.00154	0.00407	0.000241	−0.000955	0.00245
	(0.00170)	(0.00323)	(0.00276)	(0.00358)	(0.00583)	(0.00581)
Geographical Distance	−6.82e-07***	9.88e-08	-1.37e-06***	7.90e-07	2.09e-06*	4.08e-08
	(1.55e-07)	(4.80e-07)	(3.73e-07)	(7.61e-07)	(1.25e-06)	(2.57e-07)
Road accessibility	1.75e-06*	2.91e-06	4.58e-07	−1.82e-06	-6.61e-06	4.88e-06
	(9.62e-07)	(2.46e-06)	(1.06e-06)	(2.15e-06)	(4.06e-06)	(3.59e-06)
Whether has high-speed rail	0.125***	0.205*	0.335***	−0.258	-0.128	−0.515**
	(0.0458)	(0.107)	(0.101)	(0.165)	(0.185)	(0.203)
Administrative level	0.0368	0.0300	0.00110	0.0221	0.131	0.0853
	(0.0278)	(0.0518)	(0.0452)	(0.0697)	(0.0798)	(0.0859)
Intercept term	−0.870***	-0.0738	0.694	−0.120	−0.829	0.172
	(0.214)	(0.580)	(0.573)	(1.175)	(2.174)	(0.795)
Time fixed effects	YES	YES	YES	YES	YES	YES
City fixed effects	YES	YES	YES	YES	YES	YES
Sample	7257	2116	2518	1247	769	687

Note: Significant at $^{***}p < 0.01$, $^{**}p < 0.05$, $^{*}p < 0.1$.

under the 900 km distance threshold. There are two main reasons for this: firstly, there are fewer airline frequencies between cities below 900 km; secondly, it is caused by different transport mode selection mechanisms. The distance between cities in China is generally in the range of 200 to 1,200 km (Table 9.5). While within 900 km, high-speed rail, as the first choice of intercity transport travel, has a significant competitive advantage over air transport in enhancing the intensity, frequency, and efficiency of cross-city knowledge flows.

9.1.2.4 *The effect of aviation accessibility in the three megacities*

In this section, we present regression results that demonstrate the influence of aviation accessibility on the transfer of technological innovation in the three key cities of Beijing, Shanghai, and Shenzhen, which together form the radius of technology innovation transfer (Table 9.6). It shows that the effect of air accessibility is not significant below a distance of 600 km radius. This may be due to the substitution effect of HSR on air within the short transport journey. The opening of HSR in recent years has had a strong impact on air transport that some short-haul air routes were forced to stop operating. Airlines had to turn to remote areas and develop branch airlines to extend into the areas not covered by HSR in order to seek new profit growth points (Ding & Wang, 2013). The rapid development of HSR has weakened the intensity of airline influence on technology transfer to the three innovation hub cities of Beijing, Shanghai, and Shenzhen. Its increasingly convenient and dense HSR network has had a substitution effect on aviation, reducing aviation links with neighboring cities and weakening the innovation spillover effect of aviation. Glancing over the other radius ranges, the regression results are all significantly positive and all pass the 1% significance level test. Specifically, the regression coefficients are 0.0602 for 600–900 km, 0.0510 for 900–1,200 km, 0.0806 for 1,200–1,500 km, 0.132 for 1,500–1,800 km, and 0.132, and 0.091 for 1,800 km and above.

The results of the complete data set indicate a consistent pattern where the impact of air accessibility on technology transfer to core cities varies with distance. Specifically, there is a spiral pattern where

Table 9.6 Regression result for the core cities.

	Model_1	Model_2	Model_3	Model_4	Model_5	Model_6
Variables	Below 600 km	600–900 km	900–1200 km	1,200–1,500 km	1,500–1,800 km	Above 1,800 km
Flight frequency	−0.0144	0.0602***	0.0510***	0.0806***	0.132***	0.091***
	(0.0194)	(0.0163)	(0.0153)	(0.0219)	(0.0271)	(0.0189)
Economic development level	0.557**	0.497	1.958***	0.657	1.208***	2.959***
	(0.238)	(0.316)	(0.217)	(0.424)	(0.430)	(0.245)
Research and education resource	−0.564***	0.385***	−0.569***	0.0551	−0.275	−0.0313
	(0.191)	(0.141)	(0.135)	(0.205)	(0.239)	(0.298)
Innovation input	**0.362***	**0.191**	**0.474***	**−0.0140**	**0.812***	**0.865***
	(0.119)	**(0.138)**	**(0.116)**	**(0.193)**	**(0.150)**	**(0.153)**
Industrial development level	**2.333***	**0.780**	**3.876***	**2.177**	**−0.446**	**1.372**
	(0.629)	**(0.698)**	**(0.653)**	**(0.950)**	**(0.787)**	**(0.878)**
Technological similarity	0.341*	0.588***	0.541***	−0.0462	0.485**	0.580***
	(0.198)	(0.161)	(0.204)	(0.175)	(0.210)	(0.226)
Road accessibility	0.823***	0.284*	0.496***	0.282*	0.475*	0.153
	(0.145)	(0.171)	(0.146)	(0.166)	(0.244)	(0.209)
High-speed rail	0.0946***	−0.0685	-0.156	0.00369	0.316	0.122
	(0.0205)	(0.353)	(0.265)	(0.352)	(0.294)	(0.245)
Intercept term	−5.349	−9.090	-34.07***	25.58	61.95***	6.889
	(3.916)	(6.693)	(6.985)	(16.17)	(20.96)	(6.723)
Time fixed effects	YES	YES	YES	YES	YES	YES
City fixed effects	YES	YES	YES	YES	YES	YES
Sample	1,697	859	1,310	645	375	495

Note: Significant at: $^{***}p < 0.01$, $^{**}p < 0.05$, $^{*}p < 0.1$.

the effect weakens, then strengthens, and then weakens again. The regression coefficient shows this trend as it decreases from 0.0602 for distances between 600 and 900 km to 0.0510 for distances between 900 and 1,200 km, before increasing to 0.132 for distances between 1,500 and 1,800 km. The peak is reached at this point before the coefficient decreases to 0.091 for distances of 1,800 km and above.

Similar to the radiation range of the full samples, the optimal radiation range of air transport for technology innovation transfer in the three core cities is approximately between a distance of 1,500 and 1,800 km. This finding is broadly similar to that of Huang *et al.* (2012) who analyzed that the economic radiation range of cities such as Beijing and Shanghai is based on urban field strength theory. The optimal spatial thresholds are still far below the east–west or north–south longitudinal range in China, indicating that geographical distance is still a strong spatial constraint on the transfer of technological innovation. With the continuous improvement of the national "hub-and-spoke" airport system, the western aviation hub cities and the capital cities of the northeast region (e.g., Urumqi, Xi'an, Lanzhou, Xining, Yinchuan, Changchun, Harbin, etc.) have high aviation accessibility. However, even under ideal aviation accessibility conditions, the technology transfer from national innovation centers such as Beijing, Shanghai, and Shenzhen to these cities is still limited in terms of radiation and diffusion.

9.2 The Effect of High-Speed Railway

9.2.2 *Conceptual background and theoretical logic*

9.2.2.1 *Agglomeration externalities and knowledge spillover*

The introduction of HSR has shortened the distance between cities, eliminated regional barriers to economic activity, facilitated the clustering of production factors and economic activities, and overcome constraints to knowledge spillover imposed by geographic separation (Yin *et al.*, 2015; Wang *et al.*, 2017; Heuermann & Schmieder, 2018; Wang & Cai, 2020; Zhang, 2021). Specifically, it affects economic agglomeration and promotes knowledge spillover mainly through the following three mechanisms. Firstly, HSR has a direct impact on regional accessibility by reducing transportation costs, which in turn enhances the location conditions of cities and attracts the concentration of production resources such as population (Shao *et al.*, 2017). This provides a significant source of high-quality labor for urban and regional development, generating agglomeration through the "labor pool" effect (Du & Peng, 2017). Secondly, HSR has a profound impact on the location choice of different types of firms (Hodgson, 2018), while knowledge-intensive firms prefer to cluster in cities along HSR lines to reduce production costs (Wetwitoo & Kato, 2017). In this process, knowledge-intensive firms generate agglomeration economies by sharing resources, competing, and learning from each other, thus promoting tacit knowledge diffusion and knowledge spillover (Chen & Vickerman, 2017). Finally, production factors have a certain convergence, and the cities along the route connected to the HSR network will continue to attract the spatial concentration of production factors and economic activities (Pyka *et al.*, 2019), and expand the agglomeration through the "circular and accumulative effect". The externalities will further enhance the knowledge spillover and innovation performance of core cities through "circular and accumulative causation" (Fang & Yi, 2014).

9.2.2.2 *Network externalities and knowledge spillover*

As a modern transportation way, HSR facilitates the unhindered flow of innovation factors such as capital and technology, enhances economic ties between cities, and creates network externalities. This chapter delves into how the introduction of HSR leads to network externalities and shapes intercity knowledge spillovers, focusing on three key aspects: Firstly, HSR has a positive impact on talent flows by facilitating faster and more convenient transportation. It also increases face-to-face knowledge exchanges, such as attending academic conferences, which promotes cross-regional research cooperation and technology transactions. This leads to the formation of closer innovation cooperation networks among cities, creating network externalities that can facilitate knowledge spillovers (Gao & Zheng, 2020; van Meeteren *et al.*, 2016). Secondly, HSR network allows cities to access knowledge from a larger spatial scope through network embedding, sharing resources, and learning from each other through network pipelines (Glaeser *et al.*, 2016). This promotes urban networking and lays a strong foundation for knowledge spillovers and new knowledge creation (Cortinovis & Oort, 2019). In particular, the flow of innovation factors is enhanced by the establishment of a knowledge flow system that combines "local buzz" with a "global pipeline". This system enables cities to absorb external knowledge, enhance interaction with local knowledge, generate knowledge spillovers, and improve their innovation performance (Bathelt *et al.*, 2004). Finally, HSR improves the efficiency of innovation resource allocation by highlighting the comparative advantages of cities and matching resources on a larger spatial scale (Ma *et al.*, 2020). This leads to improved efficiency in production factor allocation and economies of scale, which generate network externalities and promote knowledge spillovers (Sheng *et al.*, 2020).

In summary, while scholars have conducted studies on the impact of HSR on urban innovation performance, these studies have mainly focused on how HSR can improve urban accessibility, promote factor mobility, and encourage economic agglomeration. In this way, they tend to neglect an important perspective: network externalities, which explore how cities can capture these externalities to boost innovation performance after the opening of HSR. To address this gap, this chapter provides a theoretical analysis of both agglomeration and network externalities to investigate HSR's effect on urban innovation performance. We then employ both the PSM-DID model and spatial econometric model to test our hypotheses empirically. By doing so, we aim to deepen our understanding of how HSR can impact urban innovation performance.

9.2.3 *Empirical results*

This chapter focuses on 285 prefecture-level cities from 2008 to 2016 to explore how the exogenous event of HSR opening impacts urban innovation performance. We selected cities that opened HSR between 2007 and 2016 as the treatment group, while those that did not open during the period were used as the control group. We obtained information on HSR openings mainly from the National Railway Bureau's website. To avoid potential lag effects in innovation, we removed sample values for the year in which HSR was opened since many HSR lines in the sample were opened during July–December. Moreover, China's HSR lines primarily connect hub-type cities, making it more appropriate to exclude cities at the two ends of the line due to their endogeneity. Therefore, we only included cities along the middle of the line in our sample.

9.2.3.1 *Spatial correlational analysis*

We assess the innovation performance of cities by quantifying the number of patents per 10,000 people. To explore the relationship between urban innovation and HSR, we analyze the spatial correlation between the HSR lines opened in 2016 and the average number of patents per 10,000 people in cities from 2007 to 2016. We present our findings through plotted data (Figure 9.2). When examining the spatial distribution of the number of patents, we observe a general trend of decreasing values from east to west zones. Notably, areas with high patent counts exhibit a spatial pattern characterized by significant clustering and minimal scattering. These high-value regions are primarily concentrated within the core cities of the BTH, YRD, PRD, and *Chengyu* regions, along with other urban clusters throughout the country. In addition, there is also a sporadic distribution of high-value areas in some provincial capitals in the central and western regions. The HSR network has a "four vertical and four horizontal" pattern, with more intensive lines in the eastern and central regions. Cities along HSR lines have significantly higher urban innovation performance than those along non-HSR lines, with innovation corridors on the Beijing–Shanghai,

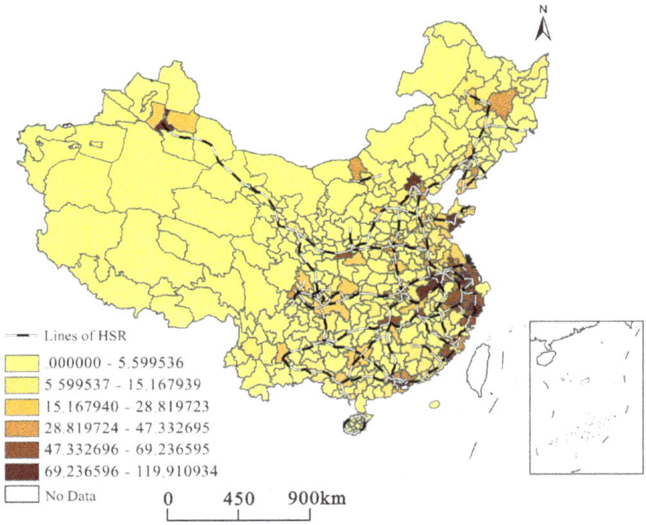

Figure 9.2 Colocational distribution between urban innovation capacity and HSR.
Source: Guan *et al.* (2021).

Beijing–Guangzhou, and Beijing–Harbin lines. Transportation hub cities with multiple interacting lines become high-value clusters of innovation levels. There is a strong spatial coupling between the spatial distribution of HSR lines and regional innovation levels.

Since knowledge spillover is influenced by the decay law of distance, the spatial autocorrelation of urban innovation performance is measured by using the spatial statistics function of ArcGIS software. The results show that the global Moran's I index of regional innovation level is 0.11 and $Z = 14.9$ ($p = 0.00$), showing a significant positive spatial correlation. Therefore, the local clustering patterns of regional innovation levels were further analyzed using LISA plots, and the spatial statistical functions were used to classify the regional innovation level clustering patterns into four types: high-high clustering, high-low clustering, low-high clustering, and low-low clustering. City clusters such as the YRD, the BTH, and the PRD attract innovation factors by virtue of their highly developed economies and produce obvious knowledge spillover effects on the surrounding areas, which become high-high agglomeration areas.

Some core cities, such as Chengdu, Xi'an, and Chongqing, have also become high concentration areas for innovation factors in a few surrounding regions by virtue of their stronger economies, which also generate certain knowledge spillover to the surrounding regions. Although previous studies have explored the relationship between HSR and innovation and found that improved transportation facilities promote regional innovation, the spatial autocorrelation of the improved innovation performance of cities after accessing the HSR network has been ignored, which affects the accuracy of the conclusions. Therefore, based on the PSM-DID model, this chapter further measures the knowledge spillover effects of HSR on regional innovation from agglomeration externality and network externality by spatial econometric model to deepen the understanding of the mechanism of HSR affecting urban innovation performance.

9.2.3.2 *Endogeneity test*

The planning and arrangement of HSR are primarily carried out by departments such as the National Development and Reform Commission. These departments often prioritize cities with significant regional advantages, such as a developed economy. In this way, this approach to site selection for HSR may result in endogenous problems that affect regional innovation. Therefore, sample data were established via PSM, to make the treatment group and control group meet the common trend hypothesis. Based on this, regression analysis was conducted via difference-in-differences model by controlling city properties including economic scale, which reduced endogenous problem caused by unobservable factors. Moreover, the current arrangement of HSR is closely linked to locational factors, such as landform. However, these factors are exogenous to modern economic development and may not be directly correlated with the innovation levels of Chinese cities. To address this issue, we extracted China's digital elevation data using GIS software and calculated the geographical gradient which was then used as an instrumental variable for HSR opening in our regression analysis to further verify its endogeneity.

9.2.3.3 *Baseline regression*

As a precondition of difference-in-differences model, the parallel trend hypothesis requires the experimental group and control group to meet the same variation trend before facing exogenous shock. We compared the innovation level changes of treatment group and control group before and after the introduction of HSR, to analyze the parallel trend hypothesis. Beijing–Tianjin HSR was opened in 2008, but the innovation level of cities with and without HSR did not change much in 2008. After 2009, the regional innovation level of cities with HSR was enhanced, showing that the influence of HSR opening on regional innovation level has a hysteresis effect over time. In addition, cities with and without HSR presented approximate regional innovation level and the variation trends were also similar. After 2009, the difference in innovation between cities with and without HSR increased gradually, and the increased speed of regional innovation level in cities with HSR was much faster than that of cities without HSR. It presents that the exogenous shock, that is, opening of HSR produces a significant influence on regional innovation. Therefore, the use of difference-in-differences model in this thesis can meet the precondition of parallel trend hypothesis (Figure 9.3).

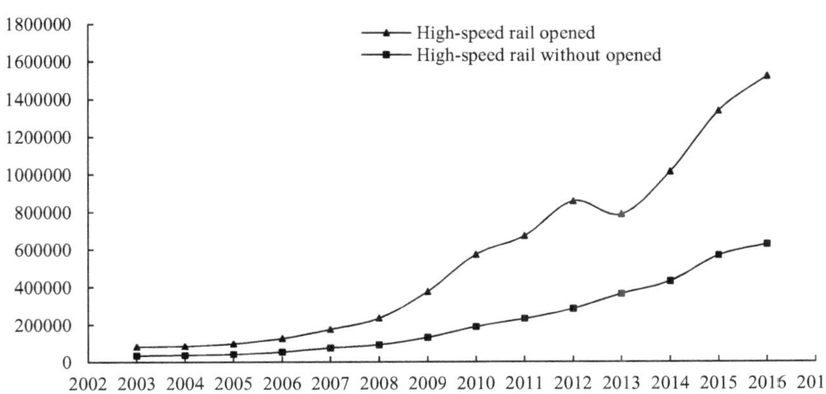

Figure 9.3 Trends in the number of urban patent applications with and without HSR in China.

According to the model, different variables were substituted into models. The inspection results show that the previous problem does not exist. Therefore, regression analysis was conducted for the models respectively. In the regression process, Models 7–10 are regression results of all samples, and Models 11–14 are regression results of cities at both ends of HSR. Models 7 and 11 are the unilabiate difference results without control variable. The variation coefficients of HSR of all models are positive, indicating cities with HSR have a higher innovation level when compared with cities without HSR. When comparing the regression results of all samples (Models 7–10) with those where node cities at both ends of HSR were removed in the treatment group (Models 11–14), it was observed that the coefficients of each model decreased somewhat. This suggests that the opening of HSR has a more significant facilitating effect on innovation in cities located at both ends, as opposed to those situated along the HSR. In other words, a policy inclination exists in the selection of cities at both ends of the line in HSR construction.

According to the above analysis, the results of Model 14 are the most accurate. The regression analysis reveals that the coefficient for HSR opening is 0.032, and it passes the significance test at the 1% level. This indicates that cities with an HSR experience a 0.32% increase in the number of patents per 10,000 people compared to those without high-speed rail. In addition, the regression coefficient of non-node city samples is significantly positive, demonstrating that HSR construction does not restrain the innovation development of cities along the line. In other words, at the level of innovation, "corridor effect" does not exist in China's HSR construction. The regression of deleting node cities at both ends has reduced the endogenous problem of the model to some extent, but it cannot avoid the endogenous problem caused by the preference for cities with obvious regional advantages. Therefore, by learning from the method of instrumental variable construction, we selected geographical gradient data of Chinese cities as an instrumental variable of HSR opening for regression, to further verify the endogeneity of the model (Table 9.7). Regression results show that the regression

Table 9.7 Regression results at national level.

Variables \ Models	Total sample					Node cities deleted		
			DID	PSM-DID		DID		PSM-DID
	Model 7	Model 8	Model 9	Model 10	Model 11	Model 12	Model 13	Model 14
Dependent variable	Lnpat	Lnpat	Lnpat	Lnpat	Lnpat	Lnpat	Lnpat	Lnpat
Variable of HSR	0.312***	0.112***	0.063***	0.047***	0.223***	0.059***	0.046***	0.032***
	(0.034)	(0.025)	(0.006)	(0.005)	(0.036)	(0.025)	(0.006)	(0.005)
Constant	2.793***	8.614***	0.952***	–0.900***	0.394***	4.973***	0.538***	–2.206***
	(0.047)	(1.191)	(0.005)	(0.305)	(0.031)	(1.073)	(0.005)	(0.213)
Control variables	NO	YES	NO	YES	NO	YES	NO	YES
Time fixed effect	YES	YES	YES	YES	YES	YES	YES	YES
City fixed effect	YES	YES	YES	YES	YES	YES	YES	YES
n	2966	2966	2918	2918	2596	2596	2450	2450
R^2	0.916	0.947	0.962	0.969	0.887	0.936	0.953	0.971

Notes: Robust standard errors are reported in parentheses, clustered at the city level; $^{*}p < 0.10$, $^{**}p < 0.05$, $^{***}p < 0.01$.

coefficient of instrumental variable and regional innovation is 1.898, and significant at the level of 1%; the standard error is 0.333, showing that the regression results of instrumental variable and benchmark regression have some similarities. Meanwhile, the *F* statistic in first-stage regression is 76.06 and greater than 10, further proving the effectiveness of the instrumental variable.

9.2.3.4 Regional heterogeneity regression

To explore the spatial heterogeneity of HSR construction on innovation, the paper divided China into the eastern, central, and western regions (including northeastern regions) according to the geographical location. DID model regression and PSM-DID model regression were conducted for the east, middle, and west (Table 9.8).

The coefficients of models are positive, and pass significance test at the level of 1%. This shows that the opening of HSR also promotes production factor flow and knowledge spillover between cities inside the region, and has a significant positive effect on the innovation level of eastern, middle, and western regions. The regression coefficients for HSR opening and regional innovation level in

Table 9.8. Regression results at regional level.

Models / Variables	Eastern region	Central region	Western region	Eastern region	Central region	Western region
		DID			PSM-DID	
	Model_7	Model_8	Model_9	Model_10	Model_11	Model_12
Dependent variable	Lnpat	Lnpat	Lnpat	Lnpat	Lnpat	Lnpat
Variable of HSR	0.114^{***}	0.064^{***}	0.097^{***}	0.028^{***}	0.013^{***}	0.029^{***}
	(0.044)	(0.032)	(0.057)	(0.007)	(0.004)	(0.006)
Constant	4.751^{***}	2.789^{***}	3.645^{***}	-3.619^{***}	-3.181^{***}	-1.722^{***}
	(1.887)	(1.887)	(1.655)	(0.398)	(0.354)	(0.172)
Control variables	YES	YES	YES	YES	YES	YES
Time fixed effect	YES	YES	YES	YES	YES	YES
City fixed effect	YES	YES	YES	YES	YES	YES
Sample size	847	957	787	840	912	696
R^2	0.954	0.912	0.913	0.964	0.986	0.984

Note: Significant at $^{***}p < 0.01$, $^{**}p < 0.05$, $^{*}p < 0.1$.

the models are similar between the eastern and western regions, but greater than that in the middle region. The eastern region has a more advanced economy and a wealth of innovation resources. The excellent transportation infrastructure has significantly facilitated knowledge flow between eastern cities and other regions, thereby intensifying the diffusion of tacit knowledge and promoting innovation activities. Moreover, the eastern regions have also attracted innovation resources from the middle and western regions due to their favorable innovation environments. In contrast, the economy in the western regions is relatively underdeveloped, and only a few core cities including Chongqing, Chengdu, and Xi'an have access to high-speed rail. Under the effect of HSR, the flow and agglomeration of properties from western regions to these core cities is accelerated. The disequilibrium of internal innovation resource distribution is enlarged, but knowledge spillover and the launching of innovation activities are also promoted. While the opening of HSR has positively affected the innovation level of eastern regions, this effect is less pronounced in middle regions where several cities have opened HSR, shortening the distance to eastern regions. This has resulted in a more balanced distribution of economic activities across the region, providing greater convenience for the flow of innovation elements to eastern regions.

9.2.3.5 *Urban heterogeneity regression*

According to the classification standard of *Green Paper for Small and Medium Cities* in 2010, cities were divided into three grades (large-sized, medium-sized, and small-sized), to explore the effect of HSR opening on regional innovation of different scales. Cities of different grades were placed in DID model and PSM-DID model for regression, and node cities at both ends of the HSR were deleted.

The coefficient in Model 9 is insignificant, and the coefficients of other models are positive (Table 9.9). This means that the opening of HSR has a significant positive effect on the innovation level of large and medium cities, but the conclusion about the positive effect on small cities is not robust in DID regression. The major reason lies

Table 9.9 Regression results at urban level.

Variables \ Models	Large-sized Cities	Medium-sized Cities	Small-sized Cities	Large-sized Cities	Medium-sized Cities	Small-sized Cities
	DID			PSM-DID		
	Model_7	Model_8	Model_9	Model_10	Model_11	Model_12
Dependent variable	Lnpat	Lnpat	Lnpat	Lnpat	Lnpat	Lnpat
Variable of HSR	0.067*** (0.036)	0.049*** (0.026)	0.025 (0.056)	0.027*** (0.007)	0.025*** (0.007)	0.021*** (0.004)
Constant	6.738*** (1.326)	3.735*** (1.758)	2.967** (1.450)	1.667*** (0.297)	−1.822*** (0.204)	−1.790*** (0.233)
Control variables	YES	YES	YES	YES	YES	YES
Time fixed effect	YES	YES	YES	YES	YES	YES
City fixed effect	YES	YES	YES	YES	YES	YES
Sample size	891	1141	559	882	1088	478
R^2	0.964	0.910	0.857	0.970	0.964	0.986

Note: Significant at $^{***}p < 0.01$, $^{**}p < 0.05$, $^{*}p < 0.1$.

in the great innovation development heterogeneity of small cities. In terms of the effect, the HSR opening variables of Models 10–12 are 0.027, 0.025, and 0.021, respectively, indicating that the effect degrees of HSR opening on the number of patents per 10,000 people in large-sized, medium-sized, and small-sized cities are 0.27%, 0.25%, and 0.21%. The effect coefficient on big cities is the largest, as the opening of HSR will help to strengthen the agglomeration of innovation resources in big cities. After agglomerating innovation resources, big cities show excellent innovation and transformation capacity.

9.2.3.6 *Decomposition of mechanisms*

Since knowledge spillover has spatial correlation, we also use spatial econometric models to analyze the impact mechanism of HSR opening on urban innovation performance in terms of agglomeration externality and network externality. The estimation results are shown in Table 9.10, and we can draw the following three conclusions.

The results of the analysis indicate a clear and significant positive relationship between the opening of HSR and economic

Table 9.10 The effect of HSR on nodal innovation capacity.

Variables \ Models	SLM		SEM		SLM		SEM	
	Dag	Lnpat	Dag	Lnpat	Degree	Lnpat	Degree	Lnpat
HSR	0.073^{**} (1.71)		0.052^{*} (1.21)		0.395^{***} (3.35)		0.294^{**} (2.46)	
Mediating variable		0.34^{***} (3.83)		0.23^{***} (2.23)		0.516^{***} (15.47)		0.469^{***} (14.27)
W*Agg	0.042^{***} (2.62)	0.081^{***} (2.94)						
W*Degree					0.176^{***} (3.04)	0.051^{**} (2.32)		
Lambda			0.724^{***} (5.38)	0.908^{***} (16.41)			0.623^{***} (3.68)	0.849^{***} (10.15)
Controls	YES	YES	YES	YES	YES	YES	YES	YES
Time fixed effect	YES	YES	YES	YES	YES	YES	YES	YES
City fixed effect	YES	YES	YES	YES	YES	YES	YES	YES
Constant	-0.16^{**} (−2.24)	-0.364^{**} (−2.53)	0.02^{***} (0.275)	0.128 (0.354)	-0.46^{***} (−2.82)	-0.22^{*} (−1.92)	0.019 (0.12)	0.067 (0.36)
R^2	0.34	0.51	0.38	0.42	0.37	0.54	0.62	0.57
obs	2966	2966	2966	2966	2966	2966	2966	2966

Note: Significant at $^{***}p < 0.01$, $^{**}p < 0.05$, $^{*}p < 0.1$.

agglomeration, as evidenced by the positive coefficients. This suggests that the introduction of HSR has contributed significantly to the concentration of production requirements, such as population, in the region. The benefits of this economic agglomeration are manifold, including the generation of externalities through mechanisms such as sharing, learning, and matching, which have proven to promote knowledge spillover and foster urban innovation performance. Furthermore, the analysis reveals that the spatial spillover effects of economic agglomeration are also significant, with neighboring areas benefiting from the presence of concentrated economic activity, thereby contributing to local innovation performance. In short, these findings demonstrate that the opening of HSR has had a considerable impact on regional innovation, both directly and indirectly through its role in promoting economic agglomeration.

The coefficient of the intensity of urban innovation linkage is also significantly positive for the opening of HSR. The analysis reveals that the opening of HSR has a significantly positive impact on the intensity of urban innovation linkage, which in turn contributes to the improvement of urban innovation performance. This finding suggests that the increased regional accessibility resulting from access to the HSR network not only facilitates the free flow of innovation factors and cross-regional innovation cooperation but also strengthens intercity socioeconomic linkages, creating network externalities that promote knowledge spillover and foster urban innovation. Moreover, the analysis highlights the significant positive impact of the spatial lag term of the strength of urban innovation linkages, indicating that the innovation linkages in geographically proximate areas influence each other and contribute significantly to the overall innovation performance of the region.

The effect of urban network externalities on urban innovation performance improvement is higher than that of agglomeration externalities. In both the SLM and SEM models, the regression coefficients of urban network externalities are higher than those of agglomeration externalities, reflecting that the improvement of urban innovation performance benefits more from the networked development of cities due to the opening of HSR, i.e., the effect of network externalities on intercity knowledge spillover is stronger. In the studies on the impact of transportation infrastructure on regional development, scholars generally focus on the regional economic agglomeration effect generated by improved accessibility, while ignoring the network externality due to increased intercity socioeconomic linkages. In the context of economic globalization, urban network externalities have become the core driving force for regional innovation improvement and its sustainable development. Therefore, it is necessary to pay attention not only to the agglomeration externality generated by the agglomeration of production factors but also to the network externality generated by the increased intercity linkages.

9.2.4 *Robustness tests*

9.2.4.1 *Parallel trend test*

The increase in regional innovation performance is the result of natural evolution caused by the economic development of the city, rather than the facilitating effect triggered by HSR. To eliminate the influence

Table 9.11 The results of placebo test.

Variables	Whole Nation	The Eastern	The Central	The Western	Large-sized Cities	Medium-sized Cities	Small-sized Cities
Dependent variable	Lnpat	Lnpat	Lnpat	Lnpat	Lnpat	Lnpat	Lnpat
D_{ummy}	0.024 (0.019)	0.064 (0.044)	−0.071 (0.027)	0.011 (0.035)	0.009 (0.028)	−0.006 (0.029)	0.067 (0.128)
Constant	5.085*** (1.06)	4.623*** (1.889)	2.863*** (1.582)	3.683*** (1.650)	7.029*** (1.311)	3.809*** (1.749)	1.472*** (2.459)
Control variable	YES	YES	YES	YES	YES	YES	YES
Time fixed effect	YES	YES	YES	YES	YES	YES	YES
City fixed effect	YES	YES	YES	YES	YES	YES	YES
Sample size	2591	847	957	787	891	1141	559
R^2	0.936	0.956	0.911	0.913	0.964	0.910	0.857

Note: Significant at $^{***}p < 0.01$, $^{**}p < 0.05$, $^{*}p < 0.1$.

of this situation, this thesis conducted a robustness test with the method of placebo test from the angle of random effect. The specific method is as follows: Suppose that the fact of HSR construction did not exist in the sample period from 2003 to 2016, and the time period of HSR construction is about three years; a dummy variable of three years before the actual opening of HSR is set as the time node of pseudo-HSR opening; if HSR has a causal effect on regional innovation, the coefficient of this variable will be insignificant in practical regression when this event does not actually happen. Placebo test was conducted for the overall regression model, regional heterogeneity regression model, and urban heterogeneity regression model. The results show that the dummy variable D_{ummy} of the pseudo-HSR opening time is insignificant (Table 9.11). In other words, the virtual nodes established have no treatment effect, indicating that the positive promoting effect of HSR opening on regional innovation has robustness.

9.2.4.2 *Temporal dynamic effect*

The node cities at the end were deleted, and cities along the line were chosen as the research sample, and the special geographical location of cities along the line, which could potentially impact the planning and layout of HSR. In order to analyze the temporal dynamic effect of HSR on regional innovation, we adopted a methodology from Giroud (2013) by establishing time dummy variables one year before the opening of HSR and the first year after its opening. If the increase of regional innovation originates from the influence of HSR opening, then the coefficient of T^{-1} should be significantly negative or insignificant in theory; the coefficients of T^1, T^2, and T^3 should be significantly positive. The coefficient of T^{-1} is insignificant, which can meet the theoretical expectation (Table 9.12). However, in the first year and second year after the opening of HSR, some coefficients are insignificant, which is mainly caused by the hysteresis effect of innovation activity. In T^3, the coefficients are significantly positive, indicating that the facilitating effect of HSR opening on regional innovation has a hysteresis period of two to three years. The hysteresis period of middle and eastern regions and big cities is relatively short (two years), while the hysteresis period of middle and western regions and small cities is relatively long (three years).

Table 9.12 Regression results of dynamic effect.

Variables	Whole Nation	The Eastern	The Central	The Western	Large-sized Cities	Medium-sized Cities	Small-sized Cities
T^{-1}	0.007	0.011	-0.094^{***}	−0.00	0.013	−0.041	0.057
	(0.029)	(0.042)	(0.041)	(0.044)	(0.047)	(0.033)	(0.057)
T^1	0.000^{***}	0.043	−0.029	0.041	0.063	−0.057	−0.036
	(0.038)	(0.050)	(0.049)	(0.062)	(0.053)	(0.055)	(0.072)
T^2	0.059^{***}	0.068^{***}	−0.036	0.063	0.057^{**}	0.054^{**}	0.030
	(0.020)	(0.031)	(0.027)	(0.041)	(0.030)	(0.030)	(0.049)
T^3	0.190^{***}	0.206^{***}	0.120^{***}	0.175^{***}	0.189^{***}	0.190^{***}	0.169^{***}
	(0.021)	(0.032)	(0.030)	(0.043)	(0.030)	(0.032)	(0.065)
Control variables	YES	YES	YES	YES	YES	YES	YES
Time fixed effect	YES	YES	YES	YES	YES	YES	YES
City fixed effect	YES	YES	YES	YES	YES	YES	YES
Sample size	2591	870	957	787	891	1141	559
R^2	0.936	0.906	0.911	0.913	0.965	0.911	0.857

Note: Significant at $^{***}p < 0.01$, $^{**}p < 0.05$, $^{*}p < 0.1$.

9.3 Summary

This chapter explores the impact of aviation and HSR transport on technology transfer in the Chinese city. On the one hand, it uses NBR, GIS spatial analysis, and other econometric regression analysis methods to examine the impact of air transport on technology transfer. Additionally, it explores the distance decay effect of air transport on technology transfer radiation by dividing the areas into circles. The study concludes with an analysis of the spatial extent of technology radiation in Beijing, Shanghai, and Shenzhen — three of China's leading innovation centers — under optimal air conditions. Meanwhile, this chapter aims to explore the interdependent relationship between aviation and innovation networks both qualitatively and quantitatively. By testing theoretical hypotheses and conducting empirical research, we hope to better understand how these two networks are coupled.

On the other hand, based on Chinese city panel data, this chapter also analyzes the impact of HSR opening on urban innovation performance and its spatial heterogeneity using PSM model (PSM-DID). On this basis, a spatial econometric model is constructed to understand the economic effects of HSR by influencing the mechanism of regional innovation from agglomeration externalities and network externalities of HSR opening.

Increased aviation accessibility has a positive effect on technology transfer at both the city node level and city-pair level. At the city node level, the more frequent the city's airline connections with other cities in the airline network, the more conducive it is to promote technology transfer between the cities. At the city-pair level, the closer the airline connections between cities, the more beneficial it is to promote technological innovation interactions between cities.

The marginal effect of aviation accessibility on technology transfer varies in different spatial scales and shows a non-linear function with geographic distance, with the optimal radiation range being between 1,500 km and 1,800 km. Within the distance range of 0 km to 900 km, aviation accessibility does not appear to have a significant impact on technology transfer. However, as the distance increases beyond 900 km and reaches 1,800 km or more, the influence of aviation accessibility on the transfer of technology gradually intensifies before eventually decreasing. It is noteworthy that the turning point at which the impact of air accessibility on technology transfer shifts occurs within the range of 1,500 km to 1,800 km, marking the inflection point of the radiation effect. In terms of the radiation range of the core cities, Beijing, Shanghai, and Shenzhen show a fluctuating change from weakening to strengthening to weakening in terms of the effect of aviation accessibility on the technology transfer within different distances. Although the optimal radiation range for technology transfer is between 1,500 km and 1,800 km under ideal aviation accessibility conditions, the radius of innovation radiation with Beijing, Shanghai, and Shenzhen as the center of the circle is still much lower than the east-west or north-south depth range of China. This suggests that geographical distance is still a strong spatial constraint on the transfer of technological innovation.

When it comes to the effect of high-speed railways, the opening of HSR significantly improves urban innovation performance, but there exists significant spatial heterogeneity in the intensity of the effect of HSR on urban innovation. In terms of enhancing innovation performance, the impact is most pronounced in economically developed eastern regions and metropolitan areas. This implies that urban innovation performance benefits substantially from advanced transportation infrastructure. However, it also accentuates the innovation gap between regions, contributing to a profound effect on the geographical pattern of uneven development.

The introduction of HSR not only facilitates the concentration of production factors such as population in cities located along the route but it also strengthens economic ties between these cities. As a result, agglomeration and network externalities are generated, which contribute to knowledge spillover and serve as a catalyst for enhancing urban innovation performance. Compared with agglomeration externalities, urban network externalities have a stronger impact on urban innovation performance and become the core driving force for sustainable regional development in the knowledge economy.

After excluding possible changes in other factors, the opening of HSR still has a significant contribution to the improvement of urban innovation performance. Through time dynamic analysis, it is found that there is a time lag effect of about three years for the impact of HSR opening on urban innovation performance, i.e., the long-term effect of transportation infrastructure improvement on regional development is more superior.

PART V
Policy Implication and Conclusion

CHAPTER 10

POLICIES FOR ENHANCING TECHNOLOGY TRANSFER NETWORK

10.1 National Perspective

10.1.1 *Optimizing national innovation system*

Innovation is a vital driver of national development and competitiveness. To accomplish innovation one needs to focus on S&T development to promote the in-depth implementation of the innovation-driven development strategy. It is crucial to address the major needs of the country while prioritizing research and development to stay at the forefront of scientific and technological advancements. By constantly investing in both basic and applied research, we can achieve breakthroughs in core technologies that propel us toward progress and success. In this way, it is imperative for nations to strategically enhance the construction of their national S&T elements. This includes systematically driving reforms within the science and technology system, as it is crucial for sustainable growth and development. Specifically, the nations need to focus on the following aspects:

(a) Cultivating innovation growth poles, and building a series of national science centers and regional innovation highlands to enhance their roles in supporting high-level scientific research;

(b) Continuously promoting the integration of science and technology with the economy and society to support high-quality economic development with science and technology innovation;
(c) Accelerating the construction and improvement of a national technology transfer system, and vigorously developing intermediaries for the socialization of technology transfer;
(d) Promoting innovation programs, reform, and improvement in the management, transforming scientific and technological achievements, and at the same time, putting forward the sharing of innovation resources and innovation evaluation and reward;
(e) Broadening the construction of transfer pipelines among enterprises, universities, and research institutions, and completing the top-level design for innovation-driven development;
(f) Strengthening the construction of talent teams for innovation development, optimizing the structure of talents, and introducing and cultivating leading talents and innovation teams in various fields of science and technology.

10.1.2 *Enhancing openness innovation*

The nation should also enhance the level of openness in S&T innovation and increase technological exchanges with developed countries. The absorption and learning of external technologies can help improve local innovation capacity. This book finds that China's transnational technology pipelines are highly concentrated and its technological dependence on developed countries such as the US and Japan remains high, making its integration into global innovation networks risky.

On the one hand, there is a need to continue to expand the breadth of transnational technology import and promote patent technology transactions with countries along the Belt and Road to enhance the diversity and balance of transnational technology pipelines. On the other hand, there is a need to increase participation in international scientific programs and projects and enhance the capacity to take the lead in organizing resources, to change the role from minor participation to significant participation so as to accelerate the launch of international scientific programs and projects led by China. It is also important to fully support scientists and researchers around the world to jointly carry out technological innovation research.

Therefore, there is a need to enhance the country's external science and technology openness, strengthen bilateral or multilateral cooperation and exchanges in science and technology between governments, establish mechanisms for innovation dialogue with a number of countries, especially technologically developed countries, and participate in and promote consultations and cooperation on science and innovation issues under multilateral mechanisms.

10.2 Regional and Local Perspective

10.2.1 *Cultivating regional innovation clusters*

Regional innovation clusters are the growth cores of innovation network and the energy sources that radiate and drive regional innovation development. Therefore, strengthening the cultivation and enhancement of regional innovation clusters to achieve the ladder linkage development of the national innovation system is crucial. The cultivation and strengthening of regional innovation clusters need to be strengthened, on the one hand, to rely on strategic industry clusters to undertake the industrial transfer, knowledge spillover, and technology transformation from international technological innovation centers and national science centers; and on the other hand, promote the absorption and transformation of knowledge and technology within innovation clusters to facilitate the development and diffusion of new products, technologies, and industries, thereby radiating and driving regional development. Specifically, this point can be achieved through the following aspects:

(a) Accelerating the construction of innovation clusters in the BTH, the YRD, and the PRD urban agglomerations, with innovative cities such as Beijing, Shanghai, and Shenzhen as the core.

(b) Establishing a sound unified technology transfer market and an efficient technology transfer mechanism to promote the integrated development of technological innovation within innovation clusters.

(c) Enhancing technology transfer and transformation among innovation clusters, as well as technology linkages and cooperation between innovation clusters and inland cities.

(d) Laying out and building several innovation city systems with different levels and functional gradient linkages.

(e) Enhancing the technology R&D strength and technology transfer as well as the transformation capacity of major innovation cities by attracting high-tech talents, encouraging technological innovation, and strengthening cooperation among industries, universities, and research institutes.

This will increase the density and flow of intercity technology corridors, and drive the whole country to realize the development of the national innovation system in a hierarchical way.

10.2.2 *Improving local innovation capacity*

Every region and city has unique strengths when it comes to S&T innovation. It is thus important to emphasize local and regional characteristics to enhance innovation capacity around areas of regional strength. To ensure sustainable development in this field, it is crucial to understand and capitalize on these strengths. By identifying the specific areas of expertise and direction of each region and city, we can effectively focus our efforts on fostering innovation that will drive progress and growth. Furthermore, there is a need to cultivate regional innovation growth poles with local characteristics and obvious advantages by prioritizing the strength of local innovation forces and advantageous industries. This will lead to the construction of a national–regional innovation system with a differentiated layout, complementary functions, and positive interaction.

Taking the findings of this book as an example, in the later stages of transnational technology import activities, Beijing and Shenzhen are mainly absorbed in the fields of audio-visual technology and digital communications, Shanghai has a higher demand for technologies in the fields of computer technology and motor-instrument-energy, while Guangzhou has been focusing on technologies in the chemistry sector for its development. Therefore, each city has distinctive technological needs and specific areas of focus when it comes to innovation. To fully leverage the potential of each region, it is vital to harness their unique strengths and characteristics. By driving innovation that is anchored in local knowledge and economic conditions, we can optimize the benefits of S&T in each city and propel growth in a sustainable manner.

10.3 Actor Perspective

10.3.1 *Increasing R&D investment*

Chinese enterprises should strive to enhance their openness and cooperation with the outside world by increasing investment in innovation and research & development (R&D). This can be achieved by establishing innovation R&D centers and industrial parks overseas, which will enable them to tap into global innovation elements and resources. Additionally, strengthening technology trade and R&D cooperation with foreign transnational companies, leading universities, and research institutions can facilitate technology transfer and broaden the scope of technology spillover.

Enterprises should also prioritize the strengthening of their in-house scientific and technological R&D capabilities by increasing financial and human resources investment in innovation, as well as the construction of innovative organizations. This will enable them to better learn, apply, and reinvent external technologies. Amidst fierce market competition, they should maintain a sense of urgency and competitiveness, thereby fostering exchanges and cooperation with other cities within the cluster. This can help promote the spillover effect of local innovation and actively enhance their R&D activities, thus avoiding reliance on external advanced technologies.

10.3.2 *Promoting industry–university–research cooperation*

To enhance the central role of universities and research institutes, it is crucial to promote in-depth integration between industry, academia, and research. Developing local innovation networks and innovation milieu can positively impact China's external technology pipelines, but university clusters must play a more prominent role in supporting local integration into global innovation networks.

Top universities and research institutions should act as a national team for knowledge production and talent cultivation. This approach can help strengthen the deep integration of knowledge innovation and industrial value chains, and deepen the integration of industry, university, and research. However, the current challenge is the poor coupling of science and technology innovation capacity of urban higher education and urban science and technology levels. To improve this situation, joint research with universities and research institutes on scientific and technological problems should be vigorously carried out. Additionally, new market-oriented and flexible R&D institutions should be actively built, and efforts to promote scientific and technological innovation and entrepreneurship should be stepped up. Continuous improvements to the incubator, accelerator, crowdsourcing space, and other business incubation systems are also necessary to cultivate new advantages for local participation in global science and technology competition and cooperation.

Intermediaries can play a critical role in strengthening the linkage among enterprises, government, universities, and research institutions. Science and technology intermediaries can become a bridge or broker in the industry–university–research network. By leveraging the help of intermediaries, successful communication and collaboration among these entities can be achieved more efficiently and effectively.

CHAPTER 11

CONCLUSIONS AND DISCUSSIONS

11.1 Conclusions

In the background of glocalization, the scope, speed, and strength of cross-regional knowledge, as well as innovation element flows such as technologies, talents, and capital have reached unprecedented levels. Emerging market countries such as China are continuously integrating into the global innovation network, catching up and leapfrogging in capacity and status, and gradually becoming one of the global scientific centers and innovative highlands. To enhance competitiveness in S&T innovation, it is important to maintain technological exchange and cooperation with the outside world, of which the data on patent transactions is a good indicator of the cross-regional flow of technological elements. The effect of regional technology transfer activities on local innovative capability is an issue that needs to be explored in depth. Therefore, based on the data on patent transactions and patent applications retrieved from the Patent Information Service Platform and Wanfang database, this book establishes a theoretical framework for "global–local" interaction of technology transfer networks. It aims at integrating transnational, national, and regional scales to study the spatial evolution of multiple-scalar technology transfer networks in Chinese cities as well as their impact mechanisms and spillover effects.

11.1.1 *Periodic dynamics and diversifications*

From the perspective of technology transfer, the breadth and intensity of multiple-scalar technology transfer activities in China had shown a phased change over time, with the breadth and intensity of the technology transfer networks increasing and its fields diversifying.

During the study period, the intensity of technology transfer in China demonstrated varying degrees of growth. The introduction of technology can be broadly classified into three phases: an initial phase from 2001 to 2005, a slow-growth phase from 2007 to 2012, and a fluctuating-growth stage from 2013 to 2018. Similarly, technology export can be divided into three distinct stages: an initial stage from 2001 to 2006, a fluctuating-growth stage from 2007 to 2012, and a rapid growth stage from 2013 to 2018.

China's technology transfer network has witnessed significant growth in recent times, marked by an unmistakable hierarchy that distinguishes the intensity of technology transfer across each city. Alongside the yearly increase in the total amount of technology import in China, the breadth and depth of the technology import network have expanded, resulting in a more complex and denser network.

The technology transfer encompasses a diverse range of categories and sectors, with new additions being made constantly. In different periods, China has shown varying preferences for foreign technology import and export. From the import side, the electrical engineering sector has consistently been in a dominant position, with a rapidly growing import volume. Meanwhile, the mechanical engineering sector has shown significant improvements. When it comes to the export side, there has been a shift from instrumentation to the electrical engineering sector as the dominant sector. However, other sectors have also demonstrated significant progress in this regard.

In terms of technology diversity, relatedness, and similarity evolution, externally imported technologies are becoming increasingly varied, while their relatedness and similarity to local technologies are also on the rise. Among all cities in China, Beijing, Shenzhen, and Shanghai stand out as top performers across these three indicators, suggesting that they possess strong capabilities to absorb various types of technologies.

11.1.2 *Spatial heterogeneity and shifts*

Technology transfer activities and urban technological innovation activities follow a high degree of macroscopic spatial agglomeration effects, with technology transfer hubs combining the triple characteristics of hierarchy, spatial polarization, as well as functional balance and complementarity. The functional organization shows a pattern of evolution from functional separation in the starting stage, functional integration in the development stage to functional synergy in the maturity stage.

The technology transfer network exhibits scale-free characteristics, with a decreasing trend over time that leads to a convergence of intercity differences. As the absolute value of the power index of intercity technology introduction activities diminishes in the later stages, the heterogeneity of intercity technology absorption decreases as well. This narrowing gap in intercity technology absorption capacity results in a more balanced distribution of technology across cities. Its hinterworld forms three differentiated hinterworld growth patterns of gradual expansion (from inside to outside expansion), jump expansion (alternating between inside and outside expansion) and gradual contraction (from outside to inside expansion).

Technology transfer activities and urban technology innovation activities highly follow the macroscopic spatial agglomeration effect. Both the size of urban technology innovation and external technology transfer, as well as the concentration of central nodes in the technology transfer network are important factors.

Technology transfer activities exhibit both path-dependence and path-breakthrough characteristics within the spatial domain. As technology absorption becomes increasingly complex, the trend is shifting from dependence toward breakthrough. From the perspective of transferred technologies correlation, most cities are currently experiencing relatively low levels of both similarity and relatedness of technology transfers. However, there is a faster growth in the number of similar technology introductions, indicating a path dependence on locally available technologies that are being learned. At the same time, the scope of intercity technology transfer activities has been expanding, and the number of technology pipelines has increased to foster path breakthrough.

11.1.3 *Spillover effect*

The impact of external technologies on local innovation output depends on their relevance to the local knowledge base. When transnational technologies are both related and similar to the local knowledge base, they positively affect local innovation output. Similarly, domestic imported technologies that align with local knowledge base have a facilitating effect on innovation output.

Examining the spillover effects from the centrality of external technology networks, we find that the in-degree centrality of the transnational technology introduction networks has a boosting effect on local innovation output. However, the weighted in-degree centrality has a suppressing effect. The degree centrality, closeness centrality, and betweenness centrality of domestic technology introduction networks all facilitate local innovation output.

11.1.4 *Determinants*

The innovation milieu of cities has a significant impact on the evolution of domestic technology transfer networks. The promotion of external technology transfer is driven by several factors, including the level of economic development, investment in innovation, public funding expenditure, degree of decentralization, and marketization. Both public financial power and market competitiveness have the efficacy of dominating innovation resources, but cities with stronger financial power are more likely to become the hubs of technology transfer networks. Technological innovation capacity and the growth of the number of high-tech enterprises positively influence technology transfer, while the administrative hierarchy matching reveals a differential effect.

However, the local innovation milieu has multiple effects on transnational technology transfer. The level of local economy, the increased number of S&T employees, and the establishment of national enterprise technology centers can effectively increase the intensity of foreign technology transfer, while the number of universities can inhibit transnational technology transfer. In this way, there is a coupling dissonance between the level of higher education in cities and their S&T production capacity. The amount of goods exported by cities can promote the transfer of transnational patents, while the amount of goods imported has a negative impact.

The expansion of domestic technology channels enhances the size of transnational technology transfer. The breadth and intensity of domestic technology transfer both have significant positive promotion effects on transnational intercity technology. At the same time, the diversity domestic technology transfer also plays a significant role in enhancing innovation performance and optimizing innovation networks. The closeness of domestic intercity technology transfer is conducive to local technology export to foreign countries.

11.1.5 *Transportation effect*

Increased airline accessibility positively promotes technology innovation transfer at both the city level and the city-pair level. At the city level, the more frequent the city's aviation connections with other cities in the aviation network, the more favorable it is to promote the transfer of proprietary technology between the city and other cities. At the city-pair level, the closer the aviation frequency linkage between cities, the more favorable it is to promote technological innovation interactions between cities. The marginal effect of aviation accessibility on the transfer of technology innovation varies across different spatial scopes and exhibits a non-linear relationship with geographical distance. Our analysis reveals that the optimal radiation range falls between 1,500 and 1,800 km, suggesting that there is an ideal distance for technology innovation transfer through air travel. However, it is important to note that geographical distance continues to exert strong spatial constraints on the transfer process, highlighting the need for strategic planning and investment in overcoming these barriers.

The introduction of HSR has a positive impact on urban innovation performance, although the strength of this effect varies significantly across different regions. While transportation infrastructure has the potential to enhance innovation capacity, it can also contribute to widening disparities between regions and shape the geographical pattern of uneven development. Our analysis suggests that the opening of HSRs generates both agglomeration externalities and network externalities, which work together to promote knowledge spillover and boost urban innovation performance. However, compared to agglomeration externalities, network externalities have a more profound impact on innovation performance within urban areas.

11.2 Discussions

While this book offers an in-depth analysis of evolutionary landscapes, spillover effects, and determinants of China's technology transfer network at transnational, national, and regional scales, there are still some shortcomings to be addressed in future research.

11.2.1 *Lack of technology flows from a global eyesight*

The book lacks analysis and validation on a global scale. Obtaining global patent transfer data including detailed transfer address and technology classification remains a challenge, so this book focuses on analyzing external technology transfer networks in Chinese cities. While this approach allows for a thorough examination of the pattern-mechanism effects within China, evaluating technology transfer patterns and spillover effects on a global scale can illuminate differences in innovation levels across countries and regions. This can enable targeted recommendations for innovation development in different regions. Additionally, since China is a rapidly developing country with unique paths and characteristics in science and technology innovation, the conclusions drawn from this book may lack validation on a global scale.

11.2.2 *Lack of effect of local buzz on global pipelines*

There is also a lack of analysis on the effect of local buzz on external pipelines. As the concept of local buzz encompasses a variety of activities and behaviors, it is difficult to define and obtain data, so this book only analyzes the spillover effect of external pipelines on local innovation capacity, i.e., the global–local interaction mechanism, but lacks the impact of local buzz on global pipelines. However, this is an important part of the "global pipeline–local buzz" research framework and therefore needs to be complemented and deepened in later studies.

11.2.3 *Lack of comprehension of innovation chains*

The comprehensive analysis of multiple data on the technology market is neglected. Technological innovation encompasses multiple processes such as R&D investment, technology production, technology application, and technology diffusion, as well as multiple approaches such as patents, products, and projects. Although patent rights transaction is an effective indicator of supply and demand in the technology market, analysis from the patent data is not comprehensive enough, as a large proportion of technological innovations are not produced and flow through patent applications and transfers. To gain a better understanding of the pattern and impact of technological diffusion, it is imperative to conduct a thorough analysis of knowledge exchange and cooperation within institutions, trade in technological products, and collaboration in research projects. By doing so, we can more accurately depict the trajectory of technological transfer and development paths in various regions.

REFERENCES

Aarstad, J., Kvitastein, O. A., & Jakobsen, S.-E. (2016). Local buzz, global pipelines, or simply too much buzz? A critical study. *Geoforum, 75*, 129–133.

Abernathy, W. J., & Utterback, J. M. (1978). Patterns of innovation in technology. *Technology Review, 80*(7), 40–46.

Acs, Z., & Preston, L. (1997). Small and medium enterprises, technology, and globalization: Introducing to a special issue on small and medium-sized enterprises in the global economy. *Small Business Economics, 9*(1), 1–6.

Afuah, A. (2013). Are network effects really all about size? The role of structure and conduct. *Strategic Management Journal, 34*, 257–273.

Aghion, P., & Howitt, P. (1992). A model of growth through creative destruction. *Econometrica, 60*(2), 323–351.

Agrawal, A., Kapur, D., & McHale, J. (2008). How do spatial and social proximity influence knowledge flows? Evidence from patent data. *Journal of Urban Economics, 64*(2), 258–269.

Ahuja, G. (2000). Collaboration networks, structural holes, and innovation: A longitudinal study. *Administrative Science Quarterly, 45*(3), 425–455.

Al-Tabbaa, O., & Ankrah, S. (2016). Social capital to facilitate 'engineered' university–industry collaboration for technology transfer: A dynamic perspective. *Technological Forecasting & Social Change, 104*, 1–15.

Albors, J., Sweeney, E., & Hidalgo, A. (2005). Transnational technology transfer networks for SMEs. A review of the state-of-the art and analysis of the European IRC network. *Production Planning & Control, 16*(4), 413–423.

Amara, N., Landry, R., & Ouimet, M. (2005). Milieux innovateurs: Determinants and policy implications. *European Planning Studies, 13*(6), 939–965.

Amiti, M., Itskhoki, O., & Konings, J. (2012). Importers, exporters, and exchange rate disconnect. *Social Science Electronic Publishing, 104*(7), 1942–1978.

Andersen, T. B., & Dalgaard, C. J. (2011). Flows of people, flows of ideas, and the inequality of nations. *Journal of Economic Growth, 16*(1), 1–32.

Anderson, P., & Tushman, M. L. (1991). Technological discontinuities and dominant designs: A cyclical model of technological change. *Research-Technology Management, 34*(3), 26–31.

Ankrah, S. N., Burgess, T. F., Grimshaw, P., & Shaw, N. E. (2013). Asking both university and industry actors about their engagement in knowledge transfer: What single-group studies of motives omit. *Technovation, 33*(2), 50–65.

Antonietti, R., & Montresor, S. (2021). Going beyond relatedness: Regional diversification trajectories and key enabling technologies (KETs) in Italian regions. *Economic Geography, 97*(2), 187–207.

Archibugi, D., & Michie, J. (1995). The globalisation of technology: A new taxonomy. *Cambridge Journal of Economics, 19*(1), 121–140.

Armanios, D. E., Eesley, C. E., Li, J., & Eisenhardt, K. M. (2017). How entrepreneurs leverage institutional intermediaries in emerging economies to acquire public resources. *Strategic Management Journal, 38*, 1373–1390.

Arora, A., Belenzon, S., Patacconi, A., & Suh, J. (2020). The changing structure of American innovation: Some cautionary remarks for economic growth. *Innovation Policy and the Economy, 20*(1), 39–93.

Arthur, W. B. (1988). Self-reinforcing mechanisms in economics. In Anderson, P., Arrow, K., & Pines, D. (Eds.) *The Economy as an Evolving, Complex System.* Reading, Mass: Addison-Wesley, 9–31.

Arthur, W. (1989). Competing technologies, increasing returns, and lock-in by historical events. *Economic Journal, 99*(394), 116–131.

Asheim, B. (1996). Industrial districts as learning regions: A condition for prosperity. *European Planning Studies, 4*, 379–400.

Asheim, B., & Coenen, L. (2005). Knowledge bases and regional innovation systems: Comparing Nordic clusters. *Research Policy, 34*(8), 1173–1190.

Asheim, B. T., & Isaksen, A. (2002). Regional innovation systems: The integration of local 'sticky' and global 'ubiquitous' knowledge. *The Journal of Technology Transfer, 27*(1), 77–86.

Asheim, B., Coenen, L., & Vang, J. (2007b). Face-to-face, buzz, and knowledge bases: Sociospatial implications for learning, innovation, and innovation policy. *Environment and Planning C, 25*(5), 655–670.

Asheim, B. T., Smith, H. L., & Oughton, C. (2011). Regional innovation systems: Theory, empirics and policy. *Regional Studies, 45*(7), 875–891.

Asheim, B., Coenen, L., Moodysson, J., & Vang, J. (2007a). Constructing knowledge-based regional advantage: Implications for regional innovation policy. *International Journal of Entrepreneurship and Innovation Management, 7*(2–5), 140–155.

Aslesen, H., Hydle, K., & Wallevik, K. (2017). Extra-regional linkages through MNCs in organizationally thick and specialized RISs: A source of new path development? *European Planning Studies, 25*(3), 443–461.

Atkinson, R. D. (2012). Why the current education reform strategy won't work to improve innovation and boost the economy, the nation needs a fundamentally new approach to education in science, technology, engineering, and mathematics. *Issues In Science and Technology, 28*(3), 29–36.

Audretsch, D. B., & Feldman, M. P. (1996a). R&D spillovers and the geography of innovation and production. *American Economic Review, 86*(3), 630–640.

Audretsch, D. B., & Feldman, M. P. (1996b). Innovative clusters and the industry life-cycle. *Review of Industrial Organization, 11*, 253–273.

Audretsch, D. B., & Feldman, M. P. (2004). Knowledge spillovers and the geography of innovation. *Handbook of Regional and Urban Economics, 4*, 2713–2739.

Audretsch, D. B., & Keilbach, M. (2004). Entrepreneurship capital and economic performance. *Regional Studies, 38*(8), 949–959.

Audretsch, D. B., Lehmann, E. E., & Wright, M. (2014). Technology transfer in a global economy. *The Journal of Technology Transfer, 39*, 301–312.

Audretsch, D. B., Lehmann, E. E., Paleari, S., & Vismara, S. (2016). Entrepreneurial finance and technology transfer. *The Journal of Technology Transfer, 41*(1), 1–9.

Aula, P., & Harmaakorpi, V. (2006). An innovative milieu: A view on regional reputation building: Case study of the Lahti urban region. *Regional Studies, 42*(4), 523–538.

Autio, E. (1998). Evaluation of RTD in regional systems of innovation. *European Planning Studies, 6*(2), 131–140.

Awate, S., Larsen, M. M., & Mudambi, R. (2015). Accessing vs sourcing knowledge: A comparative study of R&D internationalization between emerging and advanced economy firms. *Journal of International Business Studies, 46*(1), 63–86.

Baglieri, D., & Lorenzoni, G. (2014). Closing the distance between academia and market: Experimentation and user entrepreneurial processes. *The Journal of Technology Transfer, 39*(1), 52–74.

Balland, P.-A. (2012). Proximity and the evolution of collaboration networks: Evidence from research and development projects within the global navigation satellite system (GNSS) industry. *Regional Studies, 46*(6), 741–756.

Balland, P.-A., & Rigby, D. (2017). The geography of complex knowledge. *Economic Geography, 93*(1), 1–23.

Balland, P.-A., Boschma, R., Crespo, J., & Rigby, D. (2019). Smart specialization policy in the European Union: Relatedness, knowledge complexity and regional diversification. *Regional Studies, 53*(9), 1252–1268.

Balland, P. A., Jara-Figueroa, C., Petralia, S. G., Steijn, M., & Hidalgo, C. A. (2020). Complex economic activities concentrate in large cities. *Nature Human Behaviour, 4*(3), 1–7.

Barabasi, A. L., & Albert, R. (1999). Emergence of scaling in random networks. *Science, 286*(5439), 509–512.

Barnard, H., & Chaminade, C. (2011). *Global Innovation Networks: Towards a Taxonomy. Centre for Innovation, Research and Competence in the Learning Economy (CIRCLE).* Lund University. Retrieved from https://www.circle.lu.se/publications/.

Barrat, A., Barthelemy, M., Pastor-Satorras, R., & Vespignani, A. (2004). The architecture of complex weighted networks. *Proceedings of the National Academy Sciences of the United States of America, 101*(11), 3747–3752.

Basile, R., Capello, R., & Caragliu, A. (2012). Technological interdependence and regional growth in Europe: Proximity and synergy in knowledge spillovers. *Papers in Regional Science, 91*(4), 697–722.

Bassis, N. F., & Armellini, F. (2018). Systems of innovation and innovation ecosystem: A literature review in search of complementarities. *Journal of Evolutionary Economics, 28*, 1053–1080.

Basu, S., & Weil, D. N. (1998). Appropriate technology and growth. *The Quarterly Journal of Economics, 113*(4), 1025–1054.

Bathelt, H. (2005). Geographies of production: Growth regimes in spatial perspective (II)-knowledge creation and growth in clusters. *Progress in Human Geography, 29*(2), 204–216.

Bathelt, H. (2007). Buzz-and-pipeline dynamics: Towards a knowledge-based multiplier model of clusters. *Geography Compass, 1*(6), 1282–1298.

Bathelt, H. (2008). Knowledge-based clusters: Regional multiplier models and the role of "buzz" and "pipelines". In Karlsson, C. (Ed.) *Handbook of Research on Cluster Theory*. Cheltenham, UK: Edward Elgar, 78–92.

Bathelt, H., & Boggs, J. S. (2003). Toward a reconceptualization of regional development paths: Is Leipzig's media cluster a continuation of or a rupture with the past? *Economic Geography, 79*, 265–293.

Bathelt, H., & Glückler, J. (2003). Toward a relational economic geography. *Journal of Economic Geography, 3*, 117–144.

Bathelt, H., & Glückler, J. (2011). *The Relational Economy: Geographies of Knowing and Learning*. Oxford: Oxford University Press, 1–320.

Bathelt, H., & Henn, S. (2014). The geographies of knowledge transfers over distance: Toward a typology. *Environment and Planning A, 46*(6), 1403–1424.

Bathelt, H., & Li, P. (2014). Global cluster network: Foreign direct investment flows from Canada to China. *Journal of Economic Geography, 14*(1), 45–71.

Bathelt, H., & Li, P. (2020). Processes of building cross-border knowledge pipelines. *Research Policy, 49*(3), 103928.

Bathelt, H., & Li, P. (2021). Relational and evolutionary economic geography. In Fischer, M. M., & Nijkamp, P. (Eds.) *Handbook of Regional Science*. Berlin, Heidelberg: Springer, 1247–1264.

Bathelt, H., & Schuldt, N. (2008). Between luminaires and meat grinders: International trade fairs as temporary clusters. *Regional Studies, 42*(6), 853–868.

Bathelt, H., Cantwell, J., & Mudambi, R. (2018). Overcoming frictions in transnational knowledge flows: Challenges of connecting, sense-making and integrating. *Journal of Economic Geography, 18*, 1001–1022.

Bathelt, H., Kogler, D. F., & Munro, A. K. (2011). Social foundations of regional innovation and the role of university spin-offs: The case of Canada's Technology Triangle. *Industry and Innovation, 18*(5), 461–486.

Bathelt, H., Malmberg, A., & Maskell, P. (2004). Clusters and knowledge: Local buzz, global pipelines and the process of knowledge creation. *Progress in Human Geography, 28*(1), 31–56.

Batty, M., & Longley, P. A. (1994). *Fractal Cities: A Geometry of Form and Function*. London: Academic Press.

Baum, J. C., & Dutton, J. E. (1996). Introduction: The embeddedness of strategy. *Strategic Management Journal*, 20, 251–278.

Beaudry, C., & Schiffauerova, A. (2009). Who's right, Marshall or Jacobs? The localization versus urbanization debate. *Research Policy, 38*(2), 318–337

Bellais, R., & Guichard, R. (2006). Defense innovation, technology transfer and public policy. *Defence & Peace Economics, 17*(3), 273–286.

Benitez, G. B., Ayala, N. F., & Frank, A. G. (2020). Industry 4.0 innovation ecosystems: An evolutionary perspective on value cocreation. *International Journal of Production Economics, 228,* 107725.

Berends, H., & Reymen, I. (2011). External designers in product design processes of small manufacturing firms. *Design Studies, 32*(1), 86–108.

Bergman, E. M., & Maier, G. (2009). Network central: Regional positioning for innovative advantage. *The Annals of Regional Science, 43*(3), 615–644.

Bessant, J., & Rush, H. (1995). Building bridges for innovation: The role of consultants in technology transfer. *Research Policy, 24,* 97–114.

Bidault, F., & Fischer, W. A. (1994). Technology transactions: Network over markets. *R&D Management, 24*(4), 373–386.

Bilbao-Osorio, B., & Rodríguez-Pose, A. (2004). From R&D to innovation and economic growth in the EU. *Growth and Change, 35*(4), 434–455.

Binz, C., & Truffer, B. (2017). Global innovation systems: A conceptual framework for innovation dynamics in transnational contexts. *Research Policy, 46*(7), 1284–1298.

Binz, C., Truffer, B., & Coenen, L. (2014). Why space matters in technological innovation systems: Mapping global knowledge dynamics of membrane bioreactor technology. *Research Policy, 43*(1), 138–155.

Bliznets, I. A., Kartskhiya, A. A., & Smirnov, M. G. (2018). Technology transfer in digital era: Legal environment. *Journal of History Culture and Art Research, 7*(1), 354–363.

Bloom, P. N., & Dees, G. (2008). Cultivate your ecosystem. *Stanford Social Innovation Review, 6*(1), 45–53.

Borgatti, S. P., Mehra, A., Brass, D. J., & Labianca, G. (2009). Network analysis in the social sciences. *Science, 323*(5916), 892–895.

Boschma, R. A. (1997). New industries and windows of locational opportunity: A long-term analysis of Belgium. *Erdkunde, 51*(1), 1–19.

Boschma, R. (1999). The rise of clusters of innovative industries in Belgium during the industrial epoch. *Research Policy, 28,* 853–871.

Boschma, R. (2005). Proximity and innovation: A critical assessment. *Regional Studies, 39*(1), 61–74.

Boschma, R. (2017). Relatedness as driver of regional diversification: A research agenda. *Regional Studies, 51*(3), 351–364.

Boschma, R., & Capone, G. (2016). Relatedness and diversification in the European Union (EU-27) and European neighbourhood policy countries. *Environment and Planning C, 34,* 617–637.

Boscham, R., & Frenken, K. (2006). Why is economic geography not an evolutionary science? Towards an evolutionary economic geography. *Journal of Economic Geography, 6*(3), 273–302.

Boschma, R., & Frenken, K. (2010). The spatial evolution of innovation networks: A proximity perspective. In Boschma, R., & Martin, R. (Ed.) *The Handbook of Evolutionary Economic Geography*. Cheltenham: Edward Elgar, 120–135.

Boschma, R., & Iammarino, S. (2009). Related variety, trade linkages, and regional growth in Italy. *Economic Geography, 85*(3), 289–311.

Boschma, R., & Lambooy, J. G. (1999). Evolutionary economics and economic geography. *Journal of Evolutionary Economics, 9*(4), 411–429.

Boschma, R., & Ter Wal, A. L. J. (2007). Knowledge networks and innovative performance in an industrial district: The case of a footwear district in the South of Italy. *Industry and Innovation, 14*(2), 177–199

Boschma, R., Heimeriks, G., & Balland, P. A. (2014). Scientific knowledge dynamics and relatedness in biotech cities. *Research Policy, 43*(1), 107–114.

Boschma, R., Minondo, A., & Navarro, M. (2012). Related variety and regional growth in Spain. *Papers in Regional Science, 91*(2), 241–256.

Bottazzi, L., & Peri, G. (2003). Innovation and spillovers in regions: Evidence from European patent data. *European Economic Review, 47*(4), 687–710.

Bozeman, B. (2000). Technology transfer and public policy: A review of research and theory. *Research Policy, 29*(4–5), 627–655.

Bradshaw, M. (2001). Multiple proximities: Culture and geography in the transport logistics of newsprint manufactured in Australia. *Environment and Planning A, 33*(10), 1717–1739.

Brennecke, J., & Rank, O. (2017). The firm's knowledge network and the transfer of advice among corporate inventors — A multilevel network study. *Research Policy, 46*, 768–783.

Breschi, S., & Lenzi, C. (2015). The role of external linkages and gatekeepers for the renewal and expansion of US cities' knowledge base, 1990–2004. *Regional Studies, 49*(5), 782–797.

Breschi, S., & Malerba, F. (1997). Sectoral innovation systems: Technological regimes, Schumpeterian dynamics, and spatial boundaries. *Systems of Innovation: Technologies, Institutions and Organizations, 1*, 130–156.

Breschi, S., Lissoni, F., & Malerba, F. (2003). Knowledge–relatedness in firm technological diversification. *Research Policy, 32*(1), 69–87.

Brescia, F., Colombo, G., & Landoni, P. (2014). Organizational structures of knowledge transfer offices: An analysis of the world top-ranked universities. *The Journal of Technology Transfer, 41*, 132–151.

Bresnahan, T., Gambardella, A., & Saxenian, A. (2001). 'Old economy' inputs for 'new economy' outcomes: Cluster formation in the new Silicon Valleys. *Industrial and Corporate Change, 10*(4), 835–860.

Broekel, T., & Boschma, R. (2012). Knowledge networks in the Dutch aviation industry: The proximity paradox. *Journal of Economic Geography, 12*(2), 409–433.

Broekel, T., & Brachert, M. (2015). The structure and evolution of intersectoral technological complementarity in R&D in Germany from 1990 to 2011. *Journal of Evolutionary Economics, 25*, 755–785.

Broekel, T., Balland, P. A., Burger, M., & van Oort F. (2014). Modeling knowledge networks in economic geography: A discussion of four methods. *The Annals of Regional Science, 53*(2), 423–452.

Broekel, T., Lazzeretti, L., Capone, F., & Hassink, R. (2021). Rethinking the role of local knowledge networks in territorial innovation models. *Industry and Innovation, 28*(7), 805–814.

Brown, J. S., & Duguid, P. (2002). Local knowledge: Innovation in the networked age. *Management Learning, 33*(4), 427–437.

Brown, L. A. (1981). *Innovation Diffusion.* London: Methuen & Co. Ltd.

Brown, L. A., & Cox, K. R. (1971). Empirical regularities in the diffusion of innovation. *Annals of the Association of American Geographers, 61*, 551–559.

Bruneel, J., D'Este, P., & Salter, A. (2010). Investigating the factors that diminish the barriers to university-industry collaboration. *Research Policy, 39*(7), 858–868.

Buchmann, T., & Pyka, A. (2015). The evolution of innovation networks: The case of a publicly funded German automotive network. *Economics of Innovation and New Technology, 24*(1–2), 114–139.

Buesa, M., Heijs, J., Pellitero, M. M., & Baumert, T. (2006). Regional systems of innovation and the knowledge production function: The Spanish case. *Technovation, 26*(4), 463–472.

Bum, C. D., & Jung, D. D. (2013). The effects of performance management & application capabilities and activities on technology transfer from public research institutes in Korea. *Journal of Technology Innovation, 21*(2), 199–224.

Burhop, C., & Wolf, N. (2013). The German market for patents during the "Second Industrialization" 1884–1913: A gravity approach. *Business History Review, 87*(1), 69–93.

Burt, R. S. (1992). *Structural Holes: The Social Structure of Competition.* Cambridge, MA: Harvard University Press.

Burt, R. S. (2000). The network structure of social capital. *Research in Organizational Behavior, 22*, 345–423.

Burt, R. S. (2005). *Brokerage and Closure: An Introduction to Social Capital.* Oxford, UK: Oxford University Press.

Cai, H., Feng, Z., Zhou, W., Chen, J., & Chen, Z. (2022). Understanding the spatial polarization pattern of technology transfer network in the Guangdong–Hong Kong–Macao Greater Bay area. *Growth and Change*, 1–22.

Camagni, R. (1991). *Innovation Networks: Spatial Perspectives.* London: Belhaven Press, 1–9, 121–144.

Camagni, R. P. (1995). The concept of innovative milieu and its relevance for public policies in European lagging regions. *Papers in Regional Science, 74*(4), 317–340.

Cameron, A. C., & Trivedi, P. K. (1986). Econometric models based on count data. Comparisons and applications of some estimators and tests. *Journal of Applied Economics, 1*(1), 29–53.

Cannarella, C., & Piccioni, V. (2008). Innovation diffusion and architecture and dynamics of local territorial networks. *Trames, 12*(62/57), 2, 215–237.

Cao, Z., Derudder, B., Dai, L., & Peng, Z. (2022). "Buzz-and-pipeline" dynamics in Chinese science: The impact of interurban collaboration linkages on cities' innovation capacity. *Regional Studies, 56*(2), 290–306.

Capello, R. (1999). Spatial transfer of knowledge in high technology milieux: Learning versus collective learning processes. *Regional Studies, 33*(4), 353–365.

Caragliu, A., & Nijkamp, P. (2016). Space and knowledge spillovers in European regions: The impact of different forms of proximity on spatial knowledge diffusion. *Journal of Economic Geography, 16*(3), 749–774.

Carayannis, E. G., & Campbell, D. F. J. (2009). "Mode 3" and "Quadruple Helix": Toward a 21st century fractal innovation ecosystem. *International Journal of Technology Management, 46*(3–4), 201–234.

Carayannis, E. G., & Campbell, D. F. J. (2010). Triple Helix, Quadruple Helix and Quintuple Helix and how do knowledge, innovation and the environment relate to each other? A proposed framework for a trans-disciplinary analysis of sustainable development and social ecology. *International Journal of Social Ecology and Sustainable Development, 1*(1), 41–69.

Carlsson, B. (2006). Internationalization of innovation systems: A survey of the literature. *Research Policy, 35*(1), 56–67.

Carlsson, B., & Jacobsson S., Holmén, M., & Rickne, A. (2002). Innovation systems: Analytical and methodological issues. *Research Policy*, 31, 233–245.

Casetti, E., & Semple, R. K. (1969). Concerning the testing of spatial diffusion hypotheses. *Geographical Analysis*, *1*(1), 9–154.

Cassi, L., & Plunket, A. (2014). Proximity, network formation and inventive performance: In search of the proximity paradox. *The Annals of Regional Science*, *53*(2), 1–28.

Castaldi, C., Frenken, K., & Los, B. (2015). Related variety, unrelated variety and technological breakthroughs: An analysis of US state-level patenting. *Regional Studies*, *49*(5), 767–781.

Castells M. (1996). *The Rise of the Network Society*. Cambridge, MA: Blackwell.

Catalini, C., Fons-Rosen, C., & Gaule, P. (2016). *Did cheaper flights change the geography of scientific collaboration?* Sloan Working Papers. Cambridge, USA: MIT Sloan School of Management.

Chaminade, C., & Vang, J. (2008). Globalisation of knowledge production and regional innovation policy: Supporting specialized hubs in the Bangalore software industry. *Research Policy*, *37*, 1684–1696.

Chaminade, C., De Fuentes, C., Harirchi, G., & Plechero, M. (2015). The geography and structure of global innovation networks: Global scope and regional embeddedness. *Papers in Innovation Studies*. Centre for Innovation Research at Lund University.

Chen, C. (2012). Reshaping Chinese space-economy through high-speed trains: Opportunities and challenges. *Journal of Transport Geography*, *22*(5), 312–316.

Chen, C., & Vickerman, R. (2017). Can transport infrastructure change regions' economic fortunes? Some evidence from Europe and China. *Regional Studies*, *51*(1), 144–160.

Chen, Y., Wang, C., Chen, Y., Lo, W., & Chen, K. (2019). Influence of network embeddedness and network diversity on green innovation: The mediation effect of green social capital. *Sustainability*, *11*, 5736.

Cheng, K., & Wang, Y. (2013). Mechanism and evidence of the promotion of city network on technological innovation. *Studies in Science of Science*, *31*(9), 1399–1411, 1440. (in Chinese)

Chesbrough, H. (2003). The era of open innovation. *MIT Sloan Management Review*, *44*(3), 35–41.

Chew, Y. T., & Yeung, H. W. (2001). The SME advantage: Adding local touch to foreign transnational corporations in Singapore. *Regional Studies*, *35*(5), 431–448.

Chiambaretto, P., Massé, D., & Mirc, N. (2019). All for one and one for all? — Knowledge broker roles in managing tensions of internal cooperations: The Ubisoft case. *Research Policy, 48*(3), 584–600.

Choi, H., Kim, S.-H., & Lee, J. (2010). Role of network structure and network effects in diffusion of innovations. *Industrial Marketing Management, 39*, 170–177.

Clayton, P. A., Feldman, M., & Lowe, N. (2018). Behind the scenes: Intermediary organisations that facilitate science commercialization through entrepreneurship. *Academy of Management Perspectives, 32*(1), 104–124.

Coccia, M. (2008). Spatial mobility of knowledge transfer and absorptive capacity: analysis and measurement of the impact within the geoeconomic space. *The Journal of Technology Transfer, 33*, 105–122.

Coenen, L., Moodysson, J., & Asheim, B. T. (2004). Nodes, networks and proximities: On the knowledge dynamics of the Medicon Valley Biotech Cluster. *European Planning Studies, 12*(7), 1003–1018.

Cohen, W. M., & Levinthal, D. A. (1990). Absorptive capacity: A new perspective on learning and innovation. *Administrative Science Quarterly, 35*, 128–152.

Cohendet, P., Grandadam, D., & Simon, L. (2018). The local, the global and the industry common: The case of the video game industry. *Journal of Economic Geography, 18*(5), 1045–1068.

Collins, H. (2012). *Tacit and Explicit Knowledge.* Chicago: The University of Chicago Press.

Comunian, R. (2017). Temporary clusters and communities of practice in the creative economy: Festivals as temporary knowledge networks. *Space and Culture, 20*(3), 329–343.

Cooke, P. N. (1992). Regional innovation systems: Competitive regulation in the new Europe. *Geoforum, 23*(3), 365–382.

Cooke, P. N. (1996). The new wave of regional innovation networks: Analysis, characteristics and strategy. *Small Business Economics, 8*(2), 159–171.

Cooke, P. N. (2014). Systems of innovation and the learning region. In Fischer, M. M., & Nijkamp, P. (Eds.) *Handbook of Regional Science.* Berlin: Springer-Verlag, 457–474.

Cooke, P. N., & Morgan, K. (1998a). *The Associational Economy: Firms, Regions, and Innovation.* New York: Guilford Press.

Cooke, P. N., & Morgan, K. (1998b). Wales: Innovating through global–local interaction. In Cooke, P. N., & Morgan, K. *The Associational Economy: Firms, Regions, and Innovation.* New York: Guilford Press, 134–161.

Cooke, P. N., Heidenreich, M., & Braczyk, H. (2004). *Regional Innovation Systems: The Role of Governance in a Globalised World.* London: Routledge.

Cooke, P. N., Uranga M., & Etxebarria, G. (1997). Regional innovation systems: Institutional and organisational dimensions. *Research Policy, 26*(4/5), 475–491.

Corpataux, J., Crevoisier, O., & Theurillat, T. (2009). The expansion of the finance industry and its impact on the economy: A territorial approach based on Swiss pension funds. *Economic Geography, 85*(3), 313–334.

Cortinovis, N., & Oort, F. V. (2019). Between spilling over and boiling down: Network-mediated spillovers, local knowledge base and productivity in European regions. *Journal of Economic Geography, 19*(6), 1233–1260.

Crescenzi, R., Nathan, M., & Rodríguez-Pose, A. (2016). Do inventors talk to strangers? On proximity and collaborative knowledge creation. *Research Policy, 45*(1), 177–194.

Crescenzi, R., Rodríguez-Pose, A., & Storper, M. (2007). The territorial dynamics of innovation: A Europe-United States comparative analysis. *Journal of Economic Geography, 7*(6), 673–709.

Crevoisier, O. (2004). The innovative milieus approach: Toward a territorialized understanding of the economy? *Economic Geography, 80*(4), 367–379.

Crevoiser, O., & Maillat, D. (1991). Milieu, industrial organization and territorial production system: Towards a new theory of spatial development. In Camagni R. (Ed.) *Innovation Networks: Spatial Perspectives.* London: Belhaven Press.

Csomós, G., & Tóth, G. (2016). Exploring the position of cities in global corporate research and development: A bibliometric analysis by two different geographical approaches. *Journal of Informetrics, 10*(2), 516–532.

Cunningham, J. A., & O'Reilly, P. (2018). Macro, meso and micro perspectives of technology transfer. *The Journal of Technology Transfer, 43*(3), 545–557.

Cunningham, J. A., Lehmann, E. E., Menter, M., & Seitz, N. (2019). The impact of university-focused technology transfer policies on regional innovation and entrepreneurship. *The Journal of Technology Transfer, 44*(5), 1451–1475.

da Silva, V. L., Kovaleski, J. L., & Pagani, R. N. (2022). Fundamental elements in technology transfer: An in-depth analysis. *Technology Analysis and Strategic Management, 34*(2), 223–244.

Dauth, W., & Suedekum, J. (2016). Globalization and local profiles of economic growth and industrial change. *Journal of Economic Geography, 16*(5), 1007–1034.

David, P. A. (1985). Clio and the economics of QWERTY. *American Economic Review, 75*(2), 332–337.

David, P. A. (1993). Path-dependence and predictability in dynamic systems with local network externalities: A paradigm for historical economics. In Foray, D. and Freeman, C. (Eds.) *Technology and the Wealth of Nations.* London: Pinter Publishers, 208–231.

De la Mothe, J., & Foray, D. (2001). *Knowledge Management in the Innovation Process.* Boston: Kluwer.

De Silva, M., Howells, J., & Meyer, M. (2018). Innovation intermediaries and collaboration: Knowledge-based practices and internal value creation. *Research Policy, 47*(1), 70–87.

Deng, H., Lu, L., & Yu, H. (2018). Spatial evolution of air–rail integrated transport hub area: The case of Shanghai HongQiao hub. *Human Geography, 33*(4), 130–136. (in Chinese)

Deng, P., & Lu, H. (2022). Transnational knowledge transfer or indigenous knowledge transfer: Which channel has more benefits for China's high-tech enterprises? *European Journal of Innovation Management, 25*(2), 433–453.

Desmarchelier, B., & Zhang, L. (2018). Innovation networks and cluster dynamics. *The Annals of Regional Science, 61*, 553–578.

Dicken, P. (1994). The Roepke lecture in economic geography global–local tensions: Firms and states in the global space-economy. *Economic Geography, 70*(2), 101–128.

Dicken, P., & Malmberg, A. (2001). Firms in territories: A relational perspective. *Economic Geography, 77*(4), 345–363.

Ding, J., Jin F., Wang J., & Liu, D. (2013). Competition game of high-speed rail and civil aviation and its spatial effect: A case study of Beijing-Shanghai high-speed rail. *Economic Geography, 33*(5), 104–110. (In Chinese).

Dobusch, L., & Schüßler, E. (2012). Theorizing path dependence: A review of positive feedback mechanisms in technology markets, regional clusters, and organizations. *Industrial and Corporate Change, 22*(3), 617–647.

Doloreux, D. (2004). Regional networks of small and medium-sized enterprises: Evidence from the metropolitan area of Ottawa in Canada. *European Planning Studies, 12*(2), 173–189.

Dong, X., Zheng, S., & Kahn, M. E. (2020). The role of transportation speed in facilitating high skilled teamwork across cities. *Journal of Urban Economics, 115*(1), 212–218.

Dosi, G. (1984). *Technical Change and Industrial Transformation.* London: MacMillan.

Driffield, N., Love, J., & Taylor, K. (2009). Productivity and labour demand effects of inward and outward FDI on UK industry. *The Manchester School, 77*(2), 171–203.

Du, D. (2015). *Global Science and Technology Innovation Centers: Dynamics and Modes.* Shanghai: Shanghai People's Publishing House. (in Chinese)

Du, D., & Duan, D. (2021). *China–US Science and Technology Competitiveness Assessment Report (2020).* New Jersey: World Scientific.

Du, D., Duan, D., Liu C., *et al.* (2022). *Global S&T Innovation Center Index 2022.* https://wap.peopleapp.com/article/6934209/6794921.

Du, H., & Hu, B. (2015). The tempo-spatial evolution pattern of global scientific paper output and international science cooperation. *China Soft Science,* (2), 35–43. (in Chinese)

Du, X., & Peng, M. (2017). Do high-speed trains motivate the flow of corporate highly educated talents? *Business Management Journal, 39*(12), 89–107. (in Chinese)

Duan, D. (2020). *The Spatial Evolution of Technology Transfer in China City System.* Beijing: China Social Science Press. (in Chinese)

Duan, D., Du, D., & Chen, Y. (2018). Technology transfer in China's city system: Process, pattern and influencing factors. *Acta Geographica Sinica, 73*(4), 738–754. (in Chinese)

Duan, D., Du, D., & Grimes, S. (2019). The faster the better? Economic effects of the speed of inter-city technology transfer in China. *Growth and Change, 50*(3), 1085–1101.

Duan, D., Zhang, Y., Chen, Y., & Du, D. (2019). Regional integration in the inter-city technology transfer system of the Yangtze River Delta, China. *Sustainability, 11*(10), 2941.

Duan, Y., Nie, W., & Coakes, E. (2010). Identifying key factors affecting transnational knowledge transfer. *Information & Management, 47*(7), 356–363.

Ducruet, C. (2017). Multilayer dynamics of complex spatial networks: The case of global maritime flows (1977–2008). *Journal of Transport Geography, 60,* 47–58.

Ducruet, C., & Notteboom, T. (2012). The worldwide maritime network of container shipping: Spatial structure and regional dynamics. *Global Networks, 12,* 395–423.

Dunning, J. (1977). Trade, location of economic activity and the MNE: A search for an eclectic approach. In Ohlin, B. & Hesselborn, P.-O. & Wijkman, P. M. (Eds.) *The International Allocation of Economic Activity.* London: Palgrave Macmillan, 395–418.

Dunning, J. (1981). *International Production and the Multinational Enterprise.* London: George Allen & Unwin, 52–69.

Dunning, J. (2002). *Regions, Globalization, and the Knowledge-based Economy.* Oxford: Oxford University Press.

Dyba, W., Stryjakiewicz, T., & De Marchi, V. (2020). Knowledge sourcing and cluster life cycle: A comparative study of furniture clusters in Italy and Poland. *European Planning Studies, 28*(10), 1979–1998.

D'Adda, D., Iacobucci, D., & Palloni, R. (2020). Relatedness in the implementation of Smart Specialisation Strategy: A first empirical assessment. *Papers in Regional Science, 99*(3), 405–425.

D'Agostino, G., & Scarlato, M. (2015). Innovation, socio-institutional conditions and economic growth in the Italian regions. *Regional Studies, 49*(9), 514–534.

D'Este, P., & Perkmann, M. (2011). Why do academics engage with industry? The entrepreneurial university and individual motivations. *The Journal of Technology Transfer, 36,* 316–339.

D'Este, P., Guy, F., & Iammarino, S. (2013). Shaping the formation of university-industry research collaborations: what type of proximity does really matter? *Journal of Economic Geography, 13*(4), 537–558.

Eden, L., Levitas, E., & Martinez, R. J. (1997). The production, transfer and spillover of technology: Comparing large and small multinationals as technology producers. *Small Business Economics, 9,* 53–66.

Edquist C. (1997). *Systems of Innovation: Technologies, Institutions and Organizations.* London: Pinter, 1–35.

Elkan, R. V. (1996). Catching up and slowing down: Learning and growth patterns in an open economy. *Journal of International Economics, 41*(1/2), 95–111.

Engel, J. S. (2015). Global clusters of innovation: Lessons from Silicon Valley. *California Management Review, 57*(2), 36–65.

Epicoco, M. (2021). Technological revolutions and economic development: Endogenous and exogenous fluctuations. *Journal of the Knowledge Economy, 12,* 1437–1461.

Ernst, D. (2002). Global production networks and the changing geography of innovation systems. Implications for developing countries. *Economics of Innovation & New Technology, 11*(6), 497–523.

Ernst, D. (2007). Beyond the "Global Factory" model: Innovative capabilities for upgrading China's IT industry. *International Journal of Technology & Globalisation, 3*(4), 437–459.

Ernst, D. (2009). A new geography of knowledge in the electronics industry? Asia's role in global innovation networks. *Policy Studies, 11,* 54–65.

Esposito, C. R., & Rigby, D. L. (2019). Buzz and pipelines: The costs and benefits of local and nonlocal interaction. *Journal of Economic Geography, 19*, 753–773.

Essleztbichler, J. (2015). Relatedness, industrial branching and technological cohesion in US metropolitan areas. *Regional Studies, 49*(5), 752–766.

Etzkowitz, H. (2008). *The Triple Helix: University–Industry–Government Innovation in Action.* London and New York: Routledge, 15.

Etzkowitz, H., & Leydesdorff, L. (1995). The Triple Helix: University–industry–government relations: A laboratory for knowledge based economic development. *EASST Review, 14* (1), 11–19.

Etzkowitz, H., & Leydesdorff, L. (2000). The dynamics of innovation: From national systems and "Mode 2" to a Triple Helix of university-industry-government relations. *Research Policy, 29*, 109–123.

Evangelista, R., Iammarino, S., Mastrostefano, V., & Silvani, A. (2002). Looking for regional systems of innovation: Evidence from the Italian innovation survey. *Regional Studies, 36*(2), 173–186.

Fagerberg, J. (1994). Technology and international differences in growth rates. *Journal of Economic Literature, 32*, 1147–1175.

Fan, G., & Wang, X. (2001). *China Marketization Index.* Beijing: Economic Science Press. (in Chinese)

Fang, Y., & Yi, J. (2014). Study on evolution mechanism of industrial cluster based on Brusselator Model. *Applied Mechanics & Materials, 46*(11), 687–691.

Farris, G. F. (2017). Research on innovation management and technology transfer in China. *The Journal of Technol Transfer, 32*, 123–126.

Feldman, M. P. (1994). *The Geography of Innovation.* Boston: Kluwer Academic Publishers.

Feldman, M. P. (1999). The new economics of innovation, spillovers and agglomeration: A review of empirical studies. *Economics of Innovation & New Technology, 8*(1), 5–25.

Feldman, M. P., Kogler, D. F., & Rigby, D. L. (2005). rKnowledge: The spatial diffusion and adoption of rDNA methods. *Regional Studies, 49*(5), 798–817.

Fernández, J., & Gavilanes, J. C. (2017). Learning-by-importing in emerging innovation systems: Evidence from Ecuador. *The Journal of International Trade & Economic Development, 26*(1), 45–64.

Ferraro, G., & Iovanella, A. (2016). Revealing correlations between structure and innovation attitude in inter-organisational innovation networks. *International Journal of Computational Economics and Econometrics, 6*(1), 93–113.

Ferraro, G., & Iovanella, A. (2017). Technology transfer in innovation networks: An empirical study of the Enterprise Europe Network. *International Journal of Engineering Business Management, 9,* 1847979017735748.

Fini, R., Rasmussen, E., Siegel, D., & Wilkund, J. (2018). Rethinking the commercialization of public science: From entrepreneurial outcomes to societal impacts. *Academy of Management Perspectives, 32*(1), 4–20.

Florida, R. (1995). Toward the learning region. *Futures, 27,* 527–536.

Florida, R., & Kenney, M. (1988). Venture capital, high technology, and regional-development. *Regional Studies, 22,* 33–48.

Foray, D. (2004). *Economics of Knowledge.* Cambridge: MIT Press.

Fracasso, A., & Marzetti, G. V. (2014). International R&D spillovers, absorptive capacity and relative backwardness: A panel smooth transition regression model. *International Economic Journal, 28*(1), 137–160.

Franco, M., & Haase, H. (2015). University–industry cooperation: Researchers' motivations and interaction channels. *Journal of Engineering and Technology Management, 36,* 41–51.

Freeman, C. (1987). *Technology Policy and Economic Performance: Lessons from Japan.* London: Frances Pinter Publishers.

Freeman, L. C. (1978). Centrality in social networks' conceptual clarification. *Social Networks, 1*(3), 215–239.

Frenken, K., & Boschma, R. (2007). A theoretical framework for evolutionary economic geography: Industrial dynamics and urban growth as a branching process. *Papers in Evolutionary Economic Geography, 7*(5), 635–649.

Frenken, K., Van Oort, F., & Verburg, T. (2007). Related variety, unrelated variety and regional economic growth. *Regional Studies, 41*(5), 685–697.

Fritsch, M., & Storey, D. J. (2014). Entrepreneurship in a regional context: Historical roots, recent developments and future challenges. *Regional Studies, 48*(6), 939–954.

Fromhold-Eisebith, M. (2004). Innovative milieu and social capital: Complementary or redundant concepts of collaboration-based regional development? *European Planning Studies, 12*(6), 747–765.

Fu, C. (2015). *Theoretical Study of the Regional Innovation System.* Dissertation of Jilin University. (in Chinese)

Fukudaa, K., & Watanabe, C. (2008). Japanese and US perspectives on the national innovation ecosystem. *Technology in Society, 30*(1), 49–63.

Furman, J., Porter, M., & Stern, S. (2002). The determinants of national innovative capacity. *Research Policy, 31*(6), 899–933.

Gallié, E. P. (2009). Is geographical proximity necessary for knowledge spillovers within a cooperative technological network? The case of the French biotechnology sector. *Regional Studies, 43*(1), 33–42.

Gao, Y., & Zheng, J. (2020). The impact of high-speed rail on innovation: An empirical test of the companion innovation hypothesis of transportation improvement with China's manufacturing firms. *World Development, 127*, 104838.

Gardner, P. (1994). Representations of the relationship between science and technology in the curriculum. *Studies in Science Education, 24*(1), 1–28.

Gartner. (2022). Retrieving from https://www.gartner.com/en/information-technology/glossary/hype-cycle.

Garud, R., & Karnøe, P. (2001). *Path Dependence and Creation.* London: Lawrence Erlbaum Associates.

Geoghegan, W., O'Kane, C., & Fitzgerald, C. (2015). Technology transfer offices as a nexus within the triple helix: The progression of the university's role. *International Journal of Technology Management, 68*(3–4), 255–277.

Gertler, M. S. (2008). Buzz without being there? Communities of practice in context. In Amin, A., & Roberts, J. (Eds.) *Community, Economic Creativity and Organization.* Oxford, UK: Oxford University Press.

Gertler, M. S., & Levitte, Y. (2005). Local nodes in global networks: The geography of knowledge flows in biotechnology innovation. *Industry & Innovation, 12*(4), 487–507.

Gibson, D. V., & Smilor, R. (1991). Key variables in technology transfer: A field-study based on empirical analysis. *Journal of Engineering and Technology Management, 8*, 287–312.

Gilsing, V., Bekkers, R., Freitas, I. M. B., & van der Steen, M. (2011). Differences in technology transfer between science-based and development-based industries: Transfer mechanisms and barriers. *Technovation, 31*, 638–647.

Gilsing, V., Nooteboom, B., Vanhaverbeke, W., Duysters, G., & van den Oord, A. (2008). Network embeddedness and the exploration of novel technologies: Technological distance, betweenness centrality and density. *Research Policy, 37*(10), 1717–1731.

Giroud, X. (2013). Proximity and Investment: Evidence from Plant-Level Data. *The Quarterly Journal of Economics, 128*(2), 861–915.

Gittelman, M. (2007). Does geography matter for science-based firms? Epistemic communities and the geography of research and patenting in biotechnology. *Organization Science, 18*(4), 724–741.

Giuliani, E. (2005). Cluster absorptive capacity, why do some clusters forge ahead and others lag behind? *European Urban and Regional Studies, 12,* 269–288.

Giuliani, E. (2011). Role of technological gatekeepers in the growth of industrial clusters: Evidence from Chile. *Regional Studies, 45*(10), 1329–1348.

Glaeser, E. L., Ponzetto, G., & Zou, Y. (2016). Urban networks: Connecting markets, people, and ideas. *Papers in Regional Science, 95*(1), 17–59.

Glaeser, E., Kallal, H., Scheinkman, J., & Shleifer, A. (1992). Growth in cities. *Journal of Political Economy, 100*(6), 1126–1152.

Glass, A. J. & Saggi, K. (1998). International technology transfer and the technology gap. *Journal of Development Economics, 55,* 369–398.

Glückler, J. (2007). Economic geography and the evolution of networks. *Journal of Economic Geography, 7,* 619–634.

Glückler, J. (2013). Knowledge, networks and space: Connectivity and the problem of non-interactive learning. *Regional Studies, 47*(6), 880–894.

Glückler, J., & Doreian, P. (2016). Social network analysis and economic geography-positional, evolutionary and multi-level approaches. *Journal of Economic Geography, 16*(6), 1123–1134.

Gnyawali, D. R., & Srivastava, M. K. (2013). Complementary effects of clusters and networks on firm innovation: A conceptual model. *Journal of Engineering and Technology Management, 30*(1), 1–20.

Gomulka, S. (1990). *The Theory of Technological Change and Economic Growth.* London: Routledge, 1–262.

Gould, R. V., & Fernandez, R. M. (1989). Structures of mediation: A formal approach to brokerage in transaction networks. *Sociological Methodology, 19,* 89–126.

Grabher, G. (1993). The weakness of strong ties: The lock-in of regional development in the Ruhr area. In Grabher G. (Ed.) *The Embedded Firm: On the Socioeconomics of Industrial Networks.* London: Routledge.

Grabher, G. (2006). Trading routes, bypasses, and risky intersections: Mapping the travels of "networks" between economic sociology and economic geography. *Progress in Human Geography, 30,* 163–189.

Graf, H. (2006). *Networks in the Innovation Process. Local and Regional Interactions.* Cheltenham: Edward Elgar.

Graf, H. (2011). Gatekeepers in regional networks of innovators. *Cambridge Journal of Economics, 35*(1), 173–198.

Graham, S. J., Merges, R. P., Samuelson, P., & Sichelman, T. (2009). High technology entrepreneurs and the patent system: Results of the 2008 Berkeley patent survey. *Berkeley Technology Law Journal, 24*, 1255–1327.

Granovetter, M. (1973). The strength of weak ties. *American Journal of Sociology, 78*(6), 1360–1380.

Granovetter, M. (1985). Economic action and social structure: The problem of embeddedness. *American Journal of Sociology, 91*(3), 481–510.

Griliches, Z. (1998). Patent statistics as economic indicators: A survey. In Griliches, Z. (Ed.) *R&D and Productivity: The Econometric Evidence.* Chicago: University of Chicago Press, 287–343.

Grillitsch, M., & Nilsson, M. (2015). Innovation in peripheral regions: Do collaborations compensate for a lack of local knowledge spillovers? *The Annals of Regional Science, 54*(1), 299–321.

Grillitsch, M., Asheim, B., & Trippl, M. (2018). Unrelated knowledge combinations: The unexplored potential for regional industrial path development. *Cambridge Journal of Regions, Economy and Society, 11*(2), 257–274.

Grimpe, C., & Hussinger, K. (2013). Formal and informal knowledge and technology transfer from academia to industry: Complementarity effects and innovation performance. *Industry & Innovation, 20*, 683–700.

Grosse, R. (1996). International technology transfer in services. *Journal of International Business Studies, 27*(4), 781–800.

Groysberg, B., Lee, L. E., & Nanda, A. (2008). Can they take it with them? The portability of star knowledge worker's performance. *Management Science, 54*(7), 1213–1230.

Grubesic, T. H., Matisziw, T. C., Zook, M. A. (2009). Spatio-temporal fluctuations in the global airport hierarchies. *Journal of Transport Geography, 17*, 264–275.

Guan, J., Mok, C. K., Yam, R. C., Chin, K. S., & Pun, K. F. (2006). Technology transfer and innovation performance: Evidence from Chinese firms. *Technology Forecasting & Social Change, 73*(6), 666–678.

Guan, M., Wu, S., & Liu, C. (2021). Comparing China's urban aviation and innovation networks. *Growth and Change, 52*(4), 1–14.

Gui, Q., Du D., Liu C., & Duan, D. (2019a). The geography of intercity technology transfer networks in China. *Regional Studies, Regional Science, 6*(1), 395–398.

Gui, Q., Liu, C., & Du, D. (2018a). International knowledge flows and the role of proximity. *Growth and Change, 49*(3), 532–547.

Gui, Q., Liu, C., & Du, D. (2018b). Does network position foster knowledge production? Evidence from international scientific collaboration network. *Growth and Change, 49*(4), 594–611.

Gui, Q., Liu, C., & Du, D. (2019b). Globalization of science and international scientific collaboration: A network perspective. *Geoforum, 105*, 1–12.

Gui, Q., Liu, C., Du, D., & Duan, D. (2019c). The changing geography of global science. *Environment and Planning A, 51*(8), 1615–1617.

Guo, H., & Deng, Z. (2019). Research on the integrative high-quality development of Yangtze River Delta regional economy under the new normal. *Economy and Management, 33*, 22–30. (in Chinese)

Guo, H., Chen, C., Dong, X., & Jiang, C. (2021). The evolution of transport networks and the regional water environment: The case of Chinese high-speed rail. *Regional Studies, 55*(6), 1084–1110.

Görg, H., & Strobl, E. (2005). Spillovers from foreign firms through worker mobility: An empirical investigation. *Scandinavian Journal of Economics, 107*(4), 693–709.

Görg, H., & Strobl, E. (2007). The effect of R&D subsidies on private R&D. *Economica, 74*(294), 215–234.

Haasis, T. I., Liefner, I., & Garg, R. (2018). The organization of knowledge transfer in the context of Chinese cross-border acquisitions in developed economies. *Asian Business and Management, 17*(4), 286–311.

Hall, G. R., & Johnson, R. E. (1970). *The Technology Factors in International Trade.* New York: Colombia University Press.

Hansen, T. (2015). Substitution or overlap? The relations between geographical and non-spatial proximity dimensions in collaborative innovation projects. *Regional Studies, 49*(10), 1672–1684.

Hargadon, A., & Sutton, R. I. (1997). Technology brokering and innovation in a product development firm. *Administrative Science Quarterly, 42*(12), 716–749.

Hartog, M., Boschma, R., & Sotarauta, M. (2012). The impact of related variety on regional employment growth in Finland 1993–2006: High-tech versus medium/low-tech. *Industry and Innovation, 19*(6), 459–476.

Haupt, R., Kloyer, M., & Lange, M. (2007). Patent indicators for the technology life cycle development. *Research Policy, 36*(3), 387–398.

He, C., & Mao, X. (2015). Economic globalization research based on scale-construction in western human geography. *Progress in Geography, 34*(9), 1073–1083. (in Chinese)

He, C., Zhu, S., Hu, X., & Li, Y. (2018). Proximity matters: Inter-regional knowledge spillovers and regional industrial diversification in China. *Tijdschrift Voor Economische En Sociale Geografie, 110*(2), 173–190.

Henning, M., Stam, E., & Wenting, R. (2013). Path dependence research in regional economic development: Cacophony or knowledge accumulation? *Regional Studies, 47*(8), 1348–1362.

Hensengerth, O. (2018). South-south technology transfer: Who benefits? A case study of the Chinese-built Bui dam in Ghana. *Energy Policy, 114*, 499–507.

Heringa, P. W., Horlings, E, Zouwen, M., Besselaar P., & Vierssen, W. (2014). How do dimensions of proximity relate to the outcomes of collaboration? A survey of knowledge-intensive networks in Dutch water sector. *Economics of Innovation and New Technology, 23*(7), 689–716.

Hesse, K., & Fornahl, D. (2020). Essential ingredients for radical innovations? The role of (un-)related variety and external linkages in Germany. *Papers in Regional Science, 99*(5), 1165–1183.

Heuermann, D., & Schmieder, J. F. (2018). The effect of infrastructure on worker mobility: Evidence from high-speed rail expansion in Germany. *Journal of Economic Geography, 19*(2), 335–372.

Hewitt-Dundas, N., Andreosso-O'Callaghan, B., Crone, M., Murray, J., & Roper, S. (2005). Selling global, buying local? What determines the sourcing patterns of multinational plants in Ireland? *Regional Studies, 39*(2), 225–239.

Hidalgo, C. A., Klinger, B., Barabási, A. L., & Hausmann, R. (2007). The product space conditions the development of nations. *Science, 317*(5837), 482–487.

Hilbe, J. M. (2011). *Negative Binomial Regression.* Cambridge, UK: Cambridge University Press.

Hodgson, C. (2018). The effect of transport infrastructure on the location of economic activity: Railroads and post offices in the American west. *Journal of Urban Economics, 104*(3), 59–76.

Hoekman, J., Frenken, K., & Van Oort, F. (2009). The geography of collaborative knowledge production in Europe. *The Annals of Regional Science, 43*(3), 721–738.

Hsueh, J.-T., Lin, N.-P., & Li, H.-C. (2010). The effects of network embeddedness on service innovation performance. *The Service Industries Journal, 30*(10), 1723–1736.

Hu, S., Huang, L., Fan, B., & Xiao, G. (2016). Spatial spillover effects of innovative outputs of university and its impacts on regional economic growth of China: Based on spatial econometrics analysis of provincial data. *Scientia Geographica Sinica*, *36*(12), 1767–1776. (in Chinese)

Huang, X., Du, D., & Liu, C. (2021). The evolution process and growth mechanism of global cross-border M&A network. *Acta Geographica Sinica*, *76*(10), 2536–2550. (in Chinese)

Huang, Y., Hong, T., & Ma, T. (2020). Urban network externalities, agglomeration economies and urban economic growth. *Cities*, *107*, 102882.

Huang, Y., Li, R., Zou, F., Jiang, L., Porter, A. L., & Zhang, L. (2022). Technology life cycle analysis: From the dynamic perspective of patent citation networks. *Technological Forecasting & Social Change*, *181*, 121760.

Huang, J., Sun, G., Yan, M., Liu, T., & Xiao, L. (2012). The spatio-temporal evolution of urban field strength and pattern of autocorrelation among cities in China. *Geographical Research*, 2012, *31*(8), 1355–1364. (in Chinese)

Huber, F. (2011). On the role and interrelationship of spatial, social and cognitive proximity: Personal knowledge relationships of R&D workers in the Cambridge information technology cluster. *Regional Studies*, *46*(9), 1169–1182.

Hudson, R. (2005). Rethinking change in old industrial regions: Reflecting on the experiences of North East England. *Environment and Planning A*, *37*, 581–596.

Hui, M. (2017). The new connotation of national innovation system. *QiuShi*, (1), 41–45. (in Chinese)

Hur, W., & Park, J. (2016). Network patterns of inventor collaboration and their effects on innovation outputs. *Sustainability*, *8*, 295.

Hägerstrand, T. (1953). *Innovation Diffusion as A Spatial Process*. Chicago: University of Chicago Press.

Håkansson, H. (1989). *Corporate Technological Behaviour: Co-operation and Networks*. London: Routledge.

Inkinen, T., & Suorsa, K. (2010). Intermediaries in regional innovation systems: High-technology enterprise survey from Northern Finland. *European Planning Studies*, *18*(2), 169–187.

Ivarsson, I., & Alvstam, C. G. (2009). The effect of spatial proximity on technology transfer from TNCs to local suppliers in developing countries: The case of AB Volvo in Asia and Latin America. *Economic Geography*, *81*(1), 83–111.

Jacobs, J. (1969). *The Economy of Cities*. New York: Vintage.

Jaffe, A. B. (1986). Technological opportunity and spillovers of R&D: Evidence from firms' patents, profits and market value. *American Economic Review, 76*(5), 984–1001.

Javorcik, B. S. (2004). Does foreign direct investment increase the productivity of domestic firms? In search of spillovers through backward linkages. *American Economic Review, 94*(3), 605–627.

Jensen, M. B., Johnson, B., Lorenz, E., & Lundvall, B. A. (2007). Forms of knowledge and modes of innovation. *Research Policy, 36*, 680–693.

Jiang, B. (2013). Head/tail breaks: A new classification scheme for data with a heavy-tailed distribution. *The Professional Geographer, 65*(3), 482–494.

Jiang, B. (2015). Head/tail breaks for visualization of city structure and dynamics. *Cities, 43*, 69–77.

Jiang, B., & Yin, J. (2014). Ht-index for quantifying the fractal or scaling structure of geographic features. *Annals of the Association of American Geographers, 104*(3), 530–540.

Jiao, J., Wang, J., Jin, F. & Wang, H. (2016). Impact of high-speed rail on inter-city network based on the passenger train network in China, 2003–2013. *Acta Geographica Sinica, 71*(2), 265–280. (in Chinese)

Jin, Z., & Zhu, S. (2021). Technology-introduction pattern of cities in China and its mechanism of change based on technology relatedness and complexity. *Progress in Geography, 40*(6), 897–910. (in Chinese)

Jolly, V. K. (1997). *Commercialization New Technologies.* Boston, Massachusetts: Harvard Business School Press.

Jones, B. W. (2010). Blog links as pipelines to buzz elsewhere: The case of New York theater blogs. *Environment and Planning B, 37*, 99–111.

Jones, C. I. (1995). RD-based models of economic growth. *Journal of Political Economy, 103*(4), 759–784.

Jones, O., & Tang, N. (1996). Networks for technology transfer: Linking HEIs and SMFs. *International Journal of Technology Management, 12*(7–8), 820–829.

Jung, W. (1980). Barriers to technology transfer and their elimination. *The Journal of Technology Transfer, 4*(2), 15–25.

Karlsson, B., & Johansson, C. (2005). Towards a dynamic theory for the spatial knowledge economy. In Johansson, B., Karlsson, C., & Stough, R. R. (Eds.) *Entrepreneurship and Dynamics in a Knowledge Economy.* London, UK: Routledge.

Kauffeld-Monz, M., & Fritsch, M. (2013). Who are the knowledge brokers in regional systems of innovation? A multi-actor network analysis. *Regional Studies, 47*(5), 669–685.

Kaufmann, A., Lehner, P., & Tödtling, F. (2003). Effects of the Internet on the spatial structure of innovation networks. *Information Economics and Policy, 15*, 402–424.

Kay, N. M. (2013). Rerun the tape of history and QWERTY always wins. *Research Policy, 42*, 1175–1185.

Kidwell, D. K. (2013). Principal investigators as knowledge brokers: A multiple case study of the creative actions of PIs in entrepreneurial science. *Technological Forecasting and Social Change, 80*(2), 212–220.

Kim, J. B., & Stewart, C. T. (1993). The relation between technology import and domestic R&D. *The Journal of Technology Transfer, 18*(3–4), 94–103.

Kim, L. (1997). *From Imitation to Innovation.* Cambridge: Harvard Business School Press, 1–303.

Kim, S. H., Jun, B., & Lee, J.-D. (2021). Technological relatedness: How do firms diversify their technology? *SocArXiv,* DOI: 10.31219/osf.io/47ank.

King, D. R., & Nowack, M. L. (2003). The impact of government policy on technology transfer: An aircraft industry case study. *Journal of Engineering and Technology Management, 20*(4), 303–318.

Klepper, S. (2010). The origin and growth of industry clusters: The making of Silicon Valley and Detroit. *Journal of Urban Economics, 67*, 15–32.

Kogler, D., Essletzbichler, J., & Rigby, D. (2017). The evolution of specialization in the EU15 knowledge space. *Journal of Economic Geography, 17*(2), 345–373.

Kogler, D., Rigby, D., & Tucker, I. (2013). Mapping knowledge space and technological relatedness in U.S. cities. *European Planning Studies, 21*, 1374–1391.

Kostiainen, J. (2002). Learning and the "ba" in the development network of an urban region. *European Planning Studies, 10*(5), 613–631.

Krätke, S. (2010). Regional knowledge networks: A network analysis approach to the interlinking of knowledge resources. *European Urban and Regional Studies, 17*(1), 83–97.

Krivitsky, P. N. (2012). Exponential-family random graph models for valued networks. Electronic Journal of Statistics, *6*, 1100–1128.

Kumar, N. (1987). Technology imports and local research and development in Indian manufacturing. *The Developing Economies, 25*, 220–252.

Kumar, V., Kumar, U., & Persaud, A. (1999). Building technological capability through importing technology: The case of Indonesian manufacturing industry. *The Journal of Technology Transfer, 24*, 81–96.

Laberge, Y. (2011). Intelligent cities and globalisation of innovation networks. *Urban Studies, 48*(5), 1062–1064.

Lambooy, J. G. (2004). The transmission of knowledge, emerging networks, and the role of universities: An evolutionary approach. *European Planning Studies, 12*(5), 643–657.

Laursen, K., & Salter, A. (2006). Open for innovation: The role of openness in explaining innovation performance among UK manufacturing firms. *Strategic Management Journal, 27*(2), 131–150.

Lawson, C., & Lorenz, E. (1999). Collective learning, tacit knowledge and regional innovative capacity. *Regional Studies, 33*(4), 305–317.

Lazaric, N., Longhi, C., & Tohmas, C. (2008). Gatekeepers of knowledge versus platforms of knowledge: From potential to realized absorptive capacity. *Regional Studies, 42*, 837–852.

Lazer, D., & Friedman, A. (2007). The network structure of exploration and exploitation. *Administrative Science Quarterly, 52*(4), 667–694.

Lee, B. K., & Sohn, S. Y. (2019). Disparities in exploitative and exploratory patenting performance across regions: Focusing on the roles of agglomeration externalities. *Papers in Regional Science, 98*(1), 241–263.

Lee, C., Kim, J., Kwon, O., & Woo, H.-G. (2016). Stochastic technology life cycle analysis using multiple patent indicators. *Technology Forecasting and Social Change, 106*, 53–64.

Lee, K., & Lim, C. (2001). Technological regimes, catching-up and leapfrogging: Findings from the Korean industries. *Research Policy, 30*(3), 459–483.

Lee, P. C., & Su, H. N. (2015). How to analyze technology life cycle from the perspective of patent characteristics? *2015 Proceedings of PICMET'15: Management of the Technology Age*, 2079–2083.

Lee, R. P., & Johnson, J. L. (2010). Managing multiple facets of risk in new product alliances. *Decision Sciences, 41*(2), 271–300.

Lee, S. C., Lee, K. B., & Ryu, J. Y. (2003). Technology transfer of foreign direct investment in China. *Geography, 88*, 289–299.

Lee, Y. S. (2000). The sustainability of university–industry research collaboration: An empirical assessment. *The Journal of Technology Transfer, 25*(2), 111–133.

Lerner, J. (1995). Venture capitalists and the oversight of private firms. *The Journal of Finance, 50*(1), 301–318.

Levin, D. Z., & Cross, R. (2004). The strength of weak ties you can trust: The mediating role of trust in effective knowledge transfer. *Management Science, 50*(11), 1477–1490.

Levison, M., Ward R. F., & Webb, J. W. (1973). *The Settlement of Polynesia: A Computer Simulation.* Minneapolis: University of Minnesota Press.

Leydesdorff, L., & Etzkowitz, H. (1998). The Triple Helix as a model for innovation studies. *Science and Public Policy, 25*, 195–203.

Li, B., Yue, Y., & Chen, T. (2016). Export and innovation: Empirical evidence from patent data at firm level. *World Economy, 39*(12), 72–94. (in Chinese)

Li, D., Heimeriks, G., & Alkemade, F. (2020). The emergence of renewable energy technologies at country level: Relatedness, international knowledge spillovers and domestic energy markets. *Industry and Innovation, 27*(9), 991–1013.

Li, D., Wei, Y. D., & Wang, T. (2015). Spatial and temporal evolution of urban innovation network in China. *Habitat International, 49*, 484–496.

Li, S., Garces, E., & Daim, T. (2019). Technology forecasting by analogy-based on social network analysis: The case of autonomous vehicles. *Technological Forecasting and Social Change, 148*, 119731.

Li, S., & Yayavaram, S. (2021). Attenuating the negative effects of network change on innovation: A whole network level analysis of Taiwanese business groups. *Asia Pacific Journal of Management, 38*, 151–177.

Li, T., & Zhou, R. (2016). Urban hinterworld in Yangtze River Delta: Empirical comparison of two network-based methods. *Acta Geographica Sinica, 71*(2), 236–250. (in Chinese)

Li, Y., Meng, Y., & Wang, Y. (2014). Public R&D subsidies and enterprise technology innovation: An empirical analysis based on multi-dimensional industry heterogeneity. *Science of Science and Management of S. & T., 35*(1), 33–41. (in Chinese)

Lian, X., Guo, Y., & Su, J. (2021). Technology stocks: A study on the characteristics that help transfer public research to industry. *Research Policy, 50*, 104361.

Liebowitz, S. J., & Margolis, S. E. (1995). Path dependence, lock-in, and history. *Journal of Law, Economics, & Organization, 11*(1), 205–226.

Liefner, I., & Hennemann, S. (2011). Structural holes and new dimensions of distance: The spatial configuration of the scientific knowledge network of China's optical technology sector. *Environment and Planning A, 43*(4), 810–829.

Lim, U. (2007). Knowledge externalities, spatial dependence, and metropolitan economic growth in the United States. *Environment and Planning A, 39*(4), 771–788.

Lin, L., Xu, J., Xing, L., & Guan, J. (2020). Analysis on the topological structure of inter-enterprise technology transfer network: Taking Z-park as an example. *International Journal of Modern Physics B, 34*(11), 2050131.

List, F. (2011). Lloyd S. S. (trans.) *The National System of Political Economy.* London: Longmans, Green, and Co.

Liu, B., Tang, N., & Zhu, X. (2008). Transferring technology across borders: Institutional effects in Chinese context. *The Journal of Technology Transfer, 33*(6), 619–630.

Liu, C. (2017). *Spatial Complexity of Urban-rural Road Network System.* Shanghai: Shanghai Science Popularization Press. (in Chinese)

Liu, C., & Guan, M. (2018). Spatiotemporal evolution of interurban technological flow network in the Yangtze River Delta urban agglomeration: From the perspective of patent transaction network. *Geographical Research, 37*(5), 981–994. (in Chinese)

Liu, C., & Guo, Q. (2019). Technology spillover effect in China: The spatiotemporal evolution and its drivers. *Sustainability, 11*(6), 1694.

Liu, C., & Niu, C. (2019). Spatial evolution and factors of interurban technology transfer network in Northeast China from national to local perspectives. *Acta Geographica Sinica, 74*(10), 2092–2107. (in Chinese)

Liu, C., & Yan, S. (2022). Transnational technology transfer network in China: Spatial dynamics and its determinants. *Journal of Geographical Sciences, 32*(12), 2383–2414.

Liu, C., Guan, M., & Duan, D. (2018a). Spatial pattern and influential mechanism of interurban technology transfer network in China. *Acta Geographica Sinica, 73*(8), 1462–1477. (in Chinese)

Liu, C., Gui, Q., Duan, D., & Yin, M. (2017). Structural heterogeneity and proximity mechanism of global scientific collaboration network based on co-authored papers. *Acta Geographica Sinica, 72*(4), 737–752. (in Chinese)

Liu, C., Niu, C., & Han J. (2019). Spatial dynamics of intercity technology transfer networks in China's three urban agglomerations: A patent transaction perspective. *Sustainability, 11*(6), 1647.

Liu, C., Wang, J., & Zhang, H. (2018b). Spatial heterogeneity of ports in the global maritime network detected by weighted ego network analysis. *Maritime Policy and Management, 45*(1), 89–104.

Liu, C., Wang, J., Zhang, H., & Yin, M. (2018c). Mapping the hierarchical structure of the global shipping network by Weighted Ego Network

Analysis. *International Journal of Shipping and Transport Logistics, 10*(1), 63–86.

Liu, D., & Yang, J. (2006). R&D input, economic growth and technological management. *Science of Science and Management of S. & T.*, (9), 14–18. (in Chinese)

Liu, J., Chaminade, C., & Asheim, B. (2013). The geography and structure of global innovation networks: A knowledge base perspective. *European Planning Studies, 21*(9), 1465–1473.

Liu, X., Jiang, S., Chen, H., Larson, C. A., & Roco, M. C. (2015). Modeling knowledge diffusion in scientific innovation networks: An institutional comparison between China and US with illustration for nanotechnology. *Scientometrics, 105*, 1953–1984.

Liu, X., Schwaag Serger, S., Tagscherer, U., & Chang, A. Y. (2017). Beyond catch-up-Can a new innovation policy help China overcome the middle income trap? *Science and Public Policy, 44*(5), 656–669.

Liu, X., Wei, Y., & Yang, J. (2016). Research on the node selection mechanism of technological innovation network based on evolutionary perspective. *Science & Technology Progress and Policy, 33*(10), 10–13. (in Chinese)

Liu, Y., Tan, L., Li, Z., Cheng, Y., Liu, L. (2015). Theoretical model and empirical study on internationalization of national innovation system. *Studies in Science of Science, 33*(9), 1324–1339. (in Chinese)

Liu, Z., & Chen, G. (2009). Environmental regulation and the location distribution of foreign direct investment: An empirical study based on urban panel data. *Journal of International Trade*, (3), 61–69. (in Chinese)

Long Y., & Li, Y. (2016). Should venture capitalists seek far? A comparative study of regional exit rate of venture capital investments in China. *Finance & Trade Economics, 2016*(6), 129–145. (in Chinese)

Love, J. H., & Roper, S. (1999). The determinants of innovation: R&D, technology transfer and networking effects. *Review of Industrial Organization, 15*, 43–64.

Lu, J., Sun, D., Yu, J., Li, J., & Niu, F. (2020). "Local versus nonlocal" enterprise linkages of global cities: A comparison between Beijing and Shanghai, China. *Complexity, 2020*, 891534.

Lundvall, B. (1992). *National Systems of Innovation: Towards a Theory of Innovation and Interactive Learning*. London: Pinter Publisher.

Luo, X., Mao, W., Wang, B., & Liu, C. (2022). The impacts of aviation and high-speed rail on urban innovation capacity in China. *Progress in Geography, 41*(12), 2203–2217. (in Chinese)

Lusher, D., Koskinen, J., & Robins, G. (2013). *Exponential Random Graph Models for Social Network: Theory, Methods, and Applications* (Vol. 35). New York: Cambridge University Press.

Lysak, L. (2003). Globalisation versus glocalization. *Ekonomicky Casopis, 51*(8), 931–946.

Lyu, G. (2016). *The Structure of Innovation Networks of China's Equipment Manufacturing Industry*. Dissertation of East China Normal University. (in Chinese)

Lyu, K. (2013). *Knowledge Base, Learning Fields and Innovation: Case for Super-hard Material Industry*. Dissertation of Henan University. (in Chinese)

Ma, H., & Xu, X. (2022). The effects of proximities on the evolving structure of intercity innovation networks in the Guangdong–Hong Kong–Macao Greater Bay Area: Comparison between scientific and technology knowledge. *International Journal of Urban Sciences*, http://doi.org/10.1080/12265934.2022.2085154.

Ma, H., Li, Y., & Huang, X. (2020). Proximity and the evolving knowledge polycentricity of megalopolitan science: Evidence from China's Guangdong–Hong Kong–Macao Greater Bay Area, 1990–2016. *Urban Studies, 58*(12), 2405–2423.

Ma, H., Wei, Y. D., & Xu, X. (2022). The proximity and dynamics of intercity technology transfers in the Guangdong–Hong Kong–Macau Greater Bay Area: Evidence from patent transfer networks. *Environment and Planning A, 54*(7), 1432–1449.

Ma, Y., & Wang, S. (2008). Exploration on Triple Helix model of university–government–industry partnership. *Research in Higher Education of Engineering*, (5), 29–34. (in Chinese)

Mackinnon, D., Dawley, S., & Pike, A. (2019). Rethinking path creation: A geographical political economy approach. *Economic Geography, 95*(2), 113–135.

Maennig, W., & Ölschläger, M. (2011). Innovative milieux and regional competitiveness: The role of associations and chambers of commerce and industry in Germany. *Regional Studies, 45*(4), 441–452.

Mahdad, M., Minh, T. T., Bogers, M. L. A. M., & Piccaluga, A. (2020). Joint university–industry laboratories through the lens of proximity dimensions: Moving beyond geographical proximity. *International Journal of Innovation Science, 12*(4), 433–456.

Maillat, D. (1991). The innovation process and the role of the milieu. In Bergman, E., Maier, G., & Tödtling, F. (Eds.) *Regions Reconsidered:*

Economic Networks, Innovation and Local Development in Industrialized Countries. London: Mansell, 103–117.

Maillat, D. (1998a). From industrial district to innovative milieu: Contribution to an analysis of territorialized productive organizations. *Geographische Zeitschrift, 86*(1), 1–15.

Maillat, D. (1998b). *Innovative Milieu and New Generations of Regional Policies Development*. Oxford: Oxford University Press.

Makkonen, T., Widenfeld, A., & Willams A. M. (2016). Cross-border regional innovation system integration: An analytical framework. *Tijdschrift Voor Economische En Sociale Geografie, 108*(6), 805–820.

Makri, M., Hitt, M. A., & Lane, P. J. (2010). Complementary technologies, knowledge relatedness and invention outcomes in high technology mergers and acquisitions. *Strategic Management Journal, 31*, 602–628.

Mankiw, N. G., Romer, D., & Weil, D. N. (1992). A contribution to the empirics of economic growth. *The Quarterly Journal of Economics, 107*(2), 407–437.

Mansfield, E. (1975). International technology transfer: Forms, resource requirements, and policy. *American Economic Review, 65*(2), 372–376.

Mansfield, E., Teece, D., & Romeo, A. (1979). Overseas research and development by US-based firms. *Economica, 46*(182), 187–196.

Mao, X., & He, C. (2019a). Product relatedness and export specialization in China's regions: A perspective of global–local interactions. *Cambridge Journal of Regions, Economy and Society, 12*, 105–126.

Mao, X., & He, C. (2019b). A review of global–local interactions for regional development. *Progress in Geography, 38*(10), 1449–1461. (in Chinese)

Markman, G., Gianiodis, P., Phan, P., & Balkin, D. (2005). Innovation speed: transferring university technology to market. *Research Policy, 34*, 1058–1075.

Markusen, A. (1996). Sticky places in slippery space: A typology of industrial districts. *Economic Geography, 72*, 293–313.

Martin, R. (2010). Rethinking regional path dependence: Beyond lock-in to evolution. *Economic Geography, 86*(1), 1–27.

Martin, R., & Moodysson, J. (2011). Innovation in symbolic industries: The geography and organization of knowledge sourcing. *European Planning Studies, 19*(7), 1183–1203.

Martin, R., & Sunley, P. (2006). Path dependence and regional economic evolution. *Journal of Economic Geography, 6*(6), 395–437.

Martin, R., & Sunley, P. (2007). Complexity thinking and evolutionary economic geography. *Journal of Economic Geography, 7*(4), 16–45.

Martin, R., & Sunley, P. (2008). The place of path dependence in an evolutionary perspective on the economic landscape. In Boschma, R. A., & Martin, R. (Eds.) *The Handbook on Evolutionary Economic Geography*. Cheltenham: Edward Elgar.

Maskell, P., & Malmberg, A. (1999a). Localised learning and industrial competitiveness. *Cambridge Journal of Economics, 23*(2), 167–185.

Maskell, P., & Malmberg, A. (1999b). The competitiveness of firms and regions: 'Ubiquitification' and the importance of localized learning. *European Urban and Regional Studies, 6*(1), 9–25.

Maskell, P., & Malmberg, A. (2007). Myopia, knowledge development and cluster evolution. *Journal of Economic Geography, 7*(5), 603–618.

Maskell, P., Bathelt, H., & Malmberg, A. (2006). Building global knowledge pipelines: The role of temporary clusters. *European Planning Studies, 14*(8), 997–1013.

Massa, L., Ardito, L., & Petruzzelli, A. M. (2022). Brokerage dynamics in technology transfer networks: A multi-case study. *Technological Forecasting & Social Change, 183*, 121895.

Massard, N., & Mehier, C. (2009). Proximity and innovation through an "accessibility to knowledge" lens. *Regional Studies, 43*(1), 77–88.

Massey, D., Arango, J., Hugo, G., Kouaouci, A., Pellegrino, A., & Taylor, J. (1994). An evaluation of international migration theory: The North American case. *Population and Development Review, 20*(4), 699–751.

Mathews, J. A., & Hu, M. C. (2007). Enhancing the role of universities in building national innovative capacity in Asia: The case of Taiwan. *World Development, 35*(6), 1005–1020.

Matthiessen, C. W., Schwarz, A. W., & Find, S. (2010). World cities of scientific knowledge: Systems, networks and potential dynamics. An analysis based on bibliometric indicators. *Urban Studies, 47*(9), 1879–1897.

Maté-Sánchez-Val, M., & Harris, R. (2018). The paradox of geographical proximity for innovators: A regional study of the Spanish agri-food sector. *Land Use Policy, 73*, 458–467.

Maurseth, P. B., & Svensson, R. (2020). The importance of tacit knowledge: Dynamic inventor activity in the commercialization phase. *Research Policy, 49*(7), 104012.

Mayer, H., & Cowell, M. (2016). Capital cities as knowledge hubs: The economic geography of homeland security contracting. In Conventz, S., Derudder, B., Thierstein A., & Witlox F. *Hub Cities in the Knowledge Economy: Seaports, Airports, Brainports*. New York, NY: Routledge, 223–246.

Mazurkiewicz, A., Giesko, T., Poteralska, B., & Tan, K. H. (2022). Crossing the chasm: Overcoming technology transfer barriers resulting from changing technical requirements in the process of innovation development in R&D organisations. *Technology Analysis & Strategic Management, 34*(10), 1187–1201.

McCann, P., & Mudambi, R. (2004). The location behavior of the multinational enterprise: Some analytical issues. *Growth and Change, 35*(4), 491–524.

McKelvey, M., Alm, H., & Riccaboni, M. (2003). Does co-location matter for formal knowledge collaboration in the Swedish biotechnology-pharmaceutical sector? *Research Policy, 32*(3), 483–501.

Meeteren, M., Neal, Z., & Derudder, B. (2016). Disentangling agglomeration and network externalities: A conceptual typology. *Papers in Regional Science, 95*(1), 61–80.

Meng, X., Meng, X., & Cheng, G. (2004). A positive research on the impact that economic growth to technical input. *Information Science, 22*(8), 1019–1021. (in Chinese)

Metcalfe, S. (1995). The economic foundation of technology policy: Equilibrium and evolutionary perspectives. In Stoneman P. (Ed.) *Handbook of the Economics of Innovation and Technological Change.* Oxford: Basil Blackwell, 409–512.

Meyer, M. (2010). The rise of the knowledge broker. *Science Communication, 32*(1), 118–127.

Meyer, S., Schiller, D., & Diez, J. R. (2012). The localization of electronics manufacturing in the Greater Pearl River Delta, China: Do global implants put down local roots? *Applied Geography, 32*(1), 119–129.

Meyer-Krahmer, F., & Schmoch, U. (1998). Science-based technologies: University–industry interactions in four fields. *Research Policy, 27*(8), 835–851.

Miao, C. (2006). Global-local nexus and technological learning in industrial cluster: A case study of hair-goods industry in Xuchang, Henan province. *Acta Geographica Sinica, 61*(4), 91–100. (in Chinese)

Miao, C., & Ai, S. (2009). Innovation in space: A new perspective of learning field. *Economic Geography, 29*(7), 1057–1063. (in Chinese)

Miao, C., Wei, Y. D., & Ma, H. (2007). Technological learning and innovation in China in the context of globalization. *Eurasian Geography and Economics, 48*(6), 713–732.

Miguelez, E., & Moreno, R. (2018). Relatedness, external linkages and regional innovation in Europe. *Regional Studies, 52*(5), 688–701.

Miguélez, E., & Moreno, R. (2013). Do labour mobility and technological collaborations foster geographical knowledge diffusion? The case of European regions. *Growth and Change, 44*(2), 321–354.

Milo, R., Shen-Orr, S., Itzkovitz, S., Kashtan N, Chklovskii, D., & Alon, U. (2002). Network motifs: Simple building blocks of Complex Networks. *Science, 298*(5594), 824–827.

Min, S., Kim, J., & Sawng, Y.- W. (2020). The effect of innovation network size and public R&D investment on regional innovation efficiency. *Technological Forecasting & Social Change, 155*, 119998.

Moodysson, J. (2008). Principles and practices of knowledge creation: On the organization of "buzz" and "pipelines" in life science communities. *Economic Geography, 84*(4), 449–469.

Moon, H., Di Benedetto, A., & Kim, S. K. (2022). The effect of network tie position on a firm's innovation performance. *Journal of Business Research, 144*, 821–829.

Moore, J. F. (1993). Predators and prey: A new ecology of competition. *Harvard Business Review, 71*(3), 75–86.

Moore, J. F. (2006). Business ecosystems and the view from the firm. *Antitrust Bulletin, 51*(1), 31–75.

Moran, P. (2005). Structural vs. relational embeddedness: Social capital and managerial performance. *Strategic Management Journal, 26*, 1129–1151.

Moreno, R., Paci, R., & Usai, S. (2005). Spatial spillovers and innovation activity in European regions. *Environment and Planning A, 37*(10), 1793–1812.

Morrill, R. L. (1970). The shape of diffusion in space and time. *Economic Geography, 46*, 68–259.

Morrison, A. (2008). Gatekeepers of knowledge within industrial districts: Who they are, how they interact. *Regional Studies, 42*, 817–835.

Morrison, A., Rabellotti, R., & Zirulia, L. (2013). When do global pipelines enhance the diffusion of knowledge in clusters? *Economic Geography, 89*(1), 77–96.

Murmann, J. P., & Frenken, K. (2006). Toward a systematic framework for research on dominant designs, technological innovations, and industrial change. *Research Policy, 35*(7), 925–952.

Müller, K. (2010). Academic spin-off's transfer speed-Analyzing the time from leaving university to venture. *Research Policy, 39*, 189–199.

Nadvi, K., & Schmitz, H. (1999). Clustering and industrialization: Introduction. *World Development, 27*(9), 1503–1514.

Nam, Y., & Barnett, G. A. (2011). Globalization of technology: Network analysis of global patents and trademarks. *Technological Forecasting & Social Change, 78*, 1471–1485.

Narula, R. (2002). Innovation systems and 'inertia' in R&D location: Norwegian firms and the role of systemic lock-in. *Research Policy, 31*, 795–816.

Nathan, M., & Rosso, A. (2022). Innovative events: Product launches, innovation and firm performance. *Research Policy, 51*(1), 104373.

Neffke, F. (2009). *Productive Places: The Influence of Technological Change and Relatedness on Agglomeration Externalities.* Dissertation of Universiteit Utrecht.

Neffke, F., & Henning, M. (2013). Skill relatedness and firm diversification. *Strategic Management Journal, 34*(3), 297–316.

Neffke, F., Henning, M., & Boschma, R. (2011). How do regions diversify over time? Industry relatedness and the development of new growth paths in regions. *Economic Geography, 87*(3), 237–265.

Neffke, F., Henning, M., & Boschma, R. (2012). The impact of aging and technological relatedness on agglomeration externalities: A survival analysis. *Journal of Economic Geography, 12*, 485–517.

Nelson, R. R. (1993). *National Innovation Systems: A Comparative Analysis.* New York: Oxford University Press.

Nelson, R. R., & Phelps, E. S. (1966). Investment in humans, technological diffusion, and economic growth. *The American Economic Review, 56*, 69–75.

Nelson, R., & Winter, S. (1982). *An Evolutionary Theory of Economic Change.* Cambridge, MA: Harvard University Press.

Nepelski, D., & De Prato, G. (2018). The structure and evolution of ICT global innovation network. *Industry and Innovation, 25*, 940–965.

Newman, M. E. (2002). Assortative mixing in networks. *Physical Review Letters, 89*(20), 208701.

Niosi, J., & Bellon, B. (1994). The global interdependence of national innovation systems: Evidence, limits, and implications. *Technology in Society, 16*(2), 173–197.

Noh, H., & Lee, S. (2019). Where technology transfer research originated and where it is going: A quantitative analysis of literature published between 1980 and 2015. *The Journal of Technology Transfer, 44*(3), 700–740.

Nonaka, I., & Konno, N. (1998). The concept of "ba": Building a foundation for knowledge creation. *California Management Review, 40*(3), 40–54.

Nonaka, I., & Takeuchi, H. (1995). *The Knowledge-Creating Company: How Japanese Companies Create the Dynamics of Innovation.* New York: Oxford University Press, 1–284.

Nooteboom, B. (2000a). *Learning and Innovation in Organizations and Economies.* Oxford, UK: Oxford University Press.

Nooteboom, B. (2000b). Learning by interaction: Absorptive capacity, cognitive distance and governance. *Journal of Management and Governance, 4*(1–2), 69–92.

Nystuen, J. D., & Dacey, M. F. (1961). A graph theory interpretation of nodal regions. *Papers of the Regional Science Association, 7*(1): 29–42.

OECD. (1996). *The Knowledge-based Economy*. Paris: Organisation of Economic Co-operation and Development.

OECD. (1997). *National Innovation System*. Paris: Organisation of Economic Co-operation and Development.

Oerlemans, A. G., Meeus, T. H., & Boekema, W. M. (2001). On the spatial embeddedness of innovation networks: An exploration of the proximity effect. *Tijdschrift voor Economische en Sociale Geografie, 92*(1), 60–75.

Olcay, G. A., & Bulu, M. (2016). Technoparks and technology transfer offices as drivers of an innovation economy: Lessons from Istanbul's innovation spaces. *Journal of Urban Technology, 23*(1), 71–93.

Opper, S., & Nee, V. (2015). Network effects, cooperation and entrepreneurial innovation in China. *Asian Business & Management, 14*(4), 283–302.

Ortega, J. L. (2011). Collaboration patterns in patent networks and their relationship with the transfer of technology: The case study of the CSIC patents. *Scientometrics, 87*, 657–666.

Owen-Smith, J., & Powell, W. (2004). Knowledge networks as pipelines and conduits: The effects of spillovers in the Boston biotechnology community. *Organization Science, 15*(1), 5–21.

Palaco, I., Kim, S. K., Park, M. J., & Rho, J. J. (2022). Exploring capabilities of international technology transfer intermediaries between emerging and developed countries. *The Journal of Technology Transfer, 47*, 307–352.

Pan, F., Bi, W., Lenzer, J., & Zhao, S. (2017). Mapping urban networks through inter-firm service relationships: The case of China. *Urban Studies, 54*(16), 3639–3654.

Park, H., Ree, J. J., & Kim, K. (2013). Identification of promising patents for technology transfers using TRIZ evolution trends. *Expert Systems With Applications, 40*(2), 736–743.

Patel, P., & Pavitt, K. (1997). The technological competencies of the world's largest firms: Complex and path–dependent, but not much variety. *Research Policy, 26*(2), 141–156.

Pavitt, K. (1985). Patent statistics as indicators of innovative activities: Possibilities and problems. *Scientometrics*, 7(1–2), 77–99.

Pedersen, P. O. (1970). Innovation diffusion within and between national urban systems. *Geographical Analysis*, *2*(3), 203–254.

Penny, R. (1992). Understanding the technology transfer process. VITA Distribution Service, https://www.africa.upenn.edu/Comp_Articles/Technology_Transfer_12764.html.

Perez, C., & Soete, L. (1988). Catching up in technology: Entry barriers and windows of opportunity. In Dosi, G., Freeman, C. R., Nelson, R. R., & Soete, L.(Eds.). *Technical Change and Economic Theory*. London: Francis Pinter, 458–479.

Perkins, R., & Neumayer, E. (2010). The international diffusion of new technologies: A multitechnology analysis of latecomer advantage and global economic integration. *Annals of the Association of American Geographers*, *95*(4), 789–808.

Perkmann, M., & Walsh, K. (2008). Engaging the scholar: Three types of academic consulting and their impact on universities and industry. *Research Policy*, *37*(10), 1884–1891.

Perry, M., & Tan B. H. (1998). Global manufacturing and local linkages in Singapore. *Environment and Planning A*, *30*(1), 603–624.

Pham, X. (2007). Five principles of path creation. *Oeconomicus*, *8*(1), 5–17.

Phillips, N. (2016). China-Building an innovator. *Nature*, *533*, 32–33.

Phillips, R. (2002). Technology business incubators: How effective is technology transfer mechanisms? *Technology in Society*, *24* (3), 299–316.

Piazza, M., Mazzola, E., Abbate, L., & Perrone, G. (2019). Network position and innovation capability in the regional innovation network. *European Planning Studies*, *27*(9), 1857–1878.

Pierre, D., & Leppälä, S. (2011). Opening up the 'Jacobs Spillovers' black box: Local diversity, creativity and the processes underlying new combinations. *Journal of Economic Geography*, *11*(5), 843–863.

Piscitello, L. (2000). Relatedness and coherence in technological and product diversification of the world's largest firms. *Structural Change and Economic Dynamics*, *11*, 295–315.

Polanyi, M. (1962). *Personal Knowledge: Towards a Post-Critical Philosophy*. Chicago: University of Chicago Press.

Polanyi, M. (1966). *The Tacit Dimension*. London: Routledge & Kegan Paul.

Polenske, K. R. (2007). *The Economic Geography of Innovation*. New York: Cambridge University Press.

Poon, J. P., & Sajarattanochote, S. (2010). Asian transnational enterprises and technology transfer in Thailand. *European Planning Studies, 18*(5), 691–707.

Ponomariov, B. (2013). Government-sponsored university-industry collaboration and the production of nanotechnology patents in US universities. *The Journal of Technology Transfer, 38*(6), 749–767.

Porta, S., Latora, V., Wang, F., & Scellato, S. (2009). Street centrality and densities of retail and services in Bologna, Italy. *Environment and Planning B, 36*, 450–465.

Porter, M. E. (1998). Clusters and the new economics of competition. *Harvard Business Review, 76*(6), 77–90.

Posner, M. (1961). International trade and technical change. *Oxford Economic Papers, 13*(3), 323–341.

Pred, A. (1977). The choreography of existence: Comments on Hägerstrand's time-geography and its usefulness. *Economic Geography, 53*(2), 207–221.

Pudelko, M., & Mendenhall, M. E. (2009). The contingent nature of best practices in national competitiveness: The case of American and Japanese innovation processes. *European Management Journal, 27*(6), 456–466.

Puffert, D. (2008). Path dependence. In Whaples, R. *EH. Net Encyclopedia*, http://eh.net/encyclopedia/path-dependence/.

Pyka, A., Kudic, M., & Mueller, M. (2019). Systemic interventions in regional innovation systems: Entrepreneurship, knowledge accumulation and regional innovation. *Regional Studies, 53*(9), 1321–1332.

Qi, Y. (2013). Impact mechanism of technology gap on multinational technology transfer base on spillover. *Journal of Xuzhou Institute of Technology (Social Science Edition), 28*(4), 50–53. (in Chinese)

Qian, F., Hong, J. Hou, B., & Fang, T. (2022). Technology transfer channels and innovation efficiency: Empirical evidence from Chinese manufacturing industries. *IEEE Transactions on Engineering Management, 69*(5), 2426–2438.

Radosevic, S. (1999). *International Technology Transfer and Catch-up in Economic Development.* Nothampton, MA: Edward Edgar Publishing.

Ratti, R., Bramanti, A., & Gordon, R. E. (1999). The dynamics of innovative regions: The GREMI approach. *Urban Studies, 36*(8), 1409–1410.

Reagans, R., & McEvily, B. (2003). Network structure and knowledge transfer: The effects of cohesion and range. *Administrative Science Quarterly, 48*(2), 240–267.

Richard, R., & Doh, C. S. (2001). Democratization backwards: The problem of third-wave democracies. *British Journal of Political Science, 31*(2),331–354.

Richardson, H. W. (1973). *The Economics of Urban Size.* Lexington: Lexington Books.

Rigby, D. (2015). Technological relatedness and knowledge space: Entry and exit of US cities from patent classes. *Regional Studies, 49*(11), 1922–1937.

Rigby, D., & Essletzbichler, J. (1997). Evolution, process variety, and regional trajectories of technological change in U.S. manufacturing. *Economic Geography, 73*(3), 269–284.

Robertson, R. (1994). Globalisation or glocalisation? *Journal of International Communication, 1*(1), 33–52.

Rodrigue, J.-P., Comtois, C., & Slack, B. (2020). *The Geography of Transport Systems* (Fifth Edition). New York: Routledge.

Rodríguez-Pose, A. (1999). Innovation prone and innovation averse societies: Economic performance in Europe. *Growth and Change, 30*(1), 75–105.

Rodríguez-Pose, A., & Comptour, F. (2012). Do clusters generate greater innovation and growth? An analysis of European regions. *The Professional Geographer, 64*(2), 211–231.

Roe, M. (1996). Chaos and evolution in law and economics. *Harvard Law Review, 109*(3), 641–668.

Rogers, E. M. (1962). *Diffusion of Innovations* (Fifth edition). Glencoe: Free Press.

Romer, P. M. (1986). Increasing returns and long-run growth. *Journal of Political Economy, 94*(5), 1002–1037.

Rong, K., & Shi, Y. (2014). Business Ecosystems: Constructs, Configurations, and the Nurturing Process. London: Palgrave Macmillan.

Rossignolo, C. (1999). The urban milieu. Territorial identity in development processes. *International Journal of Urban and Regional Research, 23*(3), 603–604.

Rothwell, R. (1992). Developments towards the fifth generation model of innovation. *Technology Analysis and Strategic Management, 4*(1), 73–75.

Roxas, S. A., Piroli, G., & Sorrentino, M. (2011). Efficiency and evaluation analysis of a network of technology transfer brokers. *Technology Analysis & Strategic Management, 23*(1), 7–24.

Rycrof, R. W., & Kash, D. E. (2004). Self-organizing innovation networks: Implications for globalization. *Technovation, 24*, 187–197.

Ryssel, R., Ritter, T., & Georg Gemünden, H. (2004). The impact of information technology deployment on trust, commitment and value creation in business relationships. *Journal of Business & Industrial Marketing, 19*(3), 197–207.

Saggi, K. (2002). Trade, foreign direct investment, and international technology transfer: A survey. *World Bank Research Observer, 17*(2), 191–235.

Saint-Pierre, G. (1997). The globalization of knowledge: R&D in Canada. *Technology in Society, 19*(3–4), 271–280.

Saito, Y. (1979). *Theory of Technology Transfer.* Tokyo: Bookman's Academy (Bunshindo). (in Japanese)

Saito, Y. (1988). *Theory of Technology Development: Mechanism and Policy of Technology Development in Japan.* Tokyo: Bookman's Academy (Bunshindo). (in Japanese)

Sampson, R. (2007). R&D alliances and firm performance: The impact of technological diversity and alliance organization on innovation. *Academy of Management Journal, 50*(2), 364–386.

Santacreu-Vasut, F., & Teshima, K. (2016). Foreign employees as channel for technology transfer: Evidence from MNC's subsidiaries in Mexico. *Journal of Development Economics, 122,* 92–112.

Sapienza, H. J., Parhankangas, A., & Autio, E. (2004). Knowledge relatedness and post-spin-off growth. *Journal of Business Venturing, 19,* 809–829.

Sapsed, J., Grantham, A., & DeFillippi, R. (2007). Abridge over troubled waters: Bridging organisations and entrepreneurial opportunities in emerging sector. *Research Policy, 36,* 1314–1334.

Schamp, E. (2008). On the notion of co-evolution in economic geography. In Boschma, R., & Martin, R. (Eds.) *The Handbook of Evolutionary Economic Geography.* Chichester: Edward Elgar.

Schamp, E., Rentmeister, B., & Vivien, L. (2004). Dimensions of proximity in knowledge-based networks: The cases of investment banking and automobile design. *European Planning Studies, 12*(5), 607–624.

Scherngell, T., & Hu, Y. (2011). Collaborative knowledge production in China: Regional evidence from a gravity model approach. *Regional Studies, 45*(6), 755–772.

Schienstock, G. (2011). path dependence and path creation: Continuity vs. fundamental change in national economies. *Journal of Futures Studies, 15*(4), 63–75.

Schmoch, U. (2008). Concept of a technology classification for country comparisons. *Final Report to the World Intellectual Property Organisation* (WIPO).

Schoen, A., de la Potterie, B., & Henkel, J. (2014). Governance typology of universities' technology transfer processes. *The Journal of Technology Transfer, 39*, 435–453.

Schulz, M. (2001). The uncertain relevance of newness: Organizational learning and knowledge flows. *Academy of Management Journal, 44*, 661–681.

Schumpeter, J. A. (2020). *The Theory of Economic Development.* Oxford: Taylor and Francis.

Scott, A. J. (1999). The cultural economy: Geography and the creative field. *Media, Culture & Society, 21*(6), 807–817.

Seo, I. (2019). Region's technology brokerage patterns and dynamics of regional development. *Sustainability, 11*(9), 2540.

Seo, I., & Sonn, J. W. (2019). The persistence of inter-regional hierarchy in technology transfer networks: An analysis of Chinese patent licensing data. *Growth and Change, 50*(1), 145–163.

Setterfield, M. (1997). "History versus equilibrium" and the theory of economic growth. *Cambridge Journal of Economics, 21*(3), 365–378.

Shao, S., Tian, Z., & Yang, L. (2017). High-speed rail and urban service industry agglomeration: Evidence from China's Yangtze River Delta region. *Journal of Transport Geography, 64*(10), 174–183.

Sharif, M. N., & Haq, A. K. M. A. (1980). A time-level model of technology transfer. *IEEE Transactions on Engineering Management, 27*(2), 49–58.

Sheng, K., Li, Y., & Sun W. (2021). Spatial patterns and influencing factors of urban hinterworld in China. *Economical Geography, 41*(3), 66–76. (in Chinese)

Sheng, K., Zhang, J., & Zhang, H. (2020). Network embedding and urban economic growth in China: A study based on the corporate networks of top 500 public companies. *Acta Geographica Sinica, 75*(10), 818–834. (in Chinese)

Sheppard, E. (2002). The spaces and times of globalization: Place, scale, networks, and positionality. *Economic Geography, 78*,307–330.

Si, Y., Zeng, G., Cao, X., & Zhu, Y. (2016). Research progress of glocal innovation networks. *Progress in Geography, 35*(5), 600–609. (in Chinese)

Siegel, D. S., Waldman, D. A., Atwater, L. E., & Link, A. N. (2003). Commercial knowledge transfers from universities to firms: improving the effectiveness of university–industry collaboration. *The Journal of High Technology Management Research, 14*(1), 111–133.

Siegel, D. S., Waldman, D., & Link, A. (2003). Assessing the impact of organizational practices on the relative productivity of university technology transfer offices: An exploratory study. *Research Policy, 32*, 27–48.

Sigler, T. J. & Martinus, K. (2017). Extending beyond 'world cities' in World City Network (WCN) research: Urban positionality and economic linkages through the Australia-based corporate network. *Environment and Planning A, 49*, 2916–2937.

Siler, P., Wang, C. Q., & Liu, X. M. (2003). Technology transfer within multinational firms and its impact on the productivity of Scottish subsidiaries. *Regional Studies, 37*(1), 15–25.

Simmie, J. (2002). Knowledge spillovers and reasons for the concentration of innovative SMEs. *Urban Studies, 39*(5–6), 885–902.

Singh, J. (2005). Collaborative networks as determinants of knowledge diffusion patterns. *Management Science, 51*, 756–770.

Small, A. C. (1985). *Technology Transfer: Geographical, Economic and Technical Dimensions.* New York: Quorum Book.

Smith, H. L. (1998). Barriers to technology transfer: Local impediments in Oxfordshire. *Environment and Planning C, 16*(4), 433–448.

Snijders, T. A. (2002). Markov chain Monte Carlo estimation of Exponential Random Graph Models. *Journal of Social Structure, 3*(2), 1–40.

Snijders, T. A. B., Pattison, P. E., Robins, G. L., Handcock, M. S. (2006). New specifications for exponential random graph models. *Sociological Methodology, 36*, 99–153.

Sorenson O. (2010). Complexity, network and knowledge flows. In Boschma, R., Martin, R., *et al.* (Eds.) *The Handbook of Evolutionary Economic Geography.* Cheltenham: Edward Elgar.

Stolwijk, C. M., Vanhaverbeke, P. M., Ortt, J. R., Pieters, M. W., den Hartigh, E., & van Beers, C. (2012). The effect of internal and external technology sourcing on firm performance throughout the technology life cycle. *Technology Analysis & Strategic Management, 24*(10), 1013–1028.

Storper, M. (1997). *The Region World: Territorial Development in a Global Economy.* New York: Guilford Press, 16–17.

Storper, M., & Venables, A. (2004). Buzz: Face-to-face contact and the urban economy. *Journal of Economic Geography, 4*(4), 351–370.

Subramanian, A. M., Nishant, R., Van De Vrande, V., & Hang, C. C. (2022). Technology transfer from public research institutes to SMEs: A configurational approach to studying reverse knowledge flow benefits. *Research Policy, 51*, 104602.

Suire, R., & Vicente, J. (2014). Clusters for life or life cycles of clusters: In search of the critical factors of clusters' resilience. *Entrepreneurship & Regional Development, 26*(1–2), 142–164.

Sun, X., Cui, W., & Wang, L. (2014). The review on structural holes and enterprise innovation. *Science of Science and Management of S. & T., 35*(11), 142–152. (in Chinese)

Sun, Y., & Grimes, S. (2017). The actors and relations in evolving networks: The determinants of inter-regional technology transaction in China. *Technology Forecasting & Social Change, 125*, 125–136.

Sun, Y., & Liu, K. (2016). Proximity effect, preferential attachment and path dependence in inter-regional network: A case of China's technology transaction. *Scientometrics, 108*, 201–220.

Sung, T. K. (2009). Technology transfer in the IT industry: A Korean perspective. *Technological Forecasting & Social Change, 76*, 700–708.

Sunley, P. (2008). Relational economic geography: A partial understanding or a new paradigm? *Economic Geography, 84*(1), 1–26.

Sutton, J. (1998). *Technology and Market Structure: Theory and History*. Cambridge: MIT Press.

Swyngedouw, E. (1997). Neither global nor local: 'Glocalization' and the politics of scale. In Cox, K. R. (Ed.) *Spaces of Globalisation*. New York: Guilford, 137–166.

Swyngedouw, E. (2000). Territories of innovation-Innovation as a collective process and the globalisation of competition. In Lawton-Smith, H. (Ed.) *Technology Transfer and Industrial Change in Europe*, London: Macmillan.

Swyngedouw, E. (2004). Globalisation or 'glocalisation'? Networks, territories and rescaling. *Cambridge Review of International Affairs, 17*(1), 25–48.

Swyngedouw, E., & Kaka, M. (2003). The making of "glocal" urban modernities: Exploring the cracks in the mirror. *City, 7*(1), 5–21.

Sydow, J., Lerch, F., & Staber, U. H. (2010). Planning for path dependence? The case of a network in the Berlin-Brandenburg optics cluster. *Economic Geography, 86*(2), 173–195.

Sytch, M., & Tatarynowicz, A. (2014). Friends and foes: The dynamics of dual social structures. *Academy of Management Journal, 57*(2), 585–613.

Taira, A. (2020). Beyond the cluster: A case study of pipelines and buzz in the glocal relational space of the glove- related industry of Shikoku, Japan. *The Geographical Journal, 186*, 45–58.

Tamoschus, D. (2012). Geographies of open source biotechnology innovation: Buzz, pipelines, and proximity in a virtual cluster. *International Journal of Knowledge-Based Organizations, 2*(2), 21–39.

Tanriverdi, H., & Venkatraman, N. (2005). Knowledge relatedness and the performance of multibusiness firms. *Strategic Management Journal, 26*(2), 97–119.

Tavassoli, S., & Carbonara, N. (2014). The role of knowledge variety and intensity for regional innovation. *Small Business Economics, 43*(2), 493–509.

Taylor, M., & Taylor, A. (2012). The technology life cycle: Conceptualization and managerial implications. *International Journal of Production Economics, 140*(1), 541–553.

Taylor, P. (2001). Urban hinterworlds: Geographies of corporate service provision under conditions of contemporary globalization. *Geography, 86*(1), 51–60.

Taylor, P., & Derudder, B. (2006). *World-city Network: A Global Urban Analysis* (Second edition). London: Routledge.

Taylor, P., Hoyler, M., & Verbruggen, R. (2010). External urban relational process: Introducing central flow theory to complement central place theory. *Urban Studies, 47*(13), 2803–2818.

Teece, D. J., Pisano, G., & Shuen, A. (1997). Dynamic capabilities and strategic management. *Strategic Management Journal, 18*(7), 509–533.

Teirlinck, P., & Spithoven, A. (2008). The spatial organization of innovation: Open innovation, external knowledge relations and urban structure. *Regional Studies, 42*(5), 689–704.

Ter Wal, L. J. (2009). *The Structure and Dynamics of Knowledge Networks: A Proximity Approach.* Dissertation of University Utrecht.

Ter Wal, L. J. (2014). The dynamics of the inventor network in German biotechnology: Geographic proximity versus triadic closure. *Journal of Economic Geography, 14*(3), 589–620.

Ter Wal, L. J., & Boschma, R. A. (2009). Applying social network analysis in economic geography: Framing some key analytic issues. *The Annals of Regional Science, 43*(3), 739–756.

Tollefson, J. (2018). China declared largest source of research articles. *Nature, 553*, 390.

Torre, A. (2008). On the role played by temporary geographical proximity in knowledge transmission. *Regional Studies, 42*(6), 869–889.

Trippl, M., & Bergman, E. M. (2014). Clusters, local districts, and innovative milieux. In Fischer, M. M., & Nijkamp, P. (Eds.) *Handbook of Regional Science.* Berlin: Springer-Verlag, 439–456.

Trippl, M., Grillitsch, M., & Isaksen, A. (2018). Exogenous sources of regional industrial change: Attraction and absorption of non-local knowledge for new path development. *Progress in Human Geography, 42*(5), 687–705.

Trippl, M., Tödtling, F., & Lengauer, L. (2009). Knowledge sourcing beyond buzz and pipelines: Evidence from the Vienna software sector. *Economic Geography, 85*(4), 443–462.

Tura, T., & Harmaakorpi, V. (2005). Social capital in building regional innovative capability. *Regional Studies, 39*, 1111–1125.

Turkina, E., Assche, A. V., & Kali, R. (2016). Structure and evolution of global cluster networks: Evidence from the aerospace industry. *Journal of Economic Geography, 16*(6), 1211–1234.

Utterback, J. M. (1974). Innovation in industry and the diffusion of technology. *Science, 183*(4125), 620–626.

Van den Berghe, L., & Guild, P. D. (2008). The strategic value of new university technology and its impact on exclusivity of licensing transactions: An empirical study. The Journal of Technology Transfer, *33*(1), 91–103.

van der Pol, J. (2019). Introduction to network modeling using exponential random graph models (ERGM): Theory and an application using R-project. *Computational Economics, 54*(3), 845–875.

van Meeteren, M., Neal, Z., & Derudder, B. (2016). Disentangling agglomeration and network externalities: A conceptual typology. *Papers in Regional Science, 95*(1), 61–80.

Vasudeva, G., Zaheer, A., & Hernandez, E. (2013). The embeddedness of networks: Institutions, structural holes, and innovativeness in the fuel cell industry. *Organization Science, 24*(3), 645–663.

Vega-Gomez, F., & Miranda-Gonzalez, F. J. (2021). Choosing between formal and informal technology transfer channels: Determining factors among Spanish academicians. *Sustainability, 13*, 2476.

Vernon, R. (1966). International trade and international investment in the product cycle. *Quarterly Journal of Economics, 80*(2), 190–207.

Veugelers, R., & Cassiman, B. (2004). Foreign subsidiaries as a channel of international technology diffusion: Some direct firm level evidence from Belgium. *European Economic Review, 48*(2), 455–476.

Villani, E., Rasmussen, E., & Grimaldi, R. (2017). How intermediary organizations facilitate university–industry technology transfer: A proximity approach. *Technological Forecasting & Social Change, 114*, 86–102.

Vonhippel, E. (1994). Sticky information and the locus of problem-solving — Implications for innovation. *Management Science, 40*(4), 429–439.

Wagner, G. A., & Bologna Pavlik, J. (2020). Patent intensity and concentration: The effect of institutional quality on MSA patent activity. *Papers in Regional Science, 99*(4), 857–898.

Wahab, S. B. (2012). Defining the concepts of technology and technology transfer: A literature analysis. *International Business Research, 5*(1), 61–71.

Wakasugi, R., & Ito, B. (2009). The effects of stronger intellectual property rights on technology transfer: evidence from Japanese firm–level data. *The Journal of Technology Transfer, 34*(2), 145–158.

Walker, R. (2000). The geography of production. In Sheppard, E., & Barnes, T. (Eds.) *A Companion to Economic Geography*. Oxford, UK: Blackwell, 113–132.

Wang, X., & Xuan, X. (2019). Innovation culture, government support and R&D investment- Based on empirical data of listed companies in Beijing, Tianjin and Hebei. *Friends of Accounting*, (23), 28–33. (in Chinese)

Wang, B., Wang, T., & Liu, C. (2023a). Spatiotemporal dynamics of China's technology transfer hubs and their hinterworlds. *Acta Geographica Sinica, 78*(2), 293–314. (in Chinese)

Wang, H., & Sun, F. (2009). Analysis of the dynamics of regional innovation systems. *China Science and Technology Forum*, (1), 36–40. (in Chinese)

Wang, J., & Cai, S. (2020). The construction of high-speed railway and urban innovation capacity: Based on the perspective of knowledge spillover. *China Economic Review, 63*(10), 101–112.

Wang, J., Du, D., & Jin, F. (2019). Comparison of spatial structure and linkage systems and geographic constraints: A perspective of multiple traffic flows. *Acta Geographica Sinica, 74*(12), 2482–2494. (in Chinese)

Wang, J., Jin, F., & Jiao, J. (2017). Impacts of high-speed rail lines on the city network in China. *Journal of Transport Geography, 60*(4), 257–266.

Wang, J., Liang, Z., & Xue, L. (2014). Multinational R&D in China: Differentiation and integration of global R&D networks. *International Journal of Technology Management, 65*(1–4), 96–124.

Wang, L. (2014). *Research on the Structure and Synergistic Evolution of Regional Innovation Networks*. Dissertation of University of Science and Technology of China. (in Chinese)

Wang, Q., Zeng, G., Yang, W., & Ye, L. (2022). Research on structure and effect mechanism of technology transfer network in Yangtze River Economic Belt. *Resources and Environment in the Yangtze River, 31*(1), 1–12. (in Chinese)

Wang, W., & Lu, S. (2021). University–industry innovation community dynamics and knowledge transfer: Evidence from China. *Technovation, 106*, 102305.

Wang, Y., & Wang, C. (2018). Urban innovation linkage network and radiation ability in Yangtze River Delta. *Economic Geography, 38*(9), 130–137. (in Chinese)

Wang, Y., Ning, L., Li, J., & Prevezer, M. (2016). Foreign direct investment spillovers and the geography of innovation in Chinese regions: The role of regional industrial specialization and diversity. *Regional Studies, 50*(5), 805–822.

Wang, B., Mao, W., Piao, J., & Liu, C. (2023b). Does external linkage stimulate innovation capacity? The analysis based on "dual-pipelines" framework. *Papers in Regional Science*, https://doi.org/10.1111/pirs.12736.

Wasserman, S., & Faust, K. (1994). *Social Network Analysis: Methods and Applications.* New York: Cambridge University Press.

Watts, R. J., & Porter, A. L. (1997). Innovation forecasting. *Technology Forecasting and Social Change, 56*, 25–47.

Weitzman, M. (1998). Recombinant growth. *Quarterly Journal of Economics, 113*(2), 331–360.

Wen, H., & Zeng, G. (2005). Study of global value chain governance and local industrial network upgrading: Case of Pudong Integrated Circuit Industrial Network. *China Industrial Economy*, (7), 20–27. (in Chinese)

Wetwitoo, J., & Kato, H. (2017). High-speed rail and regional economic productivity through agglomeration and network externality: A case study of inter-regional transportation in Japan. *Case Studies on Transport Policy, 5*(4), 549–559.

Whittle, A., & Kogler, D. F. (2019). Related to what? Reviewing the literature on technological relatedness: Where we are now and where can we go? *Papers in Regional Science, 99*, 97–113.

William, F., & Gibson, D. V. (1990). *Technology Transfer: A Communication Perspective.* Beverly Hills, CA: Sage.

Wong, P. K. (1991). Technology development through subcontracting linkages: Lessons from Singapore. In Institute of Strategic and International Studies, Kuala Lumpur. *Enhancing IntraIndustry Linkages*, 49–59.

Wu, H. (2017). Exploring the path of accelerating the construction of Shanghai Science and Technology Innovation Center- Based on the perspective of enterprise innovation model. *Statistical Theory and Practice*, (12), 8–12. (in Chinese)

Wu, L., Zhu, H., & Chen, H. (2019). Comparing nanotechnology landscapes in the US and China: A patent analysis perspective. *Journal of Nanoparticle Research, 21*, 180.

Xiao, F., Wang, J., & Du, D. (2022). High–speed rail heading for innovation: The impact of HSR on intercity technology transfer. *Area Development and Policy*, *7*(3), 293–311.

Xiao, S., & Liu, J. (2018). The coordinated development analysis about the science-technology innovation ability of regional higher education. *Economic Geography*, *38*(8), 124–131. (in Chinese)

Xu, C. (2013). Universities' patents, technology transfer and firms' innovation in China: Based on the accumulation innovation. *Science of Science and Management of S. & T.*, *34*(12), 78–86. (in Chinese)

Xu, Q., Kang, X., & Yang, Z. (2017). Research on the characteristics of inter-provincial technology transferin China based on patent right transfer. *Journal of Intelligence*, *36*(7), 66–72. (in Chinese)

Yan, S. (2021). *Spatial Evolution Mechanism and Spillover Effects of the Technology Importing Network of Chinese Cities.* Dissertation of East China Normal University. (in Chinese)

Yang, L., & Liu, X. (2014). Is there "Matthew Effect" in regional technology transfer? The driving orientation of inter-regional technology transfer. *Studies in Science of Science*, *32*(12), 1820–1827, 1858. (in Chinese)

Yang, W., Yu, X., Zhang, B., & Huang, Z. (2021). Mapping the landscape of international technology diffusion (1994–2017): Network analysis of transnational patents. *The Journal of Technology Transfer*, 46,138–171.

Yao, W. (2019). The current situation of research output of universities and regional evolution and development from 2003–2015. *China Science & Technology Resources Review*, (4), 55–63. (in Chinese)

Yayavaram, S., & Ahuja, G. (2008). Decomposability in knowledge structures and its impact on the usefulness of inventions and knowledge-base malleability. *Administrative Science Quarterly*, *53*, 333–362.

Ye, Y., De Moortel, K., & Crispeels, T. (2020). Network dynamics of Chinese university knowledge transfer. *The Journal of Technology Transfer*, *45*, 1228–1254.

Yeung, H. W. (2005). Rethinking relational economic geography. *Transactions of the Institute of British Geographers*, *30*(1), 37–51.

Yeung, H. W. (2008). Industrial clusters and production networks in Southeast Asia: A global production networks approach. In Kuroiwa I., & Toh M. H. (Eds.) *Production Networks and Industrial Clusters: Integrating Economies in Southeast Asia.* Singapore: Institute of Southeast Asian Studies, 83–120.

Yeung, H. W. (2016). *Strategic Coupling: East Asian Industrial Transformation in the New Global Economy*. Ithaca: Cornell University Press.

Yeung, Y. M., & Li, X. J. (1999). Bargaining with transnational corporations: The case of Shanghai. *International Journal of Urban and Regional Research, 23*(3), 513–533.

Yi, W., Long, X., & Lin, Z. (2021). Does distance affect university knowledge spillover: Empirical evidence from the opening of Chinese high-speed rail. *China Industrial Economics, 38*(9), 99–117. (in Chinese)

Yin, M., Bertolini, L., & Duan, J. (2015). The effects of the high-speed railway on urban development: International experience and potential implications for China. *Progress in Planning, 98,* 1–52.

Yokura, Y., Matsubara, H., & Sternberg, R. (2013). R&D networks and regional innovation: A social network analysis of joint research projects in Japan. *Area, 45*(4), 493–503.

Young, S., & Lan, P. (1997). Technology transfer to China through foreign direct investment. *Regional Studies, 31*(7), 669–679.

Yu, C. (2009). *The Study on the Nonlinear Relationship between the Technology Gap and Economic Growth.* Dissertation of Hunan University, 24–26. (in Chinese)

Yu, H., Zhang, J., Zhang, M., & Fan, F. (2022). Cross-national knowledge transfer, absorptive capacity, and total factor productivity: the intermediary effect test of international technology spillover. *Technology Analysis & Strategic Management, 34*(6), 625–640.

Yu, L., Li, H., Wang, Z., & Duan, Y. (2019). Technology imports and self-innovation in the context of innovation quality. *International Journal of Production Economics*, 214, 44–52.

Yu, Q. (2017). "No county left behind?" The distributional impact of high-speed rail upgrades in China. *Journal of Economic Geography, 17*(3), 489–520.

Zeng, Q., & Wang, G. (2014). An empirical study on the relations between network capability of industrial network and innovation performance. *Economic Geography, 34*(10), 111–118. (in Chinese)

Zhang, G., Duan, H., & Zhou, J. (2016). Investigating determinants of inter-regional technology transfer in China: A network analysis with provincial patent data. *Review of Managerial Science, 10,* 345–364.

Zhang, H., & Lang, C. (2013). The impact of past performance and network heterogeneity on knowledge creation: A central network position isn't enough. *Studies in Science of Science, 31*(10), 1581–1589. (in Chinese)

Zhang, H., & Li, Z. (2011). Weighted ego network for forming hierarchical structure of road networks. *International Journal of Geographical Information Science, 25*(2), 255–272.

Zhang, J. (2021). Does HSR promote regional innovation growth or innovation convergence? *Technology in Society, 64*(1), 101–472.

Zhang, J., & Sun, Y. (2011). A comparative study on technology spillovers from FDI and import. *World Economy Study,* (12), 51–58, 85–86. (in Chinese)

Zhang, R. (2018). *Cultivating Tropical Rain Forest of Innovation: Study on the Evolution of Shanghai City Innovation Ecosystem.* Shanghai: East China Normal University Press. (in Chinese)

Zhang, Y., & Li, D. (2017). Evolution of innovation network of local enterprises in China from the perspective of globalization: A case study of Huawei. *World Regional Studies, 26*(6), 92–100. (in Chinese)

Zhang, Y., & Sun, G. (2016). Relation and evolution of the administrative hierarchy, urban scale and the domestic tourism development-Take Yangtze River Delta for example. *Economic Geography, 36*(4), 188–194. (in Chinese)

Zhang, Y., Zeng, G., & Cheng, J. (2013). The evolution of Zhangjiang IC industry cluster based on global pipeline and local buzz. *Areal Research and Development, 32*(3), 38–43. (in Chinese)

Zhao, W. (2020). Hub economy and its development mechanism: An empirical study taking transportation hub as an example. *Human Geography, 35*(3), 115–122. (in Chinese)

Zhao, Z., Han, C., & Wang, L. (2019). Collaborative innovation network of defence-related science and technology industry alliance: Structure and evolution. *Journal of Public Management, 16*(4), 156–167. (in Chinese)

Zheng, W. W., Wang, X., Tian, D. J., etc. (2013). Water pollutant fingerprinting tracks recent industrial transfer from coastal to inland China: A case study. *Scientific Reports, 3,* 1031.

Zheng, X., & Si, L. (2006). Research on the theoretic growth of NIS. *R&D Management, 18*(5), 1–8. (in Chinese)

Zhou, C., Cao, Z., & Zeng, G. (2019). Cluster networks and evolution path of China's electronic information industry innovation. *Geographical Research, 38*(9), 2212–2225. (in Chinese)

Zhou, K., & Li, C. (2012). How knowledge affects radical innovation: Knowledge base, market knowledge acquisition, and internal knowledge sharing. *Strategic Management Journal, 33*(9), 1090–1102.

Zhou, M., & Sun, L. (2016). The transfer of patent right, characteristics of spatial network and the way for regional collaborative innovation in the area of Beijing, Tianjin and Hebei. *Studies in Science of Science, 34*(11), 1736–1743+1760. (in Chinese)

Zhou, R. (2019). Analysis on features and influencing factors of network space structure of Chinese urban agglomerations. *Statistics and Decision Making, 35*(14), 90–94. (in Chinese)

Zhou, Y., & Tong, X. (2003). An innovative region in China: Interaction between multinational corporations and local firms in a high-tech cluster in Beijing. *Economic Geography, 79*(2), 129–152.

Zucker, L. G., Darby, M. R., Furner, J., Liu, R. C., & Ma, H. (2007). Minerva unbound: Knowledge stocks, knowledge flows and new knowledge production. *Research Policy, 36*(6), 850–863.

Zwitter, A., & Hazenberg, J. (2020). Decentralised network governance: Blockchain technology and the future of regulation. *Frontiers in Blockchain, 3*, 1–12.

INDEX

T

U

V

W

Y

East China Normal University Scientific Reports

Published (continued from page ii)

Subseries on Educational Information Technology

Published

Vol. 15 *Achieving Greater Educational Impact through Data Intelligence: Practice, Challenges and Expectations of Education*
by Bian Wu (East China Normal University, China), Yiling Hu (East China Normal University, China) and Xiaoqing Gu (East China Normal University, China)

Vol. 14 *Design and Development of a Wiki-based Collaborative Process Writing Pedagogy: Putting Technological, Pedagogical, and Content Knowledge (TPACK) in Action*
by Xuanxi Li (Huzhou University, China)

Subseries on Data Science and Engineering

Published

Vol. 13 *Load Balance for Distributed Real-time Computing Systems*
by Junhua Fang (Soochow University, China), Rong Zhang (East China Normal University, China) and Aoying Zhou (East China Normal University, China)

Vol. 12 *Biological Language Model: Theory and Application*
by Qiwen Dong (East China Normal University, China), Xiaoyang Jing (Yunnan University, China), Aoying Zhou (East China Normal University, China) and Xiuzhen Hu (Inner Mongolia University of Technology, China)

Vol. 11 *Probabilistic Approaches for Social Media Analysis: Data, Community and Influence*
by Kun Yue (Yunnan University, China), Jin Li (Yunnan University, China), Hao Wu (Yunnan University, China), Weiyi Liu (Yunnan University, China) and Zidu Yin (Yunnan University, China)

Vol. 10 *Clustering and Outlier Detection for Trajectory Stream Data*
by Jiali Mao (East China Normal University, China),
Cheqing Jin (East China Normal University, China) and
Aoying Zhou (East China Normal University, China)

Vol. 9 *Concurrency Control and Recovery in OLTP Systems: High Scalability and Availability*
by Peng Cai (East China Normal University, China),
Jinwei Guo (East China Normal University, China) and
Aoying Zhou (East China Normal University, China)

Vol. 8 *Network Data Mining and Analysis*
by Ming Gao (East China Normal University, China),
Ee-Peng Lim (Singapore Management University, Singapore) and
David Lo (Singapore Management University, Singapore)

Vol. 7 *Time-Aware Conversion Prediction for E-Commerce*
by Wendi Ji (East China Normal University, China),
Xiaoling Wang (East China Normal University, China) and
Aoying Zhou (East China Normal University, China)

Vol. 6 *Discovery and Fusion of Uncertain Knowledge in Data*
by Kun Yue (Yunnan University, China), Weiyi Liu
(Yunnan University, China), Hao Wu (Yunnan University, China),
Dapeng Tao (Yunnan University, China) and
Ming Gao (East China Normal University, China)

Vol. 5 *Review Comment Analysis for E-commerce*
by Rong Zhang (East China Normal University, China),
Aoying Zhou (East China Normal University, China),
Wenzhe Yu (East China Normal University, China),
Yifan Gao (East China Normal University, China) and
Pingfu Chao (East China Normal University, China)

Vol. 4 *Opinion Analysis for Online Reviews*
by Yuming Lin (Guilin University of Electronic Technology, China),
Xiaoling Wang (East China Normal University, China) and
Aoying Zhou (East China Normal University, China)

Vol. 3 *Querying and Mining Uncertain Data Streams*
by Cheqing Jin (East China Normal University) and
Aoying Zhou (East China Normal University

Directly Under the Series

Vol. 2 *School Mathematics Textbooks in China: Comparative Study and Beyond*
by Jianpan Wang (East China Normal University)

Vol. 1 *High-Frequency Trading and Probability Theory*
by Zhaodong Wang (East China Normal University) and Weian Zheng (East China Normal University)

East China Normal University Scientific Reports

www.ingramcontent.com/pod-product-compliance
Lightning Source LLC
LaVergne TN
LVHW010324220926
848484LV00004BB/838
* 9 7 8 9 8 1 1 2 7 4 9 5 4 *